Social Neuroscience of Psychiatric Disorders

Social Neuroscience of Psychiatric Disorders is about the role of the Social Brain in neuropsychiatry. The need to belong to social groups and interact with others has driven much of the evolution of the human brain. The relatively young field of social neuroscience has made impressive strides towards clarifying the neural correlates of the Social Brain, but, until recently, has not focused on mental and neurological disorders. Yet, the Social Brain underlies all brain-behaviour disorders, and nearly every neuropsychiatric illness involves social behavioural disturbances.

This unique and ground-breaking volume is a major step forward in deciphering the impact of the Social Brain on neuropsychiatric disorders. Investigators evaluate neuropsychiatric disorders in the context of recent advances in social neuroscience to reveal the impact of Social Brain mechanisms on neuropsychiatric disorders and allow readers to glimpse the exciting potential advances in this field in the years to come.

This book was originally published as a special issue of *Social Neuroscience*.

Facundo Manes is Professor of Behavioural Neurology and Cognitive Neuroscience at Favaloro University, Argentina. He is also Director of the Institute of Cognitive Neurology (INECO) in Buenos Aires.

Mario F. Mendez is Professor of Neurology and Psychiatry & Biobehavioral Sciences at the University of California, Los Angeles, School of Medicine, and is also Director of the Neurobehavior Unit at the Greater Los Angeles VA Medical Center, California, USA.

T0239961

Social Neuroscience of Psychiatric Disorders

A Special Issue of the Journal *Social Neuroscience*

Edited by

Facundo Manes and Mario F. Mendez

Routledge
Taylor & Francis Group

LONDON AND NEW YORK

First published 2013
by Psychology Press Ltd

Published 2017 by Routledge
2 Park Square, Milton Park, Abingdon, Oxon OX14 4RN
711 Third Avenue, New York, NY 10017, USA

First issued in paperback 2017

Routledge is an imprint of the Taylor & Francis Group, an informa business

British Library Cataloguing in Publication Data
A catalogue record for this book is available from the British Library

ISSN: 1747-0919

Typeset in Times New Roman
by Taylor & Francis Books

Publisher's Note
The publisher would like to make readers aware that the chapters in this book may be referred to as articles as they are identical to the articles published in the special issue. The publisher accepts responsibility for any inconsistencies that may have arisen in the course of preparing this volume for print.

ISBN 13: 978-1-138-10977-3 (pbk)
ISBN 13: 978-1-84872-757-1 (hbk)

Contents

CONTENTS

The emerging impact of social neuroscience on neuropsychiatry and clinical neuroscience

Mario F. Mendez[1] and Facundo Manes[2]

[1]David Geffen School of Medicine at UCLA School of Medicine, and Neurobehavior Unit, V.A. Greater Los Angeles Healthcare System, Los Angeles, CA, USA
[2]Institute of Neuroscience, Favaloro University, and Institute of Cognitive Neurology (INECO), Buenos Aires, Argentina; and World Federation of Neurology Research Group on Aphasia and Cognitive Disorders

Social neuroscience has made great strides toward clarifying the neural basis of brain–behavior relationships. In the last 25 years, social neuroscience has made contributions to many fields, including cognitive neuroscience, social psychology, ethology, economics, and even philosophy. The field has spawned a highly productive collaboration between investigators in these areas, many of whom have profited from the great leap forward in functional neuroimaging.

It is now time for social neuroscience to make similar contributions to neuropsychiatry. Social behavior is integral to all brain–behavior disorders. Neuropsychiatry, and an understanding of the topics in this special issue, depend upon the basic neural correlates of social behavior. It begins with social perception, or the ability to detect the presence of another mental agent, including the critical roles of face recognition and the fusiform face area (Kanwisher & Yovel, 2006), and the superior temporal gyrus or sulcus (STG/STS) for observed biological motion. Mechanisms for social simulation of not only others' movements but also their intentions and emotions have emerged in conjunction with discoveries in a mirror-neuron system. Areas of socioemotional significance include the anterior insula and anterior cingulate cortex, the anterior temporal cortex, and, especially pertinent to the articles in this issue, the amygdala, which plays a role in emotional salience and significance.

Our brains are primed for thinking about the minds of others. Making inferences about others' mental states, or theory of mind (ToM), produces increased activity in the medial prefrontal cortex (MPFC), temporoparietal junction (TPJ), and medial parietal cortex (Frith & Frith, 2006).

In this special edition, we view neuropsychiatry in the context of these advances in social neuroscience. Nearly every neuropsychiatric illness involves social behavioral disturbances (Adolphs, 2010), and much of the promise and justification for the investment in this burgeoning field is its possible clinical implications. To what extent do neuropsychiatric illnesses result from neural impairments in social perception, social simulation and the appreciation of social salience, mental state attribution, or disturbances in social regulation? Psychiatric illness may be a continuum with variability in the healthy population, as reflected in one article in this issue. Finally, understanding the social neurocognitive source of symptoms such as suspiciousness and apathy can be a step toward the development of therapies for these disorders.

CONTENTS OF THE SPECIAL ISSUE

The papers in this series study or review various clinical entities by varied methodology. Clinical

Correspondence should be addressed to: M. F. Mendez, Neurobehavior, VAGLA, 11301 Wilshire Blvd., Los Angeles, CA 90073, USA. E-mail: mmendez@ucla.edu entities include autism spectrum disorder (ASD) and its variant, pervasive developmental disorder (PPD); schizophrenia and schizotypal disorder; depression; attention-deficit hyperactivity disorder (ADHD); traumatic brain disease (TBI); psychopathy; frontotemporal dementia; and Parkinson's disease. The methodology includes behavioral assessments, eye-tracking, event-related potentials (ERP), and brain imaging, particularly functional magnetic resonance imaging (fMRI). Topics covered include social attention and perception, amygdalar reactivity, social attention, social and emotional recognition and competence, social approach and withdrawal, suspiciousness, immoral or corrupt behavior, and ToM.

SPECIFIC STUDIES

Three articles in this series start with an exploration of face perception in ASD, particularly as it relates to the amygdala. The amygdala is involved in guiding fixations on the eyes, a region of the face that is socially salient, and it plays a role in emotion and emotional inference from faces. Birmingham, Cerf, and Adolphs compare social attention in ASD to that in S.M., a patient with rare, bilateral, amygdalar lesions. As revealed by eye-tracking to complex social scenes that contain faces, S.M., but not ASD subjects, increased her gaze to the eyes when the task required social attention. The authors conclude that face-perception difficulty in ASD is not due to amygdalar dysfunction in directing gaze to the eyes but instead to insensitivity to socially relevant information. McPartland et al. compare ASD and non-ASD subjects on N170, an ERP marker of early face processing, and on behavior measures of face recognition. Individuals with ASD show decreased N170s, slowed face processing, and decreased sensitivity to face inversion. The authors postulate that decreased face perception in ASD is the consequence of a developmentally reduced social drive. Uono, Sato, and Toichi present a study of emotion-expression recognition in mild PPD, a variant of ASD, which reveals decreased performance on the Ekman face emotions, especially fearful ones. Fearful-expressions recognition does not improve with age, as in normally developing children, and is worse with the severity of PPD, suggesting an atypical development of facial expression recognition. In PPD, impaired recognition of fearful faces may result from abnormal amygdalar development for these socially important stimuli.

Three other studies investigate impairments in face-emotion recognition in various conditions. Among adults with ADHD, Ibáñez et al., in a similar study to McPartland et al., present faces and words to test the effects of stimulus type, valence, and face–word compatibility. The adult ADHD group shows deficits in N170 emotion discrimination, especially for positive face stimuli, in the absence of deficits in facial structural processing. In ADHD, the reduced N170 amplitude for positive stimuli suggests a specific right hemisphere impairment for the early processing of emotional faces. Among patients with depression, Derntl et al., using fMRI, investigate the neural correlates of social approach and withdrawal with facial emotional expressions. In addition to higher avoidance scores and stronger, wider activation during avoidance in depression, they report decreased amygdalar function during approach and during processing of happy faces in particular. This points to secondary effects of depression on amygdalar activity. Kumfor et al. report a study of face-emotion recognition in the three major variants of frontotemporal dementia. All patient groups have impaired overall facial emotion recognition, but particularly for negative emotions. They increase the intensity and perceptual salience of face emotions by digitally manipulating the images and show that all FTD subgroups, except semantic dementia (SD), improve with the increased emotional saliency. The authors conclude that, in SD, direct damage to the amygdala results in widespread emotion-recognition impairment, which cannot be overcome by increasing emotional salience.

A theme in several articles in this issue is mechanisms underlying schizophrenia, especially early emotion-processing deficits, inability to pick up and integrate body emotions, and susceptibility to suspicious thinking. Garrido-Vásquez, Jessen, and Kotz's review of the literature confirms the presence of emotion-perception deficits in psychiatric populations when patients are tested in dynamic and multimodal naturalistic settings and under different task (explicit and implicit task instructions) demands. The authors propose that patients with schizophrenia are impaired in a fast, pre-attentive system involving the amygdala and its surrounding network. Van den Stock et al. evaluate the perception of bodily expressions of emotions and their integration with voice in patients with schizophrenia. On static pictures of emotions, patients with schizophrenia are impaired in emotional recognition, and on videos of dynamic emotions they are significantly influenced by the auditory information. They conclude that patients with schizophrenia are impaired in recognizing whole-body expressions of emotions and have difficulty with multisensory integration of emotional information. Sasamoto et al. use the autism-spectrum quotient (ASQ) to explore social perception and the "autistic tendency" in comparison to gray matter alterations on structural MRI among patients with schizophrenia. There are significant negative correlations between the total ASQ score and gray matter volume in the area of the left STS. This region reciprocally interacts with regions, such as the amygdala, involved in attaching emotional salience to sensory input.

Another study investigates the predisposition to psychotic traits in a nonclinical population. Among Chinese students, Li et al. compared those with and without schizotypal personality traits on ERP changes elicited by suspicious thoughts. They use a clever paradigm, a novel digit-guessing task, to induce the subject's "feeling of been seen through." A "friend" is said to guess the patient's choice of digits, and they calculate difference waves (correct guess wave minus incorrect guess wave) on P3 and P3 amplitude. The amplitude is smaller for the schizotypal group, suggesting that it might be inhibited in those with higher level of paranoid ideation; the schizotypal subjects are not surprised by correct guesses of friends because schizotypicals might have more suspicious or paranoid thoughts.

A number of studies explore mental state attribution in neuropsychiatric disease. Two fMRI studies indicate that, in schizophrenia and in TBI, there is disturbed ToM related to the MPFC, a central region for ToM, with additional involvement of the TPJ in schizophrenia and of surrounding white matter in TBI. Among schizophrenics, Lee et al. examine ToM, using a specific task, a false-belief attribution condition versus false photograph for general reasoning and simple reading ability. Schizophrenics show reduced task-related activations in the TPJ and the MPFC during the false-belief condition, but not during the false-photograph condition. Among adolescents with moderate-to-severe TBI, Scheibel et al. examine ToM using an animated social attribution task. Compared to normal subjects, the TBI subjects have more diffuse and intense activation with sparing of the MPFC. Further diffusion tensor imaging for functional anisotropy as a measure of white-matter integrity shows reductions in surrounding frontal areas, indicating that white-matter changes may cause reductions in connectivity among the components of a brain network that mediates social cognition.

Although chronic social and emotional deficits are common in moderate-to-severe TBI, current techniques of assessment with ToM tasks may be insufficient. Hynes, Stone, and Kelso review existing social/emotional measures and introduce four new tasks. The Global Interpersonal Skills Test is a questionnaire that allows patients and their caregivers to identify specific social traits and behaviors that may be a source of difficulty. The Assessment of Social Context Task is a dyadic video task that examines patients' comprehension of social contextual information, including emotions, intentions, and attitudes. The Awareness of Interoception Task measures patients' sensitivity to their own heartbeat, which may underlie awareness of their emotions. Their Social Interpretations Task, a social framing task based on social geometric animation, appears less useful than the other three tasks.

Of significance to neuropsychiatry is whether there are cultural effects on mental state attribution and other social cognitive processes. As there are cultural differences in mode and development of ToM, one might expect differential brain activation patterns. Koelkebeck et al. present a paper on transcultural differences in activation of the MPFC in Japanese versus Caucasians on a ToM task of moving geometric shapes in a social pattern. The Caucasians have greater activation, possibly from a greater need to constantly distinguish between selves, others, and surroundings. The Japanese have less activation, possibly because of greater efficiency from early-life learning to be in tune with unspoken, nonexplicit social signals (ToM may take longer to develop in Japanese). Interestingly, MPFC activation in the Japanese "catches up" when there are more autistic features, probably due to compensation for reduced mentalizing abilities.

Two papers in this series address the important problem of psychopathy. Rather than more traditional regions of interest, Sato et al. analyze the MRI images of subjects with antisocial personality disorder with novel pattern classification techniques of support vector machines and maximum uncertainty linear discrimination analysis. Among the subjects with high psychopathy scores, there is decreased gray matter concentration in the STG bilaterally, especially on the right. Sobhani and Bechara review the somatic marker hypothesis as a mechanism potentially disturbed in those who manifest immoral and corrupt behavior. Their studies with the Iowa Gambling Task show that damage to ventral MPFC leads to impaired judgment and decision-making as well as failure to learn from repeated mistakes. Their review of the literature describes studies of patients with ventral MPFC lesions who show impairments in victim-based moral judgments, or acceptance of increased attempted harm to others (e.g., Young & Saxe, 2009).

This special issue extends to defining the neural circuitry for neuropsychiatric symptoms. Lawrence, Goerendt, and Brooks investigate the neuroanatomy of apathy, using a reward paradigm and positron emission tomography in patients with Parkinson's disease.

By manipulating search outcome (money reward versus valueless token) while keeping the actions of the participants constant, they examine the influence of apathy on the neural coding of money reward cues. In this study, apathy is associated with a blunted response to money in the distributed neural circuit integral to the representation of stimulus reward. Apathy may be associated with diminished high-level value representations in ventral MPFC, emotional blunting in the amygdala, and decreased reward value in the striatum and in the dopamine-dependent midbrain region of the ventral tegmental area.

THEORETICAL ISSUES HIGHLIGHTED BY THE CONTRIBUTIONS

Several of the papers in this issue demonstrate impairments at the early-input stage of social cognition—i.e., at the level of facial perception. The papers in this series advance our understanding of disturbances in this area including ASD, ADHD, and dementia. Many of these studies focus on the changes in the amygdala, a structure known to play a role in attention to the eyes and in facial emotion appreciation. Several papers on ASD indicate that insensitivity to aspects of face processing is consequent to a more fundamental insensitivity to social phenomena, and McPartland et al. further suggest that the face-perception system does not develop normally for social stimuli in ASD. In other words, this "social insensitivity" early in life leads to developmental impairments in the ability of face-perception areas and face-emotion areas to respond to social stimuli.

Face emotion is the most important information for social behavior, and studies show difficulty in face-emotion recognition in ASD/PPD, adult ADHD, depression, and semantic dementia. These studies point to more than one mechanism. For example, in ASD, recognition of fearful faces may follow the abnormal amygdalar development of responsiveness to social stimuli; in ADHD, there are right hemisphere deficits in processing of facial emotions; in depression, there is a top-down late effect on the amygdala; and in SD, there is direct injury to the amygdala with increasing intensity.

Social cognition is a key determinant of poor functioning in schizophrenia, and several studies in this series indicate difficulties with the social perception of emotion, not only from faces but also from the body, as well as the ability to integrate body, face, and other socioemotional clues. In addition, Li et al.'s study of suspicious thinking among schizotypal subjects supports the concept of a continuum from suspicion in normal subjects to delusions of reference and persecutory delusions in schizophrenia. This study is significant for the view that psychotic traits are part of a continuum from normality to psychosis.

The disturbances in ToM found in neuropsychiatric illnesses is associated with corresponding disturbances in the circuitry for mentalization (Frith & Frith, 2006). In schizophrenia, studies show reduced activation in TPJ, as well as in MPFC, and gray matter reductions in the cortical areas surrounding the left STS. Scheibel et al.'s findings in TBI indicate that these alterations in mentalizing circuits are not restricted to areas that typically mediate social cognition, as the surrounding white-matter tracts can affect this mentalizing circuit.

A final theme involves the important neuropsychiatric issue of psychopathy and the possible neural correlates of immoral or corrupt behavior. A study using novel fMRI analysis of subjects with antisocial personality disorder shows changes corresponding to psychopathy in the right STG, an area important for social perception. Sobhani and Bechara further make the case for linking disturbed somatic markers, or re-enacted social and emotional representations that affect decision-making, to the commission of immoral or corrupt acts. Usually, the ventral MPFC reactivates somatic effectors in the insula and somatosensory and related structures, and this fails to occur in psychopaths.

FUTURE DIRECTIONS

Many challenges remain for social neuroscience including its more thorough application to neuropsychiatric illnesses (Adolphs, 2010; Cacioppo et al., 2007). The studies in this special edition of *Social Neuroscience* advance our understanding of the neural underpinnings of neuropsychiatric diseases, but there remains much work to be done. As Hynes, Stone, and Kelso point out for TBI, investigators need additional social/emotional measures for clinical as well as research assessment. As Koelkebeck et al. point out, investigators need to consider the effects of culture on their results. As Sato et al. illustrate, novel fMRI computational approaches have potential applications to neuropsychiatry. Social perception, social salience and simulation of mind, social cognition and social representations in the brain, and the extent of social regulation need further clarification of their neuropsychiatric implications. The specificity of a social system in the brain and the role of specific structures (Mitchell, 2009), the primacy of social behavior and its default state, and how much social mechanisms share with nonsocial cognitive processes are all issues pertinent to understanding neuropsychiatric diseases. In sum, social neuroscience holds great promise for clarifying many aspects of brain—behavior disorders.

REFERENCES

Adolphs, R. (2010). Conceptual challenges and directions for social neuroscience. *Neuron, 65,* 752–767.

Cacioppo, J. T., Amaral, D. G., Blanchard, J. J., Cameron, J. L., Carter, C. S., Crews, D., et al. (2007). Social neuroscience: Progress and implications for mental health. *Perspectives in Psychological Science, 2,* 99–123.

Frith, D. C., & Frith, U. (2006). The neural basis of mentalizing. *Neuron, 50,* 531–534.

Kanwisher, N., & Yovel, G. (2006). The fusiform face area: A cortical region specialized for the perception of faces. *Philosophical Transactions of the Royal Society of London. Series B, Biological Sciences, 361,* 2109–2128.

Mitchell, J. P. (2009). Social psychology as a natural kind. *Trends in Cognitive Science, 13,* 246–251.

Young, L., & Saxe, R. (2009). Innocent intentions: A correlation between forgiveness for accidental harm and neural activity. *Neuropsychologia, 47,* 2065–2072.

Comparing social attention in autism and amygdala lesions: Effects of stimulus and task condition

Elina Birmingham[1,2], Moran Cerf[3], and Ralph Adolphs[2]

[1]Faculty of Education, Simon Fraser University, Burnaby, BC, Canada V5A 1S6
[2]Division of the Humanities and Social Sciences, California Institute of Technology, Pasadena, CA, 91125
[3]Division of Biology, California Institute of Technology, Pasadena, CA 91125, USA

The amygdala plays a critical role in orienting gaze and attention to socially salient stimuli. Previous work has demonstrated that SM a patient with rare bilateral amygdala lesions, fails to fixate and make use of information from the eyes in faces. Amygdala dysfunction has also been implicated as a contributing factor in autism spectrum disorders (ASD), consistent with some reports of reduced eye fixations in ASD. Yet, detailed comparisons between ASD and patients with amygdala lesions have not been undertaken. Here we carried out such a comparison, using eye tracking to complex social scenes that contained faces. We presented participants with three task conditions. In the Neutral task, participants had to determine what kind of room the scene took place in. In the Describe task, participants described the scene. In the Social Attention task, participants inferred where people in the scene were directing their attention. SM spent less time looking at the eyes and much more time looking at the mouths than control subjects, consistent with earlier findings. There was also a trend for the ASD group to spend less time on the eyes, although this depended on the particular image and task. Whereas controls and SM looked more at the eyes when the task required social attention, the ASD group did not. This pattern of impairments suggests that SM looks less at the eyes because of a failure in stimulus-driven attention to social features, whereas individuals with ASD look less at the eyes because they are generally insensitive to socially relevant information and fail to modulate attention as a function of task demands. We conclude that the source of the social attention impairment in ASD may arise upstream from the amygdala, rather than in the amygdala itself.

Keywords: Social attention; Amygdala; Autism; Scene perception; Gaze selection.

We evaluate the social meaning of scenes that contain people by directing our attention and gaze to socially relevant features. We integrate the significance of these features in the context of the entire image, and link it to our stereotypes, memory representations, and reasoning about what we see. Despite the complexity of the psychological processes and neural substrates that underlie this ability, there has been much progress regarding specific components. Healthy individuals are remarkably reliable in how they direct their gaze to important parts of a social scene. In particular, there is a strong bias to look at people's faces (Yarbus, 1967), and particularly at the eyes within faces (Birmingham, Bischof, & Kingstone, 2008a, 2008b; Henderson, Williams, & Falk, 2005; Pelphrey et al., 2002; Walker-Smith, Gale, & Findlay, 1977).

Faces and eyes attract our attention from birth (Johnson, 2005), and attract the attention of many other highly visual species (Emery, 2000) including

Correspondence should be addressed to: Elina Birmingham, Faculty of Education, Simon Fraser University, RCB 5246, 8888 University Drive, Burnaby, BC Canada V5A 1S6. E-mail: ebirming@sfu.ca

Special thanks to Brian Cheng and Catherine Holcomb for helping with subject recruitment and data analysis, to Prof. Christof Koch for use of the Eyelink 1000, and to Drs Lynn Paul and Dan Kennedy for help with diagnoses and assessments of the participants. This work was supported in part by grants from the Simons Foundation and the National Institute of Mental Health to R.A.; and a fellowship from the Natural Sciences and Engineering Research Council of Canada to E.B.

nonhuman primates (Keating & Keating, 1982). A wealth of information about other persons can be gleaned from looking at their eyes. For instance, the eyes can provide clues about people's emotional or mental state, the focus of their attention, and their communicative intent. Thus, it is not surprising that, when we look at faces, much of our gaze is directed to the eyes (Yarbus, 1967). Furthermore, face detection is speeded by the presence of eyes (Lewis & Edmonds, 2003). In fact, even 3-day-old babies seem to be tuned into the eye region of faces, preferring to look at faces with direct eye contact as opposed to faces with averted gaze (Farroni, Csibra, Simion, & Johnson, 2002). Consistent with this, it has been demonstrated that infants prefer to look at faces with eyes open rather than eyes closed (Batki, Baron-Cohen, Wheelwright, Connellan, & Ahluwalia, 2000). These findings and many others resonate with the proposal that, just as there is thought to be a module for face processing (Yovel & Kanwisher, 2004), there may be a module for detecting eyes in the social environment (Baron-Cohen, 1995). This strong interest in the eyes of faces is an important component of *social attention* (Birmingham & Kingstone, 2009).

In the following section, we provide a brief overview on what is known about the abnormalities of social attention associated with autism spectrum disorders (ASD). We then consider the role of the amygdala in directing attention to the eyes of faces, and discuss the proposal that amygdala dysfunction underpins social attention abnormalities in ASD. Finally, we note the absence of direct comparisons between patients with amygdala damage and individuals with ASD on identical tasks and stimuli—comparisons that are essential to determining the link between amygdala dysfunction and ASD. The present study sets out to provide such a comparison.

ASD AND SOCIAL ATTENTION

Many developmental disorders exhibit prominently abnormal levels of eye contact, including fragile X syndrome (Farzin, Rivera, & Hessl, 2009), Williams syndrome (Riby & Hancock, 2008), and, most notably, ASD. Both clinical and anecdotal evidence suggests that reduced or absent eye contact is a pervasive feature of ASD (Kanner, 1943; Lord, Rutter, & Le Coutier, 1994; Lord et al., 2000). Due to the fundamental role of eye contact in social communication, it is perhaps not surprising that many behavioral interventions attempt to teach individuals with ASD to make eye contact (e.g., Hwang & Hughes, 2000). Several prominent studies have documented abnormal eye gaze to

static images of faces or to films in people with ASD (Klin, Jones, Schultz, Volkmar, & Cohen, 2002; Pelphrey et al., 2002). It has also been shown that there is a failure to make use of information from the eye region of faces in order to make social judgments, in both ASD (Spezio, Adolphs, Hurley, & Piven, 2007a) and in parents of people with autism who have the broad autism phenotype (Adolphs, Spezio, Parlier, & Piven, 2008).

Despite these findings to date, two key questions remain. The first concerns the reliability of the social attention impairment in ASD. The second concerns the neural mechanisms underlying these social attention impairments.

Reliability of the social attention impairment in ASD

Although reduced eye contact in ASD is a clinically and anecdotally well-known phenomenon, and is a primary focus of early behavioral intervention, experimental studies aimed at quantifying this phenomenon have in fact yielded rather mixed results. Using eye-tracking methodology, some studies have found reduced eye contact in ASD (e.g., Klin et al., 2002; Pelphrey et al., 2002; Sterling et al., 2008), but other studies have failed to document impaired eye fixation (e.g., Neumann, Spezio, Piven, & Adolphs, 2006; van der Geest, Kemner, Camfferman, Verbaten, & van Engeland, 2002). Even more puzzling is the fact that some studies using the same subjects but different tasks or stimuli found a reduction in eye contact in one condition, but not in others (Neumann et al., 2006; Speer, Cook, McMahon, & Clark, 2007), suggesting that the deficit is only revealed in particular contexts. That is, the lack of reliability of social attention abnormalities in ASD may be attributable to variance due to individual differences in the subjects, differences in the stimuli used and their context, and differences in task demands. One major aim of our study was thus to elucidate these sources of variability, in particular the effect that different tasks would have on eye gaze when people are viewing the same stimuli.

Neural substrates of abnormal social attention in ASD

A second open question that has perhaps attracted even more attention is to understand the neural substrates that might be responsible for abnormal social attention in ASD. Two key brain regions have emerged as playing important roles: the superior temporal sulcus (Pelphrey, Morris, & McCarthy, 2005b), and the amygdala (Baron-Cohen et al., 2000; Howard et al., 2000). While these are only pieces of a larger network for social cognition that is likely to be impaired in ASD (Damasio & Maurer, 1978; Pelphrey, Adolphs, & Morris, 2005a), they have been the focus of most studies. The amygdala, in particular, has been closely tied to ASD and to abnormal eye gaze in a number of studies. For instance, morphometric studies of postmortem brains have shown that the amygdala is structurally abnormal in ASD (Kemper & Bauman, 1993; Schumann & Amaral, 2006). This structural abnormality of the amygdala is also evident from magnetic resonance imaging (MRI) quantification of living individuals (Howard et al., 2000; Schumann et al., 2004). Individuals with ASD have also been shown to have abnormal amygdala activation when fixating the eyes of faces (e.g., Dalton et al., 2005), buttressing the hypothesis that one of the key neural structures responsible for impaired eye gaze in ASD is the amygdala.

The amygdala in eye gaze

The amygdala has long been known to play critical roles in social behavior (Kluver & Bucy, 1939), emotional learning (LeDoux, 1993), aspects of face processing (Adolphs, Tranel, Damasio, & Damasio, 1994), attention (Holland & Gallagher, 1999), and reward processing (Murray, 2007). Pulling these diverse findings together into a single story has been challenging, but a key feature appears to be evaluating the social meaning of stimuli in a context-dependent way (Adolphs, 2010).

The amygdala plays a central role in processing information about eyes in faces. It is activated by eyes in functional (f)MRI studies (Kawashima et al., 1999; Whalen et al., 2004), and lesions of the amygdala impair eye gaze onto photographs of faces (Adolphs et al., 2005) as well as onto the faces of real people (Spezio, Huang, Castelli, & Adolphs, 2007b). Amygdala damage also impairs social judgments about people's faces within complex scenes (Adolphs & Tranel, 2003). Just as in autism (Baron-Cohen, Wheelwright, Hill, Raste, & Plumb, 2001), subjects with bilateral amygdala lesions are impaired in gleaning the social meaning of eyes alone (Adolphs, Tranel, & Baron-Cohen, 2002).

Most informative have been studies in a rare patient with complete bilateral lesions of the amygdala, known as SM (Adolphs, 2010; Buchanan, Tranel, & Adolphs, 2009). Recent research suggests that SM shows lack of interest in the eyes because her attention is attracted to other regions of the face, such as the nose (Kennedy & Adolphs, 2010) or mouth (Spezio et al., 2007b). In contrast, SM is able to gaze at the eyes in faces with no difficulty when instructed to do so (Adolphs et al., 2005). Kennedy and Adolphs (2010) examined the hypothesis that SM's endogenous interest in eyes is largely intact by restricting SM's visual input to the center of her gaze ("gaze-contingent" viewing), essentially eliminating competing visual information. When viewing faces with this restriction, SM looked at the eyes to the same extent as controls, suggesting that in unrestricted viewing conditions her fixations are captured away from the eyes by other regions of the face. This hypothesis is bolstered by the finding that SM's fixations to the eyes of whole faces are most abnormal in the first fixation, and become more normative in later fixations when endogenous control can take over (Kennedy & Adolphs, 2010). Taken together, these results suggest that the amygdala is normally involved in guiding fixations early on to regions of the face that are socially salient; that is, the eyes.

Comparing ASD and amygdala lesions

While there are thus considerable similarities in how people with ASD and with amygdala lesions fail to look at eyes in faces, there is also reason to suspect that they might do so for different reasons. For instance, Neumann et al. (2006) suggested that individuals with ASD show an abnormal endogenous bias away from the eyes and toward the mouth, rather than an exaggerated sensitivity to the bottom-up saliency of the features of the mouth. Specifically, individuals with ASD looked at the mouth region of "bubbles" faces even when the mouth was low contrast, suggesting an endogenous strategy for allocating visual attention to the mouth region. In contrast, as we reviewed above, it is thought that SM looks less at the eyes because she has abnormally increased exogenous attention capture by the mouth.

To elucidate the possible mechanistic role of amygdala dysfunction in the abnormal social attention of ASD, we undertook direct comparisons with SM. To test the roles of endogenously driven and exogenously driven attentional effects, we examined the effects of task (endogenous effects) and of stimuli (exogenous effects). To date, there have been no direct comparisons of social attention processes in ASD and patients with amygdala lesions. We expected to find a reduction in time spent looking at the eyes in both populations, but for different reasons. We hypothesized that SM would show normal endogenous, but abnormal exogenously driven attention to social features, whereas people with ASD would show the converse impairment. In particular, we expected that SM would fixate socially less salient regions (e.g., the mouth) in stimuli, especially early on, but show a relatively normal modulation of fixation to the eyes when driven by endogenous task demands. By contrast, we expected that individuals with ASD would show a reduction in fixations to the eyes that would be relatively stable across task conditions, due to impaired endogenous attention to social features. By this reasoning, we would expect to find the greatest abnormality in fixations to the eyes in ASD on those tasks for which the eyes are normally highly informative.

Present study

To compare SM to individuals with ASD, we chose to present participants with a variety of real-world social scenes under different task conditions, using a previously validated set of stimuli and task instructions (Birmingham et al., 2008b; Smilek, Birmingham, Cameron, Bischof, & Kingstone, 2006). In particular, we were interested in determining how fixation preferences change during a task that forces observers to process social attention information, relative to more neutral or general tasks. Previous research has validated such an approach with typically developing individuals, demonstrating that when asked to infer where other people in a scene are directing their attention, observers enhance their fixations to the eyes of those people, relative to when they are simply describing a scene (Birmingham et al., 2008b; Smilek et al., 2006). Importantly, this enhanced inspection of the eyes is driven by an understanding that eyes provide important social attention information, subjective reports revealing that observers mention "eye gaze" and "looking" more often when inferring attentional states (even though the instructions do not explicitly direct observers to report on the eye gaze of people in the scene; Smilek et al., 2006).

We chose this task manipulation to compare SM and individuals with ASD to two matched control groups, both because the task manipulation has been validated in previous work (Birmingham et al., 2008b; Smilek et al., 2006) and because it offers a rich interplay between competition for attention driven by the social saliency of stimuli shown in the scenes, and the demands made by the different tasks. Notably, fixations of typical observers to these scenes cannot be explained by visual saliency (Birmingham, Bischof, & Kingstone, 2009).

Finally, we expected a fair degree of variability in fixation patterns, as a function of both the stimulus and the participant. To explore these sources of variance,

we conducted analyses exploring viewing patterns as a function of image, and performed exploratory correlational analysis on the ASD group between time spent looking at eyes and various neuropsychological variables.

METHOD

Participants

Subject SM

SM is a 43-year-old (at the time of test) woman with bilateral amygdala damage. She suffers from a rare genetic disorder, Urbach–Wiethe disease (Hofer, 1973), which has led to complete bilateral calcification and atrophy of her amygdalae (Adolphs et al., 1994; Tranel & Hyman, 1990).

SM controls

The controls were 10 neurologically and psychiatrically healthy females of comparable age (mean age: 43.7 years; range 38–51). There was no significant difference between SM and her respective controls in age, verbal performance, or full-scale IQ ($p > .05$ for each comparison, modified two-tailed t-test; Crawford & Garthwaite, 2002).

ASD subjects

There were nine ASD subjects with high- functioning autism (mean age: 31.6 years; range 20–55). Participants with ASD were recruited from our participant database. All ASD participants met DSM-IV diagnostic criteria for autism or Asperger syndrome, and met the cutoff scores for autism or Asperger syndrome on both the Autism Diagnostic Interview–Revised (ADI–R; Lord et al. 1994) and the Autism Diagnostic Observation Schedule (ADOS, Module 4; Lord et al., 2000). There were two participants in the ASD group for whom the AD I–R was not available; for these participants, ADOS scores alone were used.

ASD controls

The ASD controls were five neurologically and psychiatrically healthy individuals with no family history of ASD (mean age: 32.6 years; range 18–48). ASD controls were recruited through local advertisement, and were screened to be similar in demographic characteristics to the ASD group. There was no significant

TABLE 1
Participant Demographic and Clinical Data

	SM (n = 1)	SM Controls (n = 10)	ASD (n = 9)	ASD Controls (n = 5)
Age (years)	43.0	43.7 (3.8) [38-51]	31.6 (12.2) [20-55]	32.6 (12.6) [18-48]
Males, Females	0,1	0,10	8,1	5,0
Full Scale IQ	88.0	107.4 (10.8)	108.8 (14.0)	97.0 (17.0)
Performance IQ	95.0	105.9 (11.6)	105.7 (14.5)	93 (22.6)
Verbal IQ	86.0	107.2 (10.9)	112.7 (17.1)	103 (15.9)
ADI-R Social (Cut-off = 10)			20.0 (4.2) [12-25]	
ADI-R Communication (Cut-off = 8)			15.0 (3.6) [10-20]	
ADI-R Stereotypy (Cut-off = 3)			5.1 (2.3) [2-9]	
ADOS Communication (Cut-off = 3/2)			4.0 (1.2) [2.0-6.0]	
ADOS Reciprocal Social Interaction (Cut-off = 6/4)			8.3 (3.5) [5.0-17.0]	
ADOS Communication + Social (Cut-off = 10/7)			12.3 (4.4) [9.0-23.0]	
ADOS Imagination/Creativity			0.6 (0.5) [0.0-1.0]	
ADOS Stereotyped behaviors and restricted interests			1.0 (1.0) [0.0-3.0]	

Data are given as mean (SD) and, for age, ADI-R and ADOS scores, ranges [in square brackets]. Higher scores on the ADOS and ADI-R are indicative of more severe impairment. For the ADOS, Cut-offs indicate cut-off values for Autism/ASD.

difference between the ASD group and ASD controls in age, verbal performance, or full-scale IQ ($p > .1$ for each comparison, Wilcoxon rank-sum test). See Table 1.

Participants had normal or corrected-to-normal vision, and all gave informed consent to participate in the studies under a protocol approved by the Institutional Review Board of the California Institute of Technology.

Apparatus

Eye movements were measured with an Eyelink 1000 eye tracker (SR Research, Ltd, Hamilton, Ontario, Canada), which has an accuracy of 0.5° of visual angle and sampling rate of 1,000 Hz. We tracked the right eye only.

Stimuli

Full color digital photos were used (Birmingham et al., 2008b). Four sets of five images (20 total) were presented. Example images can be found in Figure 1 Set 1 took place in a waiting area, in which one or three people sat on a couch or a chair; Set 2 took place in another waiting area, again with one or three people sitting on a couch and a chair. Sets 3 and 4 took place in a meeting room, in which one or three people sat at a table. Images were presented on a CRT monitor

(120 Hz), using the Psychophysics Toolbox (Brainard, 1997; Pelli, 1997) and Eyelink Toolbox (Cornelissen, Peters, & Palmer, 2002) for Matlab (Mathworks, Inc., Natick, MA, USA). Image size was 28° × 21° at the viewing distance of 80 cm, and image resolution was 1024 × 768 pixels. Four of the images were "filler" images designed to maintain participants' interest, and all contained one person doing something unusual (e.g., wearing a frisbee on his head). These four images were not included in the analysis. Additionally, two images were excluded from analysis because the mouth regions were not visible (because the actors were drinking from a cup).

Procedure

Participants were seated in a dark room, and were placed in a chin rest to ensure a fixed distance from the display computer screen. Participants were told that they would be shown several images, each one appearing for 15 s. They were informed that these images would be repeated and that they would have to complete three different tasks:

1. *Neutral task:* What kind of room is this? Explain.
2. *Describe task:* Describe the picture.
3. *Social Attention task:* Describe where people in the picture are directing their attention. How do you know?

Figure 1. Left column: Samples of the social stimuli used. Right column: the different regions of interest (eyes, mouths, heads, bodies, foreground objects, and background.). From top to bottom: sample images from Set 1, Set 2, Set 3, and Set 4.

Image and task order was randomized within task blocks of 10 images. Before the experiment, a calibration procedure was conducted. At the beginning of each trial, a fixation cross was displayed in the center of the computer screen in order to correct for drift in gaze position. Participants were instructed to fixate this cross to start the trial. Each trial required fixation calibration within 0.3° of the pre-trial fixation

cross. After image presentation, participants gave an oral response to the task instruction (e.g., describe the picture), which was recorded onto a Sony digital MP3 recorder. Participants pressed a button to proceed to the next trial.

As an additional accuracy check for eye-tracker calibration, participants were presented with an image containing nine squares, numbered 1–9. Participants were asked to fixate each square, in order from 1 to 9, as accurately as possible. This image appeared at the beginning, halfway point, and end of the experiment. The fixation plots for this image were further used to verify that there was no drift in the calibration.

Analysis

A total of 14 images per task condition (14 images × 3 tasks = 42 total trials) were analyzed. For each image, an outline was drawn around each region of interest (e.g., "eyes"), and each region's pixel coordinates were recorded. We defined the following regions in this manner: eyes, mouths, heads (excluding eyes and mouths), bodies (including arms, torso, and legs), foreground objects (e.g., tables, chairs, objects on the table), and background (e.g., walls, shelves, items on the walls) (Figure 1). Regions were pooled such that there was one composite "eye" region made up of all eye regions, one composite "head" region made up of all head regions, etc. (in cases where the image contained numerous entities).

1. Time spent looking at each region

To determine what regions were of most interest to observers, we computed the total time spent fixating each region, divided by total viewing time (*proportion of time in region*). To compare SM to her controls (*SM controls*), we used Crawford and Garthwaite's (2002) modified *t*-test for comparing an individual's score on a single test with the score of a normative or control sample. Using this approach, we compared the mean proportion of time in a region for SM to that for SM controls. Similar analyses were used to compare SM to the ASD group. To examine whether SM's preference for eyes changed between tasks, we performed paired tests on her fixation time for eyes, paired by image. Finally, to examine the influence of task on attention to eyes for SM controls, we performed an ANOVA on the proportional dwell time for eyes, using a within-subjects factor of task (Neutral, Describe, Social Attention), followed by post hoc pairwise comparisons (Tukey–Kramer multiple comparisons test; alpha = .05).

To compare the ASD group with their control group (*ASD controls*), the data were submitted to mixed 2 × 3 repeated measures ANOVAs on each region separately with Group (ASD vs. ASD control) as a between-subjects factor and Task (Neutral, Describe, Social Attention) as the within-subjects factor. Pairwise comparisons (Tukey–Kramer multiple

comparisons test, alpha = .05) were also performed post hoc to confirm differences between means.

Initial fixation proportion

To determine where observers' initial saccades landed in the visual scene, we computed the number of first, second, third, fourth, and fifth fixations that landed in a region (first five fixations). As with the proportional dwell time data, SM was compared to SM controls and the ASD group by modified *t*-tests; ASD and ASD controls were compared by mixed ANOVAs, followed by Tukey–Kramer post hoc comparisons (alpha = .05).

RESULTS

SM vs. SM controls

1. Time spent looking at each region

SM spent a considerably greater proportion of viewing time looking at the mouth region than did SM controls (SM: .21, controls: .10, Crawford's modified *t*-test, $t(9) = 1.84$; $p < .05$; see Figure 2a). There was also a clear trend for SM to spend a smaller proportion of time looking at the eyes than her controls (SM: .04, controls: .17), although this difference did not achieve significance due to the large variance in the SM controls (Crawford's modified *t*-test, $p > .10$).

In contrast, SM did increase her dwell time on the eye region of faces when the task involved inferring social attention information (Figure 2b). Paired *t*-tests confirmed that SM spent more time looking at eyes in the Social Attention task (.10) than in the Neutral task (.02), $t(14) = 2.20$, $p < .05$ (one-tailed), and the Describe task (.02), $t(14) = 1.93$, $p < .05$ (one-tailed). For SM controls, an ANOVA on the time proportion for the eye region revealed a similar effect of task, $F(2, 18) = 16.74$, $p < .0001$, with Tukey–Kramer post hoc comparisons revealing that the eyes were fixated for longer during the Social Attention task (.26) than in the Neutral (.09) and Describe (.16) tasks ($p < .05$).

2. Initial fixation proportion

SM differed most notably from SM controls in her initial fixations, which were predominantly onto the mouth rather than the eyes (see Figure 3). Analyzed individually (without correction for multiple comparisons), the second and third fixations

A

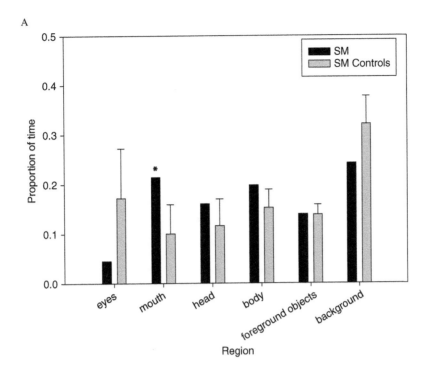

B

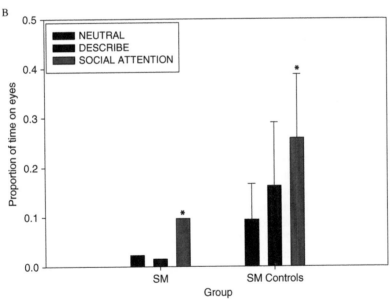

Figure 2. Fixations in patient SM. (A) Proportional dwell time for SM and her control group (means and *SD*) shown for each of the regions of interest. SM showed a trend to look less at the eyes and looked significantly more at the mouth than SM controls (*p* < .05). (B) Task effects: both SM and SM controls increased their dwell times on the eyes in the Social Attention task relative to the Describe and Neutral tasks (*p* < .05).

showed a significantly higher proportion of mouth fixations for SM than for controls; second fixation: Crawford's modified *t*-test, *t*(9) = 2.90, *p* < .01; third fixation: *t*(9) = 2.356, *p* < .05, one-tailed, but this was not significant for subsequent fixations (fixations 4 and 5, *p* > .05). There was also a trend for SM to look less at the eyes in these initial fixations, although this did not reach significance (*p* > .10).

ASD vs. ASD controls

1. Time spent looking at each region

Our Group × Task ANOVAs of proportional dwell time on each region initially revealed no differences between ASD and controls, as reflected in nonsignificant effects of group, eyes: *F*(1, 12) = 1.32; *p* > .10; heads: *F*(1, 12) = 4.57, *p* > .05; mouth:

13

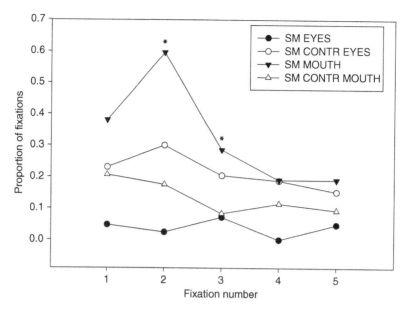

Figure 3. Proportion of initial fixations on eyes and mouth, for SM and SM controls (CONTR). SM committed significantly more of her second and third fixations to the mouth than did SM controls ($p < .05$).

$F < 1$; bodies: $F < 1$; foreground objects: $F < 1$; background: $F(1, 12) = 1.53$, $p > .10$. This suggested that the ASD and ASD control groups distributed their fixations across the scene in a broadly similar way. However, there were some hints that the ASD group was looking less at the eyes—but with a very large variance (see Figure 4a).

We also noticed a high degree of variability among the images in terms of time spent looking at the eyes, not an unexpected finding given the complexity and heterogeneity of our stimuli. To investigate fixations for each stimulus separately, we submitted the data for the ASD and ASD controls to a mixed repeated-measures ANOVA with Group, Task, and Image (random factor) as factors. This analysis revealed a significant effect of Image, $F(13, 156) = 9.27$, $p < .0001$, indicating that time spent looking at the eyes varied among the images (see Figure 4b). This effect alone was not surprising, as images varied on eye region size, and this alone could account for variations in time spent looking at the regions. However, a significant Group × Image interaction, $F(13, 156) = 2.28$, $p < .01$, could not be explained by variations in region size. Rather, it suggested that certain images produced greater group differences in time spent looking at the eyes than others. Furthermore, a significant Group × Task × Image interaction, $F(26, 312) = 1.80$, $p < .05$, suggested that this effect of Image depended on the task given to participants, as we explore below. Interestingly, these interactions with Image were absent in the comparison of SM to SM controls ($F < 1$).

In order to examine these stimulus and task effects more closely, we ranked the images according to how much time the ASD controls spent looking at the eyes (Figure 4b). We noticed that the images that tended to produce longer dwell times in the controls were from Sets 3 and 4. Interestingly, these images had in common one characteristic that differed from the images in Sets 1 and 2: The people in Sets 3 and 4 were seated around a table, whereas the people in Sets 1 and 2 were seated on couches or chairs with their full bodies visible. We constrained our analysis to images in Sets 3 and 4 in order to examine whether these images produced different gaze behavior for the ASD group. We found that for the images in Sets 3 and 4, there was a strong trend for the ASD group to look less at the eyes than ASD controls, $F(1, 12) = 3.65$, $p = .08$. Furthermore, while controls spent proportionally more time looking at the eyes in the Social Attention task than the other tasks, the ASD group did not. This observation was confirmed by a Group × Task interaction, $F(2, 24) = 7.12$, $p < .01$, and Tukey–Kramer pairwise comparisons between task conditions within each group (see Figure 4c). That is, for ASD controls, the proportional dwell time on eyes was significantly higher in the Social Attention task (.49) than in the Describe (.30) and Neutral tasks (.17), $p < 0.05$. In contrast, there was no difference between task conditions for the ASD group (Social Attention: .21; Describe: .18; Neutral: .12, $p > .05$). In other words, for this subset of images the ASD group failed to enhance their inspection of the eyes when inferring social attention information.

14

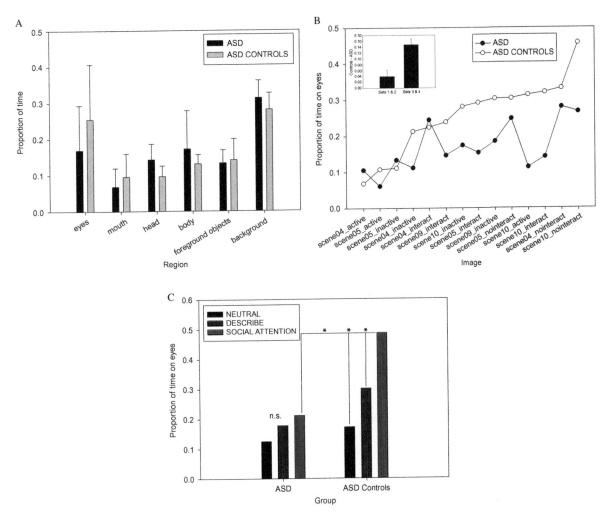

Figure 4. Fixations in ASD. (A) Proportional dwell time for each region of interest, for ASD and ASD controls (means and *SD*). (B) Dwell time broken down by individual stimuli. Sets 3 and 4 produced larger group differences than Sets 1 and 2 ($p < .05$). (C) Group and task effects for proportion of dwell time on eyes for images in Sets 3 and 4. Whereas ASD controls enhanced their fixations to the eyes in the Social Attention task ($p < .05$), the ASD group did not.

This also resulted in a significant group difference for eyes in the Social Attention task (ASD: .21 vs. ASD controls: .49 Tukey–Kramer, $p < .05$), but no significant group difference for eyes in the Describe task (ASD: .17 vs. ASD controls: .30, Tukey–Kramer, $p > .05$) or the Neutral task (ASD: .12 vs. ASD controls: .17, Tukey–Kramer, $p > .05$). This effect was specific to the eyes; that is, there were no group differences in time spent looking at mouths ($p > .10$).

2. Initial fixation proportion

A Group × Task ANOVA on the initial fixations to eyes for images in Sets 3 and 4 revealed significant group differences in the proportion of second fixations that fell on eyes. The ASD group had fewer second fixations to the eyes than did the ASD controls, ASD: .26, controls: .48; $F(1, 12) = 4.84$, $p < .05$

(see Figure 5). Further, there was a trend toward a Group × Task interaction, $F(2, 24) = 2.64$, $p = .09$. This suggested that while the ASD controls tended to commit more of their second fixations to the eyes in the Social Attention task (Social Attention: .67, Describe: .40, Neutral: .37), the ASD group did not (Social Attention: .28, Describe: .22, Neutral: .28), although this was not significant, Tukey–Kramer pairwise comparisons, $p > .05$. There were no group differences for the mouth region, $p > .10$.

Intersubject variability

Finally, having examined both effects of stimuli and of task, we explored a third and final source of variability: individual differences in the participants. We found considerable individual differences, both between the two control groups and at an individual level. For

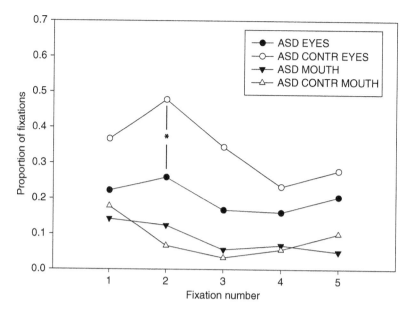

Figure 5. Proportion of initial fixations on eyes and mouths, for ASD and ASD controls (CONTR), for images in Sets 3 and 4. The ASD group looked less at the eyes than ASD controls by the second fixation ($p < .05$).

instance, the ASD control group, who was predominantly male and with a mean age of 31.6, spent slightly more time on the eyes (.25) than the SM control group (.17), who were all female and with a mean age of 43.7 (although the group difference for eyes was nonsignificant, $p > .10$). There was also large variation within each group—for instance, there were subsets of participants in each group who consistently failed to look at the eyes, and others who quite consistently showed a preference to look at the eyes. In the ASD group, we also explored correlations between individual differences in eye gaze and neuropsychological variables. Although exploratory in nature due to our small sample size, we uncovered a highly significant positive correlation between the time spent looking at eyes and scores on the communication subscale of the ADI–R (Pearson's $r = .91$, $t(6) = 4.5$, $p < .025$, two-tailed, Bonferroni corrected). That is, participants with ASD who were *more* impaired in communication (higher scores on ADI–R) spent *more* time looking at the eyes (see Supplementary Figure A). Conversely, there was a significant negative correlation in the ASD group between dwell time on the mouth and the communication subscale of the ADI–R (Pearson's $r = -.83$, $t(7) = 2.7$, $p < .025$, two-tailed, Bonferroni corrected), suggesting that individuals who were more impaired in communication looked *less* at the mouth (see Supplementary Figure B). Although we interpret these findings with caution, they are broadly consistent with previous findings of correlations between eye and mouth gaze and communication skill (e.g., Norbury et al., 2009).

SM vs. ASD

Although SM was not matched to the ASD group in terms of gender (age and full-scale, verbal, and performance IQ did not differ statistically between SM and the ASD group: Crawford's modified two-tailed t-tests, $p > .10$), the comparison of fixation patterns between SM and ASD was of particular interest, given our hypotheses about the role of the amygdala in autism. Below, we compare SM to the ASD group on time spent looking at each region, and initial placement of fixations.

1. Time spent looking at each region

SM spent considerably more time looking at the mouth region than did the ASD group, SM: .21, ASD: .07, Crawford's modified t-test, $t(8) = 2.75$, $p < .05$ (see Figure 6a). The increase in time on the mouth for SM was driven mostly by the Describe task, SM: .31, ASD: .07, $t(8) = 5.40$, $p < .001$, and less so by the Social Attention task, SM: .21, ASD: .09, $t(8) = 1.25$, $p > .10$, and the Neutral task, SM: .13, ASD: .05, $t(8) = 1.42$, $p = .10$. Although there was also a trend for SM to look less at the eyes than the ASD group, this overall difference was not significant, SM: .04, ASD: .17, $t(8) = t = -0.95$, $p > .10$, nor was it significantly different in any of the tasks: Social Attention, SM: .10; ASD: .22, $t(8) = -0.811$, $p > .10$; Describe, SM: .02; ASD: .18, $t(8) = -1.23$, $p > .10$; and Neutral, SM: .02; ASD: .11, $t(8) = -0.69$,

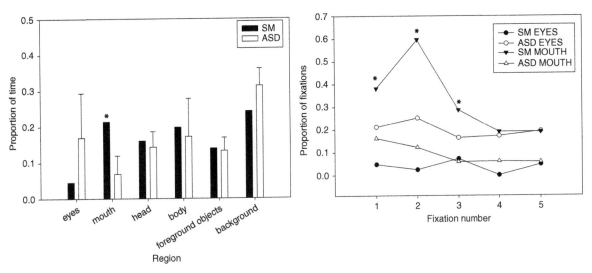

Figure 6. Comparing SM to ASD. *Left panel:* proportional dwell time for SM and the ASD group (means and *SD*) shown for each of the regions of interest. SM showed a trend to look less at the eyes and looked significantly more at the mouth than did the ASD group ($p < .05$). *Right panel:* proportion of initial fixations on eyes and mouth for SM and ASD. SM committed significantly more of her first, second, and third fixations to the mouth than did the ASD group ($p < .05$).

$p > .10$. SM did not differ from the ASD group on any other region.

2. Initial fixation proportion

The ASD group was not as likely as SM to commit their initial fixations to the mouth (see Figure 6b). Analyzed individually, the first, second, and third fixations showed a significantly higher proportion of mouth fixations for SM than for the ASD group—respectively: Crawford's modified *t*-test, $t(8) = 2.08$, $p < .05$; $t(9) = 4.15$, $p < .01$; $t(8) = 2.64$, $p < .05$, all one-tailed; and marginally more for subsequent fixations: fixation 4, $p = .06$; and fixation 5, $p = .10$. There was also a trend for SM to look less at the eyes than the ASD group in these initial fixations, although this did not reach significance ($p > .10$).

GENERAL DISCUSSION

Consistent with prior findings with the same stimuli and task (Birmingham et al., 2008b), the controls showed an expected high proportion of fixations onto the faces, and in particular the eyes, of people in the scenes. Three notable sources of variance were apparent. First, viewers looked more at the eyes in the stimuli when the task condition required more social attention. Second, there was variability due to the stimuli, with some drawing gaze more to the eyes than others. Third, there were substantial individual differences in the participants themselves.

SM looked noticeably less at the eyes and instead more at the mouth, findings consistent with prior studies in this individual (Adolphs et al., 2005). Moreover, this impairment was most noticeable on the first few fixations. By contrast, she did significantly increase her gaze to the eyes when the task required more social attention (Figure 2b). This supports our hypothesis that SM had impaired capture of attention by the social features shown in the stimuli, but intact endogenous modulation of eye gaze driven by task demands.

In the ASD group, we also found a nonsignificant, small trend towards looking less at the eye region of faces, but were struck by the very large variability. We therefore next teased apart the possible sources of this variability. First, we analyzed the results per individual stimulus, and found that some images resulted in much larger and more reliable group differences than others. Secondly, we found a significant group by stimulus by task interaction. Those stimuli showing the largest group differences in eye fixations did so because the ASD group failed to increase their eye fixations for those stimuli when the task required social attention. This finding supports our hypothesis that the ASD group had impaired endogenous modulation of eye gaze in the Social Attention task.

In addition to supporting general processing hypotheses about social attention deficits following amygdala lesions and in ASD, our findings urge caution in the interpretation of results that collapse across stimuli and/or tasks. While exploratory in nature, our analysis of interactions with image suggests that social attention deficits in ASD may be constrained to certain stimuli, a finding that warrants further investigation to establish the specific effect. We suggest that future studies pay particular attention to task demands, to differences among stimuli, and to individual differences in the participants. The large individual differences we observed were not entirely unexpected, given that we were presenting complex scenes instead of isolated faces. We were nonetheless surprised at how much variability there was in terms of time spent looking at eyes, even in our control groups, given previous reports in which the preference for eyes was striking and consistent across observers (e.g., Birmingham et al., 2008a, 2008b). It is possible that, as our control groups came from the general population (i.e., they did not come from a relatively homogeneous undergraduate subject pool), factors such as gender, age, IQ, and personality may have contributed to individual differences in preferences for looking at the eyes. Prior studies have shown that the use of eye-gaze information is influenced by the gender (Bayliss et al., 2005) and age of both participant and stimulus face (Slessor, Laird, Phillips, Bull, & Filippou, 2010), as well as aspects of social (e.g., Klin et al., 2002; Speer et al., 2007) and cognitive (Norbury et al., 2009) functioning.

Comparing ASD and amygdala lesions

Given the intense current interest in the amygdala's role in psychiatric illness, and in particular its contribution to the social impairments in autism (cf. the introduction), a particularly valuable aspect of our study was the opportunity for direct comparisons on the same task and stimuli. We found some global similarities in that both SM and the ASD group looked less at the eyes (although this did not reach statistical significance due to the large variance in the controls), but were more struck by the differences. These can be summarized as follows.

First, whereas SM showed a pronounced and significant increase in fixations to the mouth, especially at the earliest fixations, the ASD group showed no such effect. This difference is consistent with the idea that amygdala lesions impair the relative saliency of eyes compared to mouths in faces, whereas ASD has a less specific impairment in allocating social attention.

Second, whereas SM showed an intact modulation of eye gaze by the task, the ASD group showed a notable absence of such task-dependent modulation. Moreover, the effect in ASD was rather specific to certain images, resulting in a significant Group by Task by Image interaction for ASD vs. controls. This finding is consistent with the hypothesis that the amygdala is not essential for endogenous effects driven by task demands and the particular context of the stimuli,

whereas ASD shows impairments that are sensitive to precisely these factors.

Third, the amygdala does not appear to be involved in *general* social orienting, since SM had no reduction in gaze to people and faces. Instead, we suggest that the amygdala is a critical component of a "social saliency map" that specifies the relative importance of social features for allocating attention. This finding would be difficult to determine from studies presenting isolated faces to participants (e.g., Adolphs et al., 2005) and hinged on the complex scenes we used in which many features can compete for attention.

Our conclusions are qualified by our relatively small sample size and large variance. In particular, we could not identify what drove the individual variation in dwell time on the eyes in our control groups; this could have been due to personality differences, age, gender, aspects of cognitive functioning, or any combination of these factors. Future research should examine these variables and their relationship to eye gaze in more detail. We also suggest that researchers explore the effects of stimulus context on social attention. While past research has examined some aspects of scene content on eye gaze (e.g., Birmingham et al., 2008a), it remains an open question what other image characteristics (e.g., social dimensions such as trustworthiness of the people in the image, or the emotional valence and intensity of the image) may enhance the selection of eye-gaze information in neurotypical individuals. This is sure to be an exciting avenue for future research on social attention (e.g., Birmingham & Kingstone, 2009).

Implications for the amygdala's role in autism

Although the data show superficial similarities between our amygdala patient and ASD, the differences are more striking than the similarities. The findings argue that the amygdala is more critical for stimulus-driven (exogenous) social attention but not task effects (endogenous attention modulation), whereas the converse pattern of impairment is featured in ASD. One interpretation of this pattern of impairments is that the amygdala is indeed dysfunctional in autism, but that it is not the source of the dysfunction. Instead, other regions that process context- and task-dependent allocation of attention (plausibly drawing on the prefrontal cortex) may be dysfunctional, and may then pass on abnormal signals to the amygdala, which contains a saliency map driven by both exogenous and endogenous factors. This explanation would predict abnormal amygdala activation in fMRI studies of ASD, but argue that amygdala lesions by themselves do not duplicate the social dysfunction of autism—a conclusion consistent with the finding that amygdala lesions do not duplicate the social behavior of ASD either (Paul, Corsello, Tranel, & Adolphs, 2010). We thus suggest that future studies on the neural basis of the social impairments of ASD focus on the network of structures with which the amygdala is connected, rather than investigating the amygdala in isolation as the source of social cognitive impairments.

REFERENCES

Adolphs, R. (2010). What does the amygdala contribute to social cognition? *The Year in Cognitive Neuroscience, Annals of the New York Academy of Sciences, 1191*, 42–61.

Adolphs, R., Gosselin, F., Buchanan, T. W., Tranel, D., Schyns, P., & Damasio, A. R. (2005). A mechanism for impaired fear recognition after amygdala damage. *Nature, 433*(7021), 68–72.

Adolphs, R., Spezio, M. L., Parlier, M., & Piven, J. (2008). Distinct face-processing strategies in parents of autistic children. *Current Biology, 18*, 1090–1093.

Adolphs, R., & Tranel, D. (2003). Amygdala damage impairs emotion recognition from scenes only when they contain facial expressions. *Neuropsychologia, 41*(10), 1281–1289.

Adolphs, R., Tranel, D., & Baron-Cohen, S. (2002). Amygdala damage impairs recognition of social emotions from facial expressions. *Journal of Cognitive Neuroscience, 14*, 1264–1274.

Adolphs, R., Tranel, D., Damasio, H., & Damasio, A. (1994). Impaired recognition of emotion in facial expressions following bilateral damage to the human amygdala. *Nature, 372*, 669–672.

Baron-Cohen, S. (1995). *Mindblindness: An essay on autism and theory of mind.* Cambridge, MA: MIT Press.

Baron-Cohen, S., Ring, H. A., Bullmore, E. T., Wheelwright, S., Ashwin, C., & Williams, S. C. (2000). The amygdala theory of autism. *Neuroscience and Biobehavioral Reviews, 24*(3), 355–364.

Baron-Cohen, S., Wheelwright, S., Hill, J., Raste, Y., & Plumb, I. (2001). The "reading the mind in the eyes" test revised version: A study with normal adults, and adults with Asperger syndrome or high-functioning autism. *Journal of Child Psychology and Psychiatry, 42*, 241–251.

Batki, A., Baron-Cohen, S., Wheelwright, S., Connellan, J., & Ahluwalia, J. (2000). Is there an innate gaze module? Evidence from human neonates. *Infant Behavior and Development, 23*, 223–229.

Bayliss, A. P., di Pellegrino, G., & Tipper, S. P. (2005). Sex differences in eye gaze and symbolic cueing of attention. *Quarterly Journal of Experimental Psychology A, 58A*, 631–650.

Birmingham, E., Bischof, W. F., & Kingstone, A. (2008a). Social attention and real world scenes: The roles of action, competition, and social content. *Quarterly Journal of Experimental Psychology, 61*(7), 986–998.

Birmingham, E., Bischof, W. F., & Kingstone, A. (2008b). Gaze selection in complex social scenes. *Visual Cognition, 16*(2/3), 341–355.

Birmingham, E., Bischof, W. F., & Kingstone, A. (2009). Saliency does not account for fixations to eyes within social scenes. *Vision Research, 49*, 2992–3000.

Birmingham, E., & Kingstone, A. (2009). Human social attention: A new look at past, present and future investigations. *The Year in Cognitive Neuroscience, Annals of the New York Academy of Sciences 1156*, 118–140.

Brainard, D. H. (1997). *The psychophysics toolbox.* Spatial Vision, *10*, 433–436.

Buchanan, T. W., Tranel, D., & Adolphs, R. (2009). The human amygdala in social function. In P. W. Whalen & L. Phelps (Eds.), *The human amygdala* (pp. 289–320). New York, NY: Oxford University Press.

Cornelissen, F. W., Peters, E. M., & Palmer, J. (2002). The eyelink toolbox: Eye tracking within MATLAB and the psychophysics toolbox. *Behavioral Research Methods, Instrumentation and Computers, 34*, 613–617.

Crawford, J. R., & Garthwaite, P. H. (2002). Investigation of the single case in neuropsychology: Confidence limits on the abnormality of test scores and test score differences. *Neuropsychologia, 40*, 1196–1208.

Dalton, K. M., Nacewicz, B. M., Johnstone, T., Schaefer, H. S., Gernsbacher, M. A., Goldsmith, H. H., et al. (2005). Gaze fixation and the neural circuitry of face processing in autism. *Nature Neuroscience, 8*, 519–526.

Damasio, A. R., & Maurer, R. G. (1978). A neurological model for childhood autism. *Archives of Neurology, 35*, 777–786.

Emery, N. J. (2000). The eyes have it: The neuroethology, function and evolution of social gaze. *Neuroscience and Biobehavioral Reviews, 24*, 581–604.

Farroni, T., Csibra, G., Simion, F., & Johnson, M. H. (2002). Eye contact detection in humans from birth. *Proceedings of the National Academy of Sciences of the United States of America, 99*, 9603–9605.

Farzin, F., Rivera, S. M., & Hessl, D. (2009). Brief report: Visual processing of faces in individuals with fragile X syndrome: An eye tracking study. *Journal of Autism and Developmental Disorders, 39*, 946–952.

Gamer, M., & Buechel, C. (2009). Amygdala activation predicts gaze toward fearful eyes. *Journal of Neuroscience, 29*, 9123–9126.

Henderson, J. M., Williams, C. C., & Falk, R. (2005). Eye movements are functional during face learning. *Memory & Cognition, 33*(1), 98–106.

Hofer, P.-A. (1973). Urbach–Wiethe disease: A review. *Acta Dermato-Venereologica, 53*, 5–52.

Holland, P. C., & Gallagher, M. (1999). Amygdala circuitry in attentional and representational processes. *Trends in Cognitive Sciences, 3*, 65–73.

Howard, M. A., Cowell, P. E., Boucher, J., Broks, P., Mayes, A., Farrant, A., et al. (2000). Convergent neuroanatomical and behavioural evidence of an amygdala hypothesis of autism. *Neuroreport, 11*(13), 2931–2935.

Hwang, B., & Hughes, C. (2000). The effects of social interactive training on early social communicative skills of children with autism. *Journal of Autism and Developmental Disorders, 30*, 331–343.

Johnson, M. H. (2005). Subcortical face processing. *Nature Reviews Neuroscience, 6*, 766–774.

Kanner, L. (1943). Autistic disturbances of affective contact. *Nervous Child*, *3*, 217–250.

Kawashima, R., Sugiura, M., Kato, T., Nakamura, A., Natano, K., Ito K., et al. (1999). The human amygdala plays an important role in gaze monitoring. *Brain*, *122*, 779–783.

Keating, C., & Keating, E. G. (1982). Visual scan patterns of rhesus monkeys viewing faces. *Perception*, *11*, 211–219.

Kemper, T. L., & Bauman, M. L. (1993). The contribution of neuropathologic studies to the understanding of autism. *Neurology Clinics*, *11*, 175–187.

Kennedy, D. P., & Adolphs, R. (2010). Impaired fixation to eyes following amygdala damage arises from abnormal bottom-up attention. *Neuropsychologia*, *48*(12), 3392–3398.

Klin, A., Jones, W., Schultz, R., Volkmar, F., & Cohen, D. (2002). Visual fixation patterns during viewing of naturalistic social situations as predictors of social competence in individuals with autism. *Archives of General Psychiatry*, *59*(9), 809–816.

Kluver, H., & Bucy, P. C. (1939). Preliminary analysis of functions of the temporal lobes in monkeys. *Archives of Neurology and Psychiatry*, *42*, 979–997.

Lewis, M. B., & Edmonds, A. J. (2003). Face detection: Mapping human performance. *Perception*, *32*, 903–920.

LeDoux, J. E. (1993). Emotional memory: In search of systems and synapses. *Annals of the New York Academy of Sciences*, *702*, 149–157.

Lord, C., Risi, S., Lambrecht, L., Cook, E. H., Leventhal, B. L., DiLavore, P. C., et al. (2000). The Autism Diagnostic Observation Schedule–Generic: A standard measure of social and communication deficits associated with the spectrum of autism, *Journal of Autism and Developmental Disorders*, *30*, 205–223.

Lord, C., Rutter, M., & Le Coutier, A. (1994). Autism-Diagnostic Interview–Revised: A revised version of a diagnostic interview for caregivers of individuals with possible pervasive developmental disorders. *Journal of Autism and Developmental Disorders*, *24*, 659–685.

Murray, E. A. (2007). The amygdala, reward and emotion. *Trends in Cognitive Sciences*, *11*, 489–497.

Neumann, D., Spezio, M. L., Piven, J., & Adolphs, R. (2006). Looking you in the mouth: Abnormal gaze in autism resulting from impaired top-down modulation of visual attention. *Social Cognitive Affective Neuroscience*, *1*(3), 194–202.

Norbury, C. F., Brock, J., Cragg, L., Einay, S., Griffiths, H., & Nation, K. (2009). Eye-movement patterns are associated with communicative competence in autistic spectrum disorders. *Journal of Child Psychology and Psychiatry*, *50*(7), 834–842.

Paul, L. K., Corsello, C., Tranel, D., & Adolphs, R. (2010). Does bilateral damage to the human amygdala produce autistic symptoms? *Journal of Neurodevelopmental Disorders*, *2*, 165–173.

Pelli, D. G. (1997). The VideoToolbox software for visual psy- chophysics: Transforming numbers into movies. *Spatial Vision*, *10*, 437–442.

Pelphrey, K. A., Adolphs, R., & Morris, J. P. (2005a). Neuroanatomical substrates of social cognition dysfunction in autism. *Mental Retardation and Developmental Disabilities Research Reviews*, *10*, 259–271.

Pelphrey, K. A., Morris, J. P., & McCarthy, G. (2005b). Neural basis of eye gaze processing deficits in autism. *Brain*, *128*, 1038–1048.

Pelphrey, K. A., Sasson, N. J., Reznick, J. S., Paul, G., Goldman, B. D., & Piven, J. (2002). Visual scanning of faces in autism. *Journal of Autism and Developmental Disorders*, *32*, 249–261.

Riby, D. M., & Hancock, P. J. B. (2008). Viewing it differently: Social scene perception in Williams syndrome and autism. *Neuropsychologia*, *46*, 2855–2860.

Schumann, C. M., & Amaral, D. G. (2006). Stereological analysis of amygdala neuron number in autism. *Journal of Neuroscience*, *26*, 7674–7679.

Schumann, C. M., Hamstra, J., Goodlin-Jones, B. L., Lotspeich, L., Kwon, H., Buonocore, M. H., et al. (2004). The amygdala is enlarged in children but not adolescents with autism; the hippocampus is enlarged at all ages. *Journal of Neuroscience*, *24*, 6392–6401.

Slessor, G., Laird, G., Phillips, L. H., Bull, R., & Filippou, D. (2010). Age-related differences in gaze following: Does the age of the face matter? *The Journals of Gerontology. Series B, Psychological Sciences and Social Sciences*, *65B*(5), 536–541.

Smilek, D., Birmingham, E., Cameron, D., Bischof, W. F., & Kingstone, A. (2006). Cognitive ethology and exploring attention in real world scenes. *Brain Research*, *1080*, 101–119.

Speer, L. L., Cook, A. E., McMahon, W. M., & Clark, E. (2007). Face processing in children with autism: Effects of stimulus contents and type. *Autism*, *11*, 265–277.

Spezio, M. L., Adolphs, R., Hurley, R. S., & Piven, J. (2007a). Abnormal use of facial information in high-functioning autism. *Journal of Autism and Developmental Disorders*, *37*(5), 929–939.

Spezio, M. L., Huang, P.-Y. S., Castelli, F., & Adolphs, R. (2007b). Amygdala damage impairs eye contact during conversations with real people. *Journal of Neuroscience*, *27*, 3994–3997.

Sterling, L., Dawson, G., Webb, S., Murias, M., Munson, J., Panagiotides, H., et al. (2008). The role of face familiarity in eye tracking of faces by individuals with autism spectrum disorders. *Journal of Autism and Developmental Disorders*, *38*(9), 1666–1675.

Tranel, D., & Hyman, B. T. (1990). Neuropsychological correlates of bilateral amygdala damage. *Archives of Neurology*, *47*, 349–355.

van der Geest, J. N., Kemner, C., Camfferman, G., Verbaten, M. N., & van Engeland, H. (2002). Looking at images with human figures: Comparison between autistic and normal children. *Journal of Autism and Developmental Disorders*, *32*, 69–75.

Walker-Smith, G., Gale, A. G., & Findlay, J. M. (1977). Eye movement strategies involved in face perception. *Perception*, *6*(3), 313–326.

Wechsler, D. (1999). *Wechsler Abbreviated Scale of Intelligence*. San Antonio, TX: The Psychological Corporation.

Whalen, P. J., Kagan, J., Cook, R. G., Davis, F. C., Kim, H., Polis, S., et al. (2004). Human amygdala responsivity to masked fearful eye whites. *Science*, *306*, 2061.

Yarbus, A. L. (1967). *Eye movements and vision* (B. Haigh, Trans.). New York, NY: Plenum Press (original work published 1965).

Yovel, G., & Kanwisher, N. (2004). Face perception: Domain specific, not process specific. *Neuron*, *44*, 889–898.

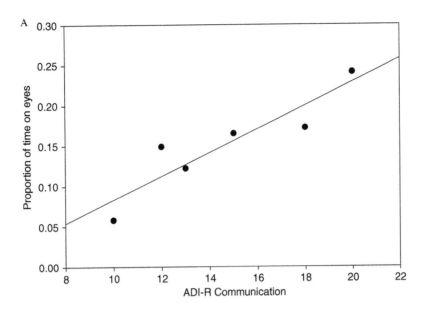

Supplementary Figure A. Positive correlation between dwell time on the eyes and scores on the ADI–R communication subscale, $r = .91$, $p < .025$, for the ASD group (with one outlier removed), suggesting that individuals with ASD who were more impaired on communication spent more time looking at the eyes.

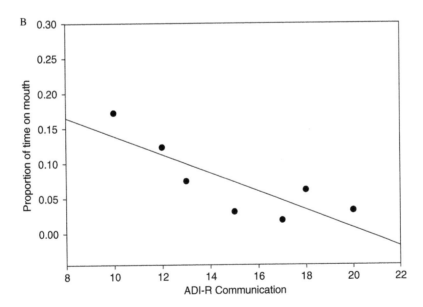

Supplementary Figure B. Negative correlation between dwell time on the mouth and scores on the ADI–R communication subscale for the ASD group, $r = -.83$, $p < .025$, suggesting that individuals with ASD who were more impaired on communication spent less time looking at the mouth.

Atypical neural specialization for social percepts in autism spectrum disorder

James C. McPartland[1], Jia Wu[1], Christopher A. Bailey[1], Linda C. Mayes[1], Robert T. Schultz[2,3], and Ami Klin[4,5]

[1]Yale Child Study Center, New Haven, CT, USA

[2]Department of Pediatrics, University of Pennsylvania, Philadelphia, PA, USA

[3]Center for Autism Research, Children's Hospital of Philadelphia, PA, USA

[4]Marcus Autism Center, Atlanta, GA, USA

[5]Department of Pediatrics, Emory University School of Medicine, Atlanta, GA, USA

The social motivation hypothesis posits that aberrant neural response to human faces in autism is attributable to atypical social development and consequently reduced exposure to faces. The specificity of deficits in neural specialization remains unclear, and alternative theories suggest generalized processing difficulties. The current study contrasted neural specialization for social information versus nonsocial information in 36 individuals with autism and 18 typically developing individuals matched for age, race, sex, handedness, and cognitive ability. Event-related potentials elicited by faces, inverted faces, houses, letters, and pseudoletters were recorded. Groups were compared on an electrophysiological marker of neural specialization (N170), as well as behavioral performance on standardized measures of face recognition and word reading/decoding. Consistent with prior results, individuals with autism displayed slowed face processing and decreased sensitivity to face inversion; however, they showed comparable brain responses to letters, which were associated with behavioral performance in both groups. Results suggest that individuals with autism display atypical neural specialization for social information but intact specialization for nonsocial information. Findings concord with the notion of specific dysfunction in social brain systems rather than nonspecific information-processing difficulties in autism.

Keywords: Perceptual expertise; N170; Event-related potential (ERP/EEG); Face perception; Autism spectrum disorder.

PRESERVED NEURAL SPECIALIZATION FOR NONSOCIAL INFORMATION IN AUTISM

The ability to efficiently perceive the human face is a crucial and early-emerging social ability. Specialized processing for faces emerges in the first days of life (Bushnell, Sai, & Mullin, 1989; Goren, Sarty, & Wu, 1975; Johnson, Dziurawiec, Ellis, & Morton, 1991; Meltzoff & Moore, 1977) and is honed by developmental experience (Nelson, 2001). Faces come to be encoded by configural processing mechanisms (Farah, Tanaka, & Drain, 1995), reflected in disproportionate impairments in recognizing both upside-down faces (the inversion effect; Yin, 1970) and facial features out of context (Tanaka & Farah, 1993). Functional neuroimaging studies show that faces elicit selective, right-lateralized hemodynamic activity in

Correspondence should be addressed to: James McPartland, Yale Child Study Center, 230 South Frontage Road, New Haven, CT 06520, USA. E-mail: james.mcpartland@yale.edu

This research was supported by NIMH R03 MH079908, NIMH K23 MH086785, NICHD PO1HD003008, a NARSAD Atherton Young Investigator Award, and CTSA Grant Number UL1 RR024139 from the National Center for Research Resources (NCRR), a component of the National Institutes of Health (NIH), and NIH Roadmap for Medical Research (USA). Its contents are solely the responsibility of the authors and do not necessarily represent the official view of NCRR or NIH.

www.psypress.com/socialneuroscience   http://dx.doi.org/10.1080/17470919.2011.586880

a portion of occipitotemporal cortex, the fusiform gyrus (Haxby et al., 1994; Kanwisher, McDermott, & Chun, 1997; Puce, Allison, Gore, & McCarthy, 1995), and intracranial electrophysiological recordings reveal face-related negative electrical activity originating from this portion of cortex (Allison, McCarthy, Nobre, Puce, & Belger, 1994). Likewise, event-related potentials (ERPs) recorded from corresponding scalp regions show a negative-going electrical deflection approximately 170 ms after viewing a face (N170; Bentin, Allison, Puce, Perez, & McCarthy, 1996). The N170 putatively reflects structural encoding, an early stage of face processing preceding higher-order processes like recognition (Bentin, Deouell, & Soroker, 1999), and is sensitive to perturbations in face configuration, including inversion (Rossion et al., 2000). Neural generators of the N170 have been localized to occipitotemporal sites including the fusiform gyrus (Itier & Taylor, 2002; Rossion, Joyce, Cottrell, & Tarr, 2003; Shibata et al., 2002), as well as the superior temporal sulcus (Itier & Taylor, 2004), lingual gyrus (Shibata et al., 2002), and posterior inferotemporal gyrus (Schweinberger, Pickering, Jentzsch, Burton, & Kaufmann, 2002; Shibata et al., 2002).

These processing strategies and brain regions are also observed in the perception of visual stimuli with which viewers have extensive experience (Diamond & Carey, 1986; Gauthier, 2000). Experts at perceiving and discriminating among exemplars within a visually homogeneous class (e.g., greebles, birds, or cars; Gauthier, Skudkarski, Gore, & Anderson, 2000; Gauthier, Williams, Tarr, & Tanaka, 1998) develop face-like patterns of brain activity, in terms of both hemodynamic (Tarr & Gauthier, 2000) and electrophysiological response, as indexed by the N170 (Rossion, Gauthier, Goffaux, Tarr, & Crommelinck, 2002). According to this model, these brain regions subserve a processing style rather than specific content, and face-related brain activity reflects, in large part, human beings' extensive experience processing human faces during development (Gauthier & Nelson, 2001).

Analogous specialization through developmental experience occurs in brain mechanisms subserving letter and word processing. Perception of printed letters (James, James, Jobard, Wong, & Gauthier, 2005) and words (McCandliss, Cohen, & Dehaene, 2003) selectively activates left fusiform gyrus and elicits a left-lateralized N170 in literate children as young as 8 years (Bentin, Mouchetant-Rostaing, Giard, Echallier, & Pernier, 1999; Maurer et al., 2006). A maturational course independent of higher-order phonological or semantic processes (Grossi, Coch, Coffey-Corina, Holcomb, & Neville, 2001; Holcomb, Coffey, & Neville, 1992) and an early time course

suggest that this "letter N170" marks prelinguistic processes related to visual perception of form (Bentin, Mouchetant-Rostaing et al., 1999) and, like the N170 elicited by faces, automatic perceptual categorization within a domain of expertise (Maurer, Brem, Bucher, & Brandeis, 2005). Neural specialization for letters is revealed by enhanced N170 amplitude to familiar alphabets but not to foreign alphabets or nonsensical letter approximations (pseudoletters; Wong, Gauthier, Woroch, DeBuse, & Curran, 2005). Converging evidence from neuroimaging studies and ERP source localization suggest left-lateralized sources in the fusiform gyrus and the inferior occipitotemporal cortex (Cohen et al., 2000; Maurer et al., 2005; Rossion et al., 2003). Though letter-related brain activity is typically contralateral to face-processing areas (Rossion et al., 2003), there is some degree of functional overlap; under special circumstances, such as precocious reading ability, right fusiform gyrus is recruited for letter and word recognition (Turkeltaub et al., 2004).

Because face perception is a well-studied social behavior, it has been employed as an avenue to understand social development in autism spectrum disorder (ASD). In ASD, decreased attention to human faces is evident by 6–12 months (Maestro et al., 2002; Osterling & Dawson, 1994), and abnormalities in face perception and recognition have been observed throughout the life span (Hobson, 1986; Hobson, Ouston, & Lee, 1988; Klin et al., 1999; Langdell, 1978; Schultz, 2005; Wolf et al., 2008). Individuals with ASD often exhibit abnormal viewing patterns to faces (Jones, Carr, & Klin, 2008; Klin, Jones, Schultz, Volkmar, & Cohen, 2002) and hypoactivation in face-related brain areas (Schultz, 2005; Schultz et al., 2000). Studies of electrophysiological markers of face perception suggest delayed N170 to human faces and decreased sensitivity to face inversion in individuals with ASD, as well as first-degree relatives (Dawson et al., 2002; Dawson, Webb, Wijsman et al., 2005; McCleery, Akshoomoff, Dobkins, & Carver, 2009; McPartland, Dawson, Webb, Panagiotides, & Carver, 2004; O'Connor, Hamm, & Kirk, 2005, 2007; Webb, Dawson, Bernier, & Panagiotides, 2006), although some studies suggest at least partially preserved face perception in some subgroups of individuals with ASD (Webb et al., 2009, 2010).

One theoretical explanation for these observed differences in face perception in ASD focuses on the role of developmental exposure to faces. The social motivation hypothesis (Dawson, Webb, & McPartland, 2005; Schultz, 2005) posits that, due to abnormalities in social drive very early in childhood, children with ASD do not attend to faces during sensitive developmental periods. Consequently, people with

ASD fail to develop typical proficiency in face processing and associated patterns of behavioral and brain specialization (Behrmann, Thomas, & Humphreys, 2006). Because the social motivation hypothesis implicates attenuated social drive as the dysfunction from which face perception difficulties originate (rather than specific dysfunction of brain regions subserving face perception), it presumes that individuals with ASD, given appropriate exposure to and interest in a stimulus class, should develop both behavioral and brain specialization (Sasson, 2006). This notion is supported by a single-case study revealing behavioral and neural indices of specialization in a child with ASD during perception of cartoon characters associated with a circumscribed interest (Grelotti et al., 2005). Though others have attempted to investigate brain response associated with experience in this population (Boeschoten, Kenemans, van Engeland, & Kemner, 2007), research has been stymied by difficulty in finding shared areas of expertise in ASD; whereas groups of study participants experienced in perceiving faces are common, groups of individuals with ASD who share a common non-face area of expertise are rare.

The current work circumvented this difficulty by examining brain activity reflecting neural specialization for letters of the alphabet. As described above, development of specialization for letters and words has been well studied, and it elicits brain activity similar to faces in terms of temporal characteristics and scalp topography (despite lateralization differences). This is a novel and uniquely appropriate comparison because, despite developmental disinterest in faces and characteristic weakness in language, facility with reading has been a noted strength in ASD since Kanner's original account (Kanner, 1943). High-functioning individuals on the autism spectrum display age-appropriate skills in single word-reading and word-decoding ability (Huemer & Mann, 2009; Nation, Clarke, Wright, & Williams, 2006; Newman et al., 2006), and a subgroup possesses precocious interest and proficiency in reading, or *hyperlexia* (Burd, Kerbeshian, & Fisher, 1985; Grigorenko et al., 2002). In this study, electrophysiological and behavioral methods were applied to compare neural specialization for faces and letters in individuals with ASD. Experiments contrasted neural response to faces versus houses, faces versus inverted faces, and letters versus pseudoletters, and compared these parameters to behavioral measures assessing proficiency in face recognition and letter and word perception. Consistent with previous work, it was hypothesized that individuals with ASD would exhibit impaired face recognition and delayed brain response to faces in the right hemisphere, as well as decreased sensitivity to face inversion. In keeping with the notion

that these atypicalities reflect developmental sequelae of social deficits, it was predicted that similar anomalies would not be observed for nonsocial stimuli; individuals with ASD would show typical skills in terms of letter and word perception and comparably enhanced response to letter stimuli with respect to unfamiliar pseudoletters. As prior work has revealed relationships among neural correlates of face perception and behavioral measures of face recognition (McPartland et al., 2004), exploratory analyses examined relationships among neural and behavioral measures of face and letter perception.

METHODS

Participants

Two groups participated in the study: individuals with ASD and medically and neuropsychiatrically healthy individuals with typical development. Exclusionary criteria for participants with ASD included seizures, neurological disease, history of serious head injury, sensory or motor impairment that would impede completion of the study protocol, active psychiatric disorder (other than ASD; screened with the Child Symptom Inventory, 4th edition; Gadow & Sprafkin, 1994), or medication known to affect brain electrophysiology. Additional exclusionary criteria for typical participants included the above as well as learning/language disability or family history of ASD. From an existing pool of subjects involved in research at the Yale Child Study Center, participants were selected based on having a Full Scale IQ in the average range or higher (Standard Score of 80 or above; Differential Ability Scales, 2nd edition, Elliott, 2007; Wechsler Intelligence Scale for Children, 4th edition, Wechsler, 2003; Wechsler Adult Intelligence Scale, 3rd edition, Wechsler, 1997). All individuals with ASD had a pre-existing diagnosis that was confirmed with reference standard diagnostic assessments for research: combination of parent interview (Autism Diagnostic Interview-Revised—ADI-R; Lord, Rutter, & Le Couteur, 1994); semistructured social behavior and communication assessment (Autism Diagnostic Observation Schedule; Lord et al., 2000); and clinical diagnosis based on DSM-IV-TR (American Psychiatric Association, 2000) criteria by an expert clinician. The ADI was not administered to one subject because a parent was unavailable for interviewing, and two individuals were included in the sample who failed to meet ADI-R onset criteria; for both of these high-functioning verbal individuals, problems were not detected until enrolled

	Typical (n=18)	ASD (n=36)
Number male (%)	15 (83.3)	32 (88.9)
Number White (%)	15 (83.3)	34 (94.4)
Number right-handed (%)	16 (88.9)	31 (86.1)
Mean age (SD)	12.6 (2.4)	11.2 (3.4)
Mean Full Scale IQ (SD)	112.9 (13.4)	105.2 (17.3)

in school with peers. In addition to the aforementioned exclusionary criteria, typical participants were recruited to match the ASD sample in terms of sex, ethnicity, handedness (Edinburgh Handedness Inventory (Oldfield, 1971), chronological age, and Full Scale IQ (Wechsler Abbreviated Scale of Intelligence; Psychological Corporation, 1999). Groups did not significantly differ on any of these variables. Behavioral assessments could not be administered to one typical participant due to time limitations. All procedures were approved by the Human Investigation Committee at the Yale School of Medicine and were carried out in accordance with the Declaration of Helsinki (1975/1983). Of an initial sample of 57 individuals with ASD and 25 typically developing participants, adequate, artifact-free data were obtained from 36 and 18 participants, respectively, in the letter/pseudoletter experiment, and 32 and 17, respectively, in the face/house experiment. Table 1 displays demographic data for the complete sample; variation in sample between experiments did not introduce differences on demographic characteristics.

EEG procedures

Stimuli

Stimuli were administered in pseudorandom sequence in two counterbalanced blocks. The first block consisted of gray-scale, digitized images of neutral faces, houses, inverted faces, and inverted houses (not included in current analyses), all displayed from a direct frontal perspective. The second block included letters and a confabulated alphabet of pseudoletters (Wong et al., 2005). Example stimuli are displayed in the legends of Figures 1 and 2. Subjects were presented with 23 stimuli from each category four times, for a total of 92 stimuli per category. Stimuli were standardized in terms of size (approximately 5° of visual angle), background color (gray), and average luminance. To maximally engage attention to individual stimuli, participants were asked to press a button whenever a stimulus repeated (nine times

for each stimulus category). Because this behavioral task was confounded with face recognition, attention to task was monitored in real time through closed-circuit video, enabling pausing of data collection and redirection of attention to stimulus presentation if needed.

Data collection

Stimuli were presented on a Pentium 4 computer controlling a 51-cm color monitor (75-Hz, 1024 × 768 resolution) running E-Prime 2.0 software (Schneider, Eschman, & Zuccolotto, 2002). Displays were viewed at a distance of 90 cm in a sound-attenuated room with low ambient illumination. EEG was recorded with Net Station 4.3. A 256-lead Geodesic sensor net (Electrical Geodesics, Inc., Eugene, OR; Tucker, 1994) was dampened with potassium chloride electrolyte solution, placed on the participant's head, and fitted according to the manufacturer's specifications. Impedances were kept below 40 kilo-ohms. ERP was recorded continuously throughout each stimulus-presentation trial, consisting of a fixation cross (randomly varying from 250 to 750 ms), stimulus (500 ms), and blank screen (500 ms). The EEG signal was amplified (x1000) and filtered (0.1-Hz high-pass filter and 100-Hz elliptical low-pass filter) via a preamplifier system (Electrical Geodesics, Inc., Eugene, OR). The conditioned signal was multiplexed and digitized at 250 Hz, using an analog-to-digital converter (National Instruments, Inc., Austin, TX) and a dedicated Macintosh computer. The vertex electrode was used as a reference, and data were re-referenced to an average reference after data collection.

Data editing and reduction

Data were averaged for each subject by stimulus type across trials. Averaged data were digitally filtered with a 30-Hz low-pass filter and transformed to correct for baseline shifts. The window for segmentation of the ERP was set from 100 ms before and 500 ms after stimulus onset. Net Station artifact detection settings were set to 200 μV for bad channels, 150 μV for eye blinks, and 150 μV for eye movements. Channels with artifacts on more than 50% of trials were marked as bad channels and replaced through spline interpolation. Segments that contained eye blinks or eye movement and those with more than 20 bad channels were also excluded. Participants with less than 46 good trials for any stimulus category were excluded from analysis. Electrodes of interest were selected by maximal observed amplitude of the N170 to faces and letters in grand-averaged data and to conform to those used in previous research. Data were averaged across eight electrodes over the left (95, 96, 97, 106, 107, 108, 116, 117) and right (151, 152, 153, 160, 161, 162, 170, 171) lateral posterior scalp. The time windows for P1 and N170 analysis, extending from 108 to 327 ms and 56 to 206 ms post-stimulus onset, respectively, were chosen by visual inspection of grand-averaged data and then customized for each subject to confirm that the component of interest was captured at each electrode. Peak amplitude and latency to peak were averaged across each electrode group within the specified time window and were extracted for each participant for each stimulus category.

Data analysis

P1 and N170 amplitudes and latencies to peak were separately analyzed by univariate repeated-measure analysis of variance (ANOVA) with two within-subject factors, Condition (face/house; face/inverted face, letter/pseudoletter) and Hemisphere (left/right). The between-subjects factor was Group (ASD/typical). A planned comparison using one-way ANOVA was employed to test the specific hypothesis that N170 to faces would be delayed in the right hemisphere in the ASD group relative to the typical group.

Behavioral procedures

Face perception

Face recognition was measured with the Benton Facial Recognition Test (Benton, Sivan, Hamsher, Varney, & Spreen, 1994). Participants viewed a gray-scale image of a face and specified one or three matches from an array of six faces, varying in shadowing and orientation.

Letter perception

The Letter-Word Identification and Word Attack subtests of the Woodcock-Johnson Tests of Achievement, 3rd edition (Woodcock, McGrew, & Mather, 2001) required the participant to read words aloud, with the former using genuine English words and the latter using novel words. Both subtests yielded a standard score ($M = 100$, $SD = 15$) derived from an age-based standardization sample.

Data analysis

Between-group differences in behavioral measures were analyzed with independent-samples t-tests. Interrelationships among behavioral measures and ERP parameters (N170 latency, amplitude) were computed with Pearson product-moment correlations.

RESULTS

Electrophysiological measures

Faces versus houses: P1 amplitude

Figure 1 displays waveforms depicting ERPs to faces and houses, and Table 2 displays mean amplitudes and SDs for both groups across both hemispheres and all conditions. Faces elicited smaller P1 amplitudes across hemisphere and group—main effect of Condition: $F(1, 47) = 18.01$, $p \leq .01$. No other significant effects were observed (all $Fs < 2.44$; all $ps > .05$).

Faces versus houses: P1 latency

Table 3 displays mean latencies and SDs for both groups across both hemispheres and all conditions. Faces elicited shorter P1 latencies across hemisphere and group—main effect of Condition: $F(1, 47) = 63.92$, $p \leq .01$. No other significant effects were observed (all $Fs < 3.10$; all $ps > .05$).

Faces versus houses: N170 amplitude

Faces elicited N170s with larger amplitudes—main effect of Condition: $F(1, 47) = 49.77$, $p \leq .01$—across hemispheres for both groups. N170 amplitude to houses was reduced in the left hemisphere across groups—Hemisphere by Condition interaction: $F(1, 47) = 6.80$, $p \leq .01$—and, relative to typical individuals, bilaterally in the ASD group—Condition by Group interaction: $F(1, 47) = 5.17$, $p \leq .05$. Right-lateralization was evident only in typically developing individuals—Hemisphere by Group interaction: $F(1, 47) = 7.22$, $p \leq .01$. No other significant effects were observed (all $Fs < 0.90$; all $ps > .05$).

Faces versus houses: N170 latency

Faces elicited N170s with shorter latencies—main effect of Condition: $F(1, 47) = 30.10$, $p \leq .01$— across

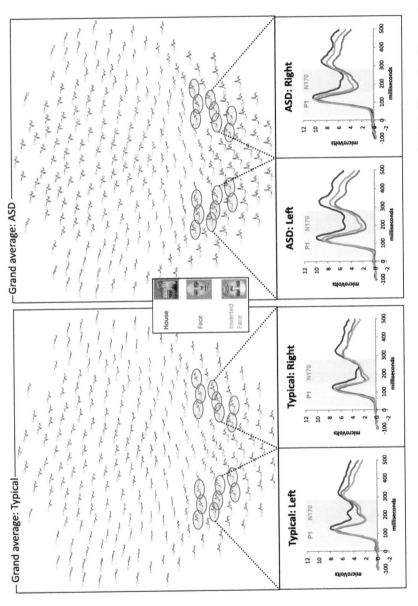

Figure 1. Grand averaged waveforms across entire scalp for faces, houses, and inverted faces for typical participants and those with ASD. Electrodes of interest in right and left hemisphere are highlighted. Subpanels display the averaged waveform across the eight specified electrodes in each hemisphere for both groups.

TABLE 2
P1 and N170 amplitude

Hemisphere	Condition	P1 M (μV)	SD	N170 M (μV)	SD
Typical group					
Left	Faces	7.77	4.4	0.39	1.7
	Houses	8.60	5.7	2.49	3.5
	Inverted faces	7.70	4.8	0.24	2.8
	Letters	6.28	3.3	−2.27	3.0
	Pseudoletters	5.89	3.4	−1.51	3.1
Right	Faces	6.79	5.0	−0.16	2.0
	Houses	7.94	5.8	0.89	2.6
	Inverted faces	7.50	5.5	−0.58	2.8
	Letters	5.03	2.5	−3.03	4.2
	Pseudoletters	5.10	3.0	−2.53	3.9
ASD group					
Left	Faces	8.94	5.6	−0.32	3.4
	Houses	10.49	6.2	3.16	4.0
	Inverted faces	9.64	5.4	0.59	3.6
	Letters	6.26	3.8	−3.09	3.6
	Pseudoletters	6.31	3.6	−2.22	3.3
Right	Faces	10.10	6.5	0.76	3.6
	Houses	11.11	6.7	3.44	4.2
	Inverted faces	11.17	6.7	1.52	3.5
	Letters	6.40	3.3	−3.57	3.1
	Pseudoletters	6.59	3.7	−2.29	2.9

TABLE 3
P1 and N170 latency

Hemisphere	Condition	P1 M (ms)	SD	N170 M (ms)	SD
Typical group					
Left	Faces	115.27	10.9	193.24	37.3
	Houses	134.71	20.6	215.59	43.4
	Inverted faces	119.44	9.7	194.82	35.3
	Letters	106.67	22.0	177.25	26.9
	Pseudoletters	111.58	25.6	179.56	26.8
Right	Faces	118.44	12.2	181.38	31.6
	Houses	133.97	15.0	217.50	36.4
	Inverted faces	121.12	9.6	188.85	32.7
	Letters	104.72	25.8	181.19	25.9
	Pseudoletters	106.89	24.7	176.03	28.5
ASD group					
Left	Faces	118.73	10.3	200.13	24.3
	Houses	134.05	11.3	221.67	31.2
	Inverted faces	125.97	10.5	205.42	27.7
	Letters	102.68	13.4	189.21	26.7
	Pseudoletters	106.29	15.1	178.83	19.3
Right	Faces	117.59	10.8	201.78	27.2
	Houses	130.25	14.4	219.38	27.8
	Inverted faces	122.78	15.4	204.66	31.3
	Letters	101.53	14.1	188.42	28.0
	Pseudoletters	103.97	14.9	183.51	23.6

hemispheres for both groups. A planned comparison confirmed the predicted differences in latency between groups in the right hemisphere, and N170 latency to faces was significantly faster (a difference of approximately 20.4 ms) in typically developing individuals than those with ASD—$F(1, 47) = 5.57$; $p \leq .05$. Figure 2 displays N170 amplitudes for faces and houses, highlighting this difference. No other significant effects were observed (all $Fs < 3.09$; all $ps > .05$).

Face versus inverted faces: P1 amplitude

Waveforms depicting ERPs to faces and inverted faces are displayed in Figure 1. Inverted faces elicited larger P1 amplitudes across hemisphere and group—main effect of Condition: $F(1, 47) = 12.41, p \leq .01$. P1 amplitude to inverted faces was larger relative to upright faces in the right hemisphere across groups—Hemisphere by Condition interaction: $F(1, 47) = 5.88$, $p \leq .05$. No other significant effects were observed (all $Fs < 3.25$; all $ps > .05$).

Face versus inverted faces: P1 latency

Faces elicited shorter P1 latencies across hemisphere and group—main effect of Condition: $F(1, 47) = 19.80, p \leq .01$. No other significant effects were observed (all $Fs < 3.31$; all $ps > .05$).

Face versus inverted faces: N170 amplitude

Across faces and inverted faces, typically developing individuals displayed enhanced amplitude in the right hemisphere, while those with ASD exhibited equivalent amplitude in both hemispheres—Hemisphere by Group interaction: $F(1, 47) = 7.70$, $p \leq .01$. Across hemisphere, typically developing individuals displayed an inversion effect in the expected direction, with larger amplitude to inverted relative to upright faces, whereas individuals with ASD displayed attenuated N170 amplitudes to inverted faces relative to upright faces—Condition by Group interaction: $F(1, 47) = 5.84, p \leq .05$. No other significant effects were observed (all $Fs < 1.40$; all $ps > .05$).

Face versus inverted faces: N170 latency

Inverted faces elicited N170s with longer latencies than upright faces—main effect of Condition: $F(1, 47) = 4.66, p \leq .05$—across hemispheres for both

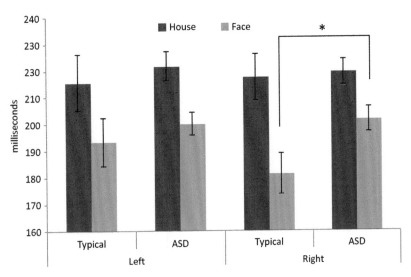

Figure 2. Mean latency of the N170 component (in ms) elicited by faces and houses for both groups in both hemispheres. Error bars represent ±1 *SE*. Significance at the $p \leq .05$ level is indicated by *.

groups. No other significant effects were observed (all *F*s < 2.69; all *p*s > .05).

Letters versus pseudoletters: P1 amplitude

Figure 3 displays waveforms depicting ERPs to letters and pseudoletters. No significant effects were observed (all *F*s < 3.08; all *p*s > .05).

Letters versus pseudoletters: P1 latency

Across hemisphere and group, letters elicited shorter P1 latency than pseudoletters—main effect of Condition: $F(1, 52) = 16.28$, $p \leq .01$. Across group and condition, P1 latency was shorter in the right hemisphere—main effect of Hemisphere: $F(1, 52) = 4.15$, $p \leq .05$. No other significant effects were observed (all *F*s < 1.07; all *p*s > .05).

Letters versus pseudoletters: N170 amplitude

For both groups, letters elicited N170s with larger amplitudes than pseudoletters across hemispheres—main effect of Condition; $F(1, 52) = 14.67$, $p \leq .01$. As displayed in Figure 4, paired-samples *t*-tests revealed that this effect was carried by significantly enhanced amplitude to letters versus pseudoletters in the typical group in the left hemisphere, $t(1, 17) = 2.12$, $p \leq .05$, and in the ASD group in both left, $t(1, 35) = 2.90$, $p \leq .01$, and right hemispheres, $t(1, 35) = 3.34$,

$p \leq .01$. No other significant effects were observed (all *F*s < 2.68; all *p*s > .05).

Letters versus pseudoletters: N170 latency

No significant effects were observed (all *F*s < 3.58; all *p*s > .05).

Behavioral measures

Face perception

Table 4 displays mean score and *SD* on behavioral measures for both groups. Individuals with ASD obtained significantly lower face recognition scores than typically developing individuals, $t(1, 51) = 3.29$, $p \leq .01$. For both groups, N170 latency to faces in the right hemisphere was correlated with face recognition skill; individuals with faster N170s displayed better face recognition performance (ASD: $r = -.39$, $p \leq .05$; Typical: $r = -.53, p \leq .05$). Among individuals with ASD, N170 amplitude to inverted faces was correlated with face-recognition performance; those with better face-recognition abilities were more likely to display an enhanced N170 associated with inversion $(r = -.47, p \leq .01)$.

Letter perception

Groups performed comparably and in the average range on word reading and decoding tasks. Among

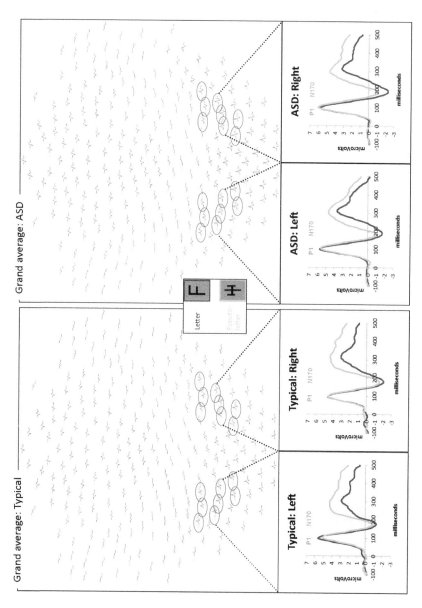

Figure 3. Grand averaged waveforms across entire scalp for letters and pseudoletters for typical participants and those with ASD. Electrodes of interest in right and left hemisphere are highlighted. Subpanels display the averaged waveform across the eight specified electrodes in each hemisphere for both groups.

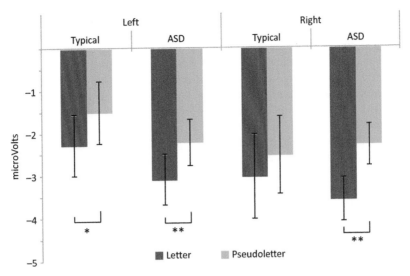

Figure 4. Amplitude of the N170 component (in μV) elicited by letters and pseudoletters for both groups in both hemispheres. Error bars represent ±1 *SE*. Significance at the $p \leq .05$ level is indicated by *, and significance at the $p \leq .01$ level is indicated by **.

TABLE 4
Performance on behavioral measures

Measure	Typical (n = 17)		ASD (n = 36)	
	M	*SD*	*M*	*SD*
Benton Face Recognition raw score	41.41	3.6	37.11	4.8
Letter-Word Identification standard score	108.41	9.9	105.67	15.0
Word Attack standard score	101.41	9.7	103.86	11.6

typically developing individuals, longer N170 latency to letters in the right hemisphere was correlated with word-reading score; those with longer latencies tended to perform better on the measure of single-word reading ($r = .64, p \leq .01$).

GENERAL DISCUSSION

The current study contrasted neural specialization for social and nonsocial information in individuals with ASD and a cohort of typically developing individuals of comparable age, ethnicity, sex, handedness, and cognitive ability. A critical social stimulus with which most adults possess great experience, the human face, was contrasted with a comparably complex visual stimulus without interpersonal relevance, houses. Consistent with predictions and with prior research (McPartland et al., 2004), individuals with ASD displayed a selective processing delay for human faces in the right hemisphere relative to typical counterparts. Individuals with ASD also showed

reduced hemispheric specialization compared to typical counterparts, who showed a marked right lateralization effect for faces. The inversion effect, a marker of neural specialization and processing experience for faces, was evident in typically developing individuals but not those with ASD. On a behavioral measure of face recognition, individuals with ASD, despite comparable intellectual ability, performed significantly worse than typically developing counterparts. Face recognition performance was associated with processing efficiency for faces; in both groups, individuals with better face-recognition abilities displayed faster N170 response. Among individuals with ASD, increased inversion effects, as reflected by a stronger response to inverted faces, were associated with better face-recognition performance. This pattern of anomalies, that is, decreased efficiency of processing, insensitivity to inversion, and impaired face recognition, is hypothesized to reflect underdeveloped specialization for faces, a downstream effect of decreased attention to faces during childhood secondary to reduced social drive from infancy (Dawson et al., 2005). Indeed, the observed correlation between neural response to face inversion and recognition performance suggests that, in this case, development of expertise and processing proficiency are related.

Though differences related to condition and lateralization were observed in an earlier component, the P1, no between-group differences were detected. Given the role of the P1 in basic visual attention and low-level sensory perception (Key, Dove, & Maguire, 2005), this pattern of results indicates that children with ASD did not differ from typically developing peers in terms of fundamental sensory perception. Indeed, despite observed differences in the basic sensory response to different classes of stimuli, groups responded similarly. These findings suggest normative sensory perception for visual information in ASD, with differences emerging at subsequent processing stages related to social perception.

Current findings concord with prior work describing deviant social development in ASD. However, scant evidence to date has informed the specificity of the observed neural processing anomalies to social information. By measuring responses to nonsocial expert stimuli, the current study demonstrated the selectivity of perceptual deficits in ASD. Following up on work showing N170-related expertise effects for letters, ERP response to letters of the Roman alphabet were compared to a confabulated alphabet of pseudoletters (Wong et al., 2005). In contrast to the discrepancies observed during perception of social stimuli, individuals with ASD displayed neural responses comparable to typical counterparts; both groups showed enhanced N170 to familiar letters. Similar results were obtained on behavioral measures, with individuals with ASD obtaining word-reading and word-decoding scores that were comparable to the typical participants in this study and within the average range. These results suggest intact behavioral and brain specialization for nonsocial information in individuals with ASD. These results are consistent with prior behavioral work demonstrating preserved word-reading and word-decoding ability in cognitively able individuals with ASD (Huemer & Mann, 2009; Nation et al., 2006; Newman et al., 2006) despite complications with reading comprehension and more sophisticated aspects of language (O'Connor & Klein, 2004). They are consistent with the notion that the social element of communication, rather than language more generally, may be most directly impacted in ASD (Paul, 2003).

The current findings have significant implications for understanding the neuropathology of ASD. Two prevailing classes of theories attribute autistic impairments to dysfunctional brain structures supporting social information processing (Dawson et al., 2005) or altered connectivity among distributed brain regions (Minshew & Williams, 2007). The former emphasizes the import of the content that is processed, and the latter accentuates the nature of processing itself. Social brain theories posit that social information is qualitatively unique and that specific brain systems have evolved to support this type of information processing. Connectivity theories, in contrast, have traditionally argued that social information is relevant only insofar as it relies on complex or cortically distributed processing mechanisms. The current work demonstrates, for the first time in a substantial sample of children with ASD, preserved specialization for a cognitive process subserved by distributed cortical regions. Specialization of letter processing develops over time and requires elaborate communication of anterior and posterior cortical regions (Krigolson, Pierce, Holroyd, & Tanaka, 2009). The demonstration of preserved neural specialization for this type of "expert" processing in ASD is not consistent with models of nonspecific, brain-wide dysfunction. Taking into account considerable evidence for atypical patterns of connectivity in ASD (Minshew & Williams, 2007), current findings emphasize the potential value of studying connectivity *within specific brain systems in a developmental context*. The observed latency delays may reflect reduced connectivity, and their specificity to social information compared to nonsocial information may indicate system-specificity in terms of atypical connectivity. By studying connectivity within specific neural circuits, scientists may also extricate atypical connectivity as a potential cause or consequence of autistic dysfunction; it is likely that origins of dysfunction in functionally specific brain systems would, through developmental maturation, lead to broader connectivity problems. Such research may also clarify to what degree problems with connectivity uniquely differentiate autism from the diversity of developmental and psychiatric disorders also manifesting atypical connectivity, such as obsessive-compulsive disorder (Garibotto et al., 2010), schizophrenia (Friston, 2002), attention deficit hyperactivity disorder (Murias, Swanson, & Srinivasan, 2007), and intellectual impairment (Zhou et al., 2008).

This work yields clinically relevant implications for the detection and treatment of ASD. Results are supportive of the broad class of interventions designed to direct the attention of children with ASD to relevant social information. When children are appropriately engaged and attuned to information—in this case, letters—typical patterns of neural specialization develop; given the right input, the brain of a person with autism can function like that a of a typical peer, without ostensible reliance on compensatory mechanisms or alternative processing strategies. Findings add to a body of evidence that electrophysiological brain activity to faces represents a viable biobehavioral risk marker for ASD, as temporal anomalies in neural correlates of face perception have been observed in children with ASD (Dawson, Webb, Wijsman et al., 2005) and infants at risk of ASD (McCleery et al., 2009).

Though the current work replicates initial findings of temporal anomalies to faces (McPartland et al., 2004), these findings have not been fully replicated in all samples (Kemner, Schuller, & van Engeland, 2006; Senju, Tojo, Yaguchi, & Hasegawa, 2005; Webb et al., 2009). Some of this variability may reflect methodological inconsistencies in terms of electrode selection (e.g., Webb et al., 2009) or employment of reference-standard diagnostic procedures (e.g., Grice et al., 2005); however, varied results may accurately reflect the phenotypic heterogeneity evident in ASD. Despite the unifying characteristic of social impairment, ASD has a remarkable diversity of manifestations, likely representing multiple etiologic pathways and developmental experiences (Jones & Klin, 2009). Considering the manner in which face processing (especially in older children and adults) has been actively shaped by experience, it is intuitive that anomalies might emerge in different ways or might not emerge universally (Jemel, Mottron, & Dawson, 2006). In this regard, like any of the symptoms characterizing autism, anomalous face perception is neither necessary nor specific. It is one potential manifestation of atypical social development that, by virtue of a deep understanding of behavioral and brain bases in typical social development, is a viable avenue for investigating social disability. Variability in electrophysiological studies of face perception may also relate to differences in visual attention (Webb et al., 2009), a trend observed in hemodynamic studies (Dalton, Nacewicz, Alexander, & Davidson, 2007; Dalton et al., 2005). Our employment of a pre-stimulus fixation crosshair reduces the likelihood that between-group differences are attributable to differences in visual attention. Furthermore, given comparable N170 latencies to visual fixations to eyes and mouths in typical development (McPartland, Cheung, Perszyk, & Mayes, 2010) and delayed processing in ASD irrespective of point of gaze on the face (McPartland, Perszyk, Crowley, Naples, & Mayes, 2011), it is unlikely that variation in visual attention alone could account for observed latency differences; resolution of this matter will ultimately require co-registration of eye-tracking and EEG.

There are several aspects of the current work that are being revisited and improved upon in ongoing research. Limiting the sample to high-functioning individuals was a necessary first step toward addressing the research questions posed in this study, but it limits generalizability to the broader range of individuals with ASD. Given that even many nonverbal children with ASD are capable of reading, these types of experiments offer a window into domains of strength and preserved neural functions of children on the autism spectrum, important goals for tailoring interventions and prescribing specific treatments. The sample in the current study focused on pre-adolescence, a time of rapid maturation of brain systems subserving face perception. Additional research in younger and older children and adults will elucidate the protracted maturational course of specialization for face perception in ASD and of letter expertise in both typical and atypical development. Of note, many participants in the current study displayed the bifid waveform morphology characteristic of pre-adult face responses (Taylor, Batty, & Itier, 2004); however, this was not evident for letter N170s. Exploiting the dense spatial sampling afforded by the 256-electrode sensor net, analyses in progress are using individual-specific three-dimensional head models (computed with sensor registration images acquired with the Geodesic Photogrammetry System; Electrical Geodesics, Inc., Eugene OR) to localize potentially distinct neural sources for these facets of the developing N170 (Perszyk et al., 2010).

Our current results reveal a different relationship between performance on the behavioral face-processing task and brain responses in individuals with ASD than that observed in a prior study (McPartland et al., 2004). Previously, results indicated slowed processing to be associated with improved face recognition. In the current study, however, individuals with ASD, like the typically developing counterparts in the current and prior study, displayed an association between faster processing and better face recognition. We hypothesize that this reflects age-related differences in the application of compensatory strategies over time. In this younger sample, more normative brain responses correlated with more normative face-recognition ability in ASD. In the older sample studied previously, the opposite trend was observed in the hemisphere contralateral to that typically associated with face perception; we interpret this as reflective of effective compensatory processing strategies. In the approximately 10 years between age 11 (current study) and age 21 (prior study), increased reliance on compensatory strategies may "overtake" weakened default processing mechanisms, ultimately resulting in better performance associated with these compensatory strategies. Though we see the pattern of results as supportive of this interpretation, it is also possible that the differences observed between studies simply reflect task effects, as the prior work relied on a visual recall task and the current study utilized a visual discrimination task with reduced memory demands. Face-recognition performance in ASD is demonstrated to vary with task characteristics (McPartland, Webb, Keehn, & Dawson, 2011).

Understanding developmental factors is particularly important in the current context in that neural specialization for letters is clearly a distinct phenomenon from face expertise, occurring over a relatively compressed period of time rather than from birth. Moreover, it is likely that qualitatively different types of experience are associated with the accrual of proficiency in letter versus face processing. It will thus be essential to examine development of specialized processing mechanisms for a greater variety of stimuli. Though it has been proposed that, like faces, letters are encoded by a holistic processing strategy (Martelli, Majaj, & Pelli, 2005), unlike faces, letters are processed at a basic rather than subordinate level of identification (James et al., 2005). The N170 has been posited to denote specialization at this basic level of identification, while later components, such as the N250, index expertise at the subordinate level of identification (Scott, Tanaka, Sheinberg, & Curran, 2006). Similar mechanisms underlying neural specialization for both faces and letters exist at early processing stages as indexed by the N170, but study of a broader range of electrophysiological components and expert stimuli will paint a clearer picture of the development and potential limitations of neural specialization in ASD.

REFERENCES

Allison, T., McCarthy, G., Nobre, A., Puce, A., & Belger, A. (1994). Human extrastriate visual context and the perception of faces, words, numbers, and colors. *Cerebral Cortex, 5*, 544–554.

American Psychiatric Association (2000). *Diagnostic and statistical manual of mental disorders: DSM-IV-TR* (4th ed.). Washington, DC: American Psychiatric Association.

Behrmann, M., Thomas, C., & Humphreys, K. (2006). Seeing it differently: Visual processing in autism. *Trends in Cognitive Sciences, 10*(6), 258–264.

Bentin, S., Allison, T., Puce, A., Perez, E., & McCarthy, G. (1996). Electrophysiological studies of face perception in humans. *Journal of Cognitive Neuroscience, 8*(6), 551–565.

Bentin, S., Deouell, L. Y., & Soroker, N. (1999). Selective visual streaming in face recognition: Evidence from developmental prosopagnosia. *Neuroreport, 10*(4), 823–827.

Bentin, S., Mouchetant-Rostaing, Y., Giard, M. H., Echallier, J. F., & Pernier, J. (1999). ERP manifestations of processing printed words at different psycholinguistic levels: Time course and scalp distribution. *Journal of Cognitive Neuroscience, 11*(3), 235–260.

Benton, A., Sivan, A., Hamsher, K., Varney, N., & Spreen, O. (1994). *Contributions to neuropsychological assessment*. New York, NY: Oxford University Press.

Boeschoten, M. A., Kenemans, J. L., van Engeland, H., & Kemner, C. (2007). Face processing in pervasive developmental disorder (PDD): The roles of expertise and spatial frequency. *Journal of Neural Transmission, 114*(12), 1619–1629.

Burd, L., Kerbeshian, J., & Fisher, W. (1985). Inquiry into the incidence of hyperlexia in a statewide population of children with pervasive developmental disorder. *Psychological Reports, 57*(1), 236–238.

Bushnell, I., Sai, F., & Mullin, J. (1989). Neonatal recognition of the mother's face. *British Journal of Developmental Pscyhology, 7*, 3–15.

Cohen, L., Dehaene, S., Naccache, L., Lehericy, S., Dehaene-Lambertz, G., Henaff, M. A., et al. (2000). The visual word form area: Spatial and temporal characterization of an initial stage of reading in normal subjects and posterior split-brain patients. *Brain, 123*(2), 291–307.

Dalton, K. M., Nacewicz, B. M., Alexander, A. L., & Davidson, R. J. (2007). Gaze-fixation, brain activation, and amygdala volume in unaffected siblings of individuals with autism. *Biological Psychiatry, 61*(4), 512–520.

Dalton, K. M., Nacewicz, B. M., Johnstone, T., Schaefer, H. S., Gernsbacher, M. A., Goldsmith, H. H., et al. (2005). Gaze fixation and the neural circuitry of face processing in autism. *Nature Neuroscience, 8*(4), 519–526.

Dawson, G., Carver, L., Meltzoff, A. N., Panagiotides, H., McPartland, J., & Webb, S. J. (2002). Neural correlates of face and object recognition in young children with autism spectrum disorder, developmental delay, and typical development. *Child Development, 73*(3), 700–717.

Dawson, G., Webb, S. J., & McPartland, J. (2005). Understanding the nature of face processing impairment in autism: Insights from behavioral and electrophysiological studies. *Developmental Neuropsychology, 27*(3), 403–424.

Dawson, G., Webb, S. J., Wijsman, E., Schellenberg, G., Estes, A., Munson, J., et al. (2005). Neurocognitive and electrophysiological evidence of altered face processing in parents of children with autism: Implications for a model of abnormal development of social brain circuitry in autism. *Development and Psychopathology, 17*(3), 679–697.

Diamond, R., & Carey, S. (1986). Why faces are and are not special: An effect of expertise. *Journal of Experimental Psychology, 115*(2), 107–117.

Elliott, C. (2007). *The Differential Ability Scales* (2nd ed.). San Antonio, TX: Harcourt Assessment.

Farah, M. J., Tanaka, J. W., & Drain, H. M. (1995). What causes the face inversion effect? *Journal of Experimental Psychology: Human Perception and Performance, 21*(3), 628–634.

Friston, K. J. (2002). Dysfunctional connectivity in schizophrenia. *World Psychiatry, 1*(2), 66–71.

Gadow, K., & Sprafkin, J. (1994). *Child Symptom Inventories manual*. Stony Brook, NY: Checkmate Plus.

Garibotto, V., Scifo, P., Gorini, A., Alonso, C. R., Brambati, S., Bellodi, L., et al. (2010). Disorganization of anatomical connectivity in obsessive compulsive disorder: A multi-parameter diffusion tensor imaging study in a subpopulation of patients. *Neurobiology of Disease, 37*(2), 468–476.

Gauthier, I. (2000). What constrains the organization of the ventral temporal cortex? *Trends in Cognitive Sciences, 4*(1), 1–2.

Gauthier, I., & Nelson, C. (2001). The development of face expertise. *Current Opinion in Neurobiology, 11*, 219–224.

Gauthier, I., Skudkarski, P., Gore, J., & Anderson, A. (2000). Expertise for cars and birds recruits brain areas involved in face recognition. *Nature Neuroscience, 3*(2), 191–197.

Gauthier, I., Williams, P., Tarr, M. J., & Tanaka, J. (1998). Training 'greeble' experts: A framework for studying expert object recognition processes. *Vision Research, 38*(15–16), 2401–2428.

Goren, C. C., Sarty, M., & Wu, P. Y. (1975). Visual following and pattern discrimination of face-like stimuli by newborn infants. *Pediatrics, 56*(4), 544–549.

Grelotti, D. J., Klin, A. J., Gauthier, I., Skudlarski, P., Cohen, D. J., Gore, J. C., et al. (2005). fMRI activation of the fusiform gyrus and amygdala to cartoon characters but not to faces in a boy with autism. *Neuropsychologia, 43*(3), 373–385.

Grice, S. J., Halit, H., Farroni, T., Baron-Cohen, S., Bolton, P., & Johnson, M. (2005). Neural correlates of eye-gaze detection in young children with autism. *Cortex, 41*(3), 342–353.

Grigorenko, E. L., Klin, A., Pauls, D. L., Senft, R., Hooper, C., & Volkmar, F. (2002). A descriptive study of hyperlexia in a clinically referred sample of children with developmental delays. *Journal of Autism and Developmental Disorders, 32*(1), 3–12.

Grossi, G., Coch, D., Coffey-Corina, S., Holcomb, P. J., & Neville, H. J. (2001). Phonological processing in visual rhyming: A developmental ERP study. *Journal of Cognitive Neuroscience, 13*(5), 610–625.

Haxby, J. V., Grady, C. L., Horwitz, B., Ungerleider, J. M., Maisog, M., & Pietrini, P. (1994). The functional organization of human extrastriate cortex: A pet-rCBFstudy of selective attention to faces and locations. *Journal of Neuroscience, 14*, 6336–6353.

Hobson, R. (1986). The autistic child's appraisal of expressions of emotion. *Journal of Child Psychology and Psychiatry, 27*(3), 321–342.

Hobson, R., Ouston, J., & Lee, A. (1988). What's in a face? The case of autism. *British Journal of Psychology, 79*, 441–453.

Holcomb, P., Coffey, S., & Neville, H. (1992). Visual and auditory sentence processing: A developmental analysis using event-related brain potentials. *Developmental Neuropsychology, 8*(2–3), 203–241.

Huemer, S. V., & Mann, V. (2009). A comprehensive profile of decoding and comprehension in autism spectrum disorders. *Journal of Autism and Developmental Disorders, 40*(4), 485–493.

Itier, R. J., & Taylor, M. J. (2002). Inversion and contrast polarity reversal affect both encoding and recognition processes of unfamiliar faces: A repetition study using ERPs. *NeuroImage, 15*(2), 353–372.

Itier, R. J., & Taylor, M. J. (2004). Source analysis of the N170 to faces and objects. *Neuroreport, 15*(8), 1261–1265.

James, K. H., James, T. W., Jobard, G., Wong, A. C., & Gauthier, I. (2005). Letter processing in the visual system: Different activation patterns for single letters and strings. *Cognitive, Affective, and Behavioral Neuroscience, 5*(4), 452–466.

Jemel, B., Mottron, L., & Dawson, M. (2006). Impaired face processing in autism: Fact or artifact? *Journal of Autism and Developmental Disorders, 36*(1), 91–106.

Johnson, M. H., Dziurawiec, S., Ellis, H., & Morton, J. (1991). Newborns' preferential tracking of face-like stimuli and its subsequent decline. *Cognition, 40*(1–2), 1–19.

Jones, W., Carr, K., & Klin, A. (2008). Absence of preferential looking to the eyes of approaching adults predicts level of social disability in 2-year-old toddlers with autism spectrum disorder. *Archives of General Psychiatry, 65*(8), 946–954.

Jones, W., & Klin, A. (2009). Heterogeneity and homogeneity across the autism spectrum: The role of development. *Journal of the American Academy of Child and Adolescent Psychiatry, 48*(5), 471–473.

Kanner, L. (1943). Autistic disturbances of affective contact. *Nervous Child, 2*, 217–250.

Kanwisher, N., McDermott, J., & Chun, M. M. (1997). The fusiform face area: A module in human extrastriate cortex specialized for face perception. *Journal of Neuroscience, 17*(11), 4302–4311.

Kemner, C., Schuller, A. M., & van Engeland, H. (2006). Electrocortical reflections of face and gaze processing in children with pervasive developmental disorder. *Journal of Child Psychology and Psychiatry, 47*(10), 1063–1072.

Key, A. P., Dove, G. O., & Maguire, M. J. (2005). Linking brainwaves to the brain: An ERP primer. *Developmental Neuropsychology, 27*(2), 183–215.

Klin, A., Jones, W., Schultz, R., Volkmar, F., & Cohen, D. (2002). Visual fixation patterns during viewing of naturalistic social situations as predictors of social competence in individuals with autism. *Archives of General Psychiatry, 59*(9), 809–816.

Klin, A., Sparrow, S., De Bildt, A., Cicchetti, D., Cohen, D., & Volkmar, F. (1999). A normed study of face recognition in autism and related disorders. *Journal of Autism and Developmental Disorders, 29*(6), 499–508.

Krigolson, O. E., Pierce, L. J., Holroyd, C. B., & Tanaka, J. W. (2009). Learning to become an expert: Reinforcement learning and the acquisition of perceptual expertise. *Journal of Cognitive Neuroscience, 21*(9), 1834–1841.

Langdell, T. (1978). Recognition of faces: An approach to the study of autism. *Journal of Child Psychology and Psychiatry, 19*(3), 255–268.

Lord, C., Risi, S., Lambrecht, L., Cook, E. H., Leventhal, B. L., DiLavore, P. C., et al. (2000). The Autism Diagnostic Observation Schedule–Generic: A standard measure of social and communication deficits associated with the spectrum of autism. *Journal of Autism and Developmental Disorders, 30*(3), 205–223.

Lord, C., Rutter, M., & Le Couteur, A. (1994). Autism Diagnostic Interview-Revised: A revised version of a diagnostic interview for caregivers of individuals with possible pervasive developmental disorders. *Journal of Autism and Developmental Disorders, 24*(5), 659–685.

Maestro, S., Muratori, F., Cavallaro, M. C., Pei, F., Stern, D., Golse, B., et al. (2002). Attentional skills during the first 6 months of age in autism spectrum disorder. *Journal of the American Academy of Child and Adolescent Psychiatry, 41*(10), 1239–1245.

Martelli, M., Majaj, N. J., & Pelli, D. G. (2005). Are faces processed like words? A diagnostic test for recognition by parts. *Journal of Vision, 5*(1), 58–70.

Maurer, U., Brem, S., Bucher, K., & Brandeis, D. (2005). Emerging neurophysiological specialization for letter strings. *Journal of Cognitive Neuroscience, 17*(10), 1532–1552.

Maurer, U., Brem, S., Kranz, F., Bucher, K., Benz, R., Halder, P., et al. (2006). Coarse neural tuning for print peaks when children learn to read. *NeuroImage, 33*(2), 749–758.

McCandliss, B. D., Cohen, L., & Dehaene, S. (2003). The visual word form area: Expertise for reading in the fusiform gyrus. *Trends in Cognitive Sciences, 7*(7), 293–299.

McCleery, J. P., Akshoomoff, N., Dobkins, K. R., & Carver, L. J. (2009). Atypical face versus object processing and hemispheric asymmetries in 10-month-old infants at risk for autism. *Biological Psychiatry, 66*(10), 950–957.

McPartland, J. C., Cheung, C. H., Perszyk, D., & Mayes, L. C. (2010). Face-related ERPs are modulated by point of gaze. *Neuropsychologia, 48*(12), 3657–3660.

McPartland, J. C., Dawson, G., Webb, S. J., Panagiotides, H., & Carver, L. J. (2004). Event-related brain potentials reveal anomalies in temporal processing of faces in autism spectrum disorder. *Journal of Child Psychology and Psychiatry, 45*(7), 1235–1245.

McPartland, J. C., Perszyk, D., Crowley, M., Naples, A. J., & Mayes, L. (2011, August). Visual attention modulates neural response to faces in autism. Paper accepted for presentation at the American Psychological Association Annual Convention.

McPartland, J. C., Webb, S. J., Keehn, B., & Dawson, G. (2011). Patterns of visual attention to faces and objects in autism spectrum disorder. *Journal of Autism and Developmental Disorders, 41*(2), 148–157.

Meltzoff, A. N., & Moore, M. K. (1977). Imitation of facial and manual gestures by human neonates. *Science, 198*(4312), 74–78.

Minshew, N. J., & Williams, D. L. (2007). The new neurobiology of autism: Cortex, connectivity, and neuronal organization. *Archives of Neurology, 64*(7), 945–950.

Murias, M., Swanson, J. M., & Srinivasan, R. (2007). Functional connectivity of frontal cortex in healthy and ADHD children reflected in EEG coherence. *Cerebral Cortex, 17*(8), 1788–1799.

Nation, K., Clarke, P., Wright, B., & Williams, C. (2006). Patterns of reading ability in children with autism spectrum disorder. *Journal of Autism and Developmental Disorders, 36*(7), 911–919.

Nelson, C. (2001). The development and neural bases of face recognition. *Infant and Child Development, 10*, 3–18.

Newman, T. M., Macomber, D., Naples, A. J., Babitz, T., Volkmar, F., & Grigorenko, E. L. (2006). Hyperlexia in children with autism spectrum disorders. *Journal of Autism and Developmental Disorders, 37*(4), 760–774.

O'Connor, K., Hamm, J. P., & Kirk, I. J. (2005). The neurophysiological correlates of face processing in adults and children with Asperger's syndrome. *Brain and Cognition, 59*(1), 82–95.

O'Connor, K., Hamm, J. P., & Kirk, I. J. (2007). Neurophysiological responses to face, facial regions and objects in adults with Asperger's syndrome: An ERP investigation. *International Journal of Psychophysiology, 63*(3), 283–293.

O'Connor, I. M., & Klein, P. D. (2004). Exploration of strategies for facilitating the reading comprehension of high-functioning students with autism spectrum disorders. *Journal of Autism and Developmental Disorders, 34*(2), 115–127.

Oldfield, R. (1971). The assessment and analysis of handedness: The Edinburgh inventory. *Neuropsychologia, 9*, 97–113.

Osterling, J. A., & Dawson, G. (1994). Early recognition of children with autism: A study of first birthday home videotapes. *Journal of Autism and Developmental Disorders, 24*(3), 247–257.

Paul, R. (2003). Promoting social communication in high functioning individuals with autistic spectrum disorders. *Child & Adolescent Psychiatric Clinics of North America, 12*(1), 87–106.

Perszyk, D., Molfese, P., Kilroy, E., Mayes, L., Klin, A., & McPartland, J. (2010). Developmental brain bases of face perception in autism as revealed by ERPs. Paper presented at the International Meeting for Autism Research, Philadelphia, PA.

Psychological Corporation (1999). *Wechsler Abbreviated Scale of Intelligence (WASI) manual*. San Antonio, TX: Psychological Corporation.

Puce, A., Allison, T., Gore, J., & McCarthy, G. (1995). Face-sensitive regions in human extrastriate cortex studied by functional MRI. *Journal of Neurophysiology, 74*(3), 1192–1199.

Rossion, B., Gauthier, I., Goffaux, V., Tarr, M. J., & Crommelinck, M. (2002). Expertise training with novel objects leads to left-lateralized facelike electrophysiological responses. *Psychological Science, 13*(3), 250–257.

Rossion, B., Gauthier, I., Tarr, M. J., Despland, P., Bruyer, R., Linotte, S., et al. (2000). The N170 occipito-temporal component is delayed and enhanced to inverted faces but not to inverted objects: An electrophysiological account of face-specific processes in the human brain. *Neuroreport, 11*(1), 69–74.

Rossion, B., Joyce, C. A., Cottrell, G. W., & Tarr, M. J. (2003). Early lateralization and orientation tuning for face, word, and object processing in the visual cortex. *NeuroImage, 20*(3), 1609–1624.

Sasson, N. J. (2006). The development of face processing in autism. *Journal of Autism and Developmental Disorders, 36*(3), 381–394.

Schneider, W., Eschman, A., & Zuccolotto, A. (2002). *E-prime user's guide*. Pittsburgh, PA: Psychology Software Tools, Inc.

Schultz, R. T. (2005). Developmental deficits in social perception in autism: The role of the amygdala and fusiform face area. *International Journal of Developmental Neuroscience, 23*(2–3), 125–141.

Schultz, R. T., Gauthier, I., Klin, A., Fulbright, R. K., Anderson, A. W., Volkmar, F., et al. (2000). Abnormal ventral temporal cortical activity during face discrimination among individuals with autism and Asperger syndrome. *Archives of General Psychiatry, 57*(4), 331–340.

Schweinberger, S. R., Pickering, E. C., Jentzsch, I., Burton, A. M., & Kaufmann, J. M. (2002). Event-related brain potential evidence for a response of inferior temporal cortex to familiar face repetitions. *Cognitive Brain Research, 14*(3), 398–409.

Scott, L. S., Tanaka, J. W., Sheinberg, D. L., & Curran, T. (2006). A reevaluation of the electrophysiological correlates of expert object processing. *Journal of Cognitive Neuroscience, 18*(9), 1453–1465.

Senju, A., Tojo, Y., Yaguchi, K., & Hasegawa, T. (2005). Deviant gaze processing in children with autism: An ERP study. *Neuropsychologia, 43*(9), 1297–1306.

Shibata, T., Nishijo, H., Tamura, R., Miyamoto, K., Eifuku, S., Endo, S., et al. (2002). Generators of visual evoked potentials for faces and eyes in the human brain as determined by dipole localization. *Brain Topography, 15*(1), 51–63.

Tanaka, J. W., & Farah, M. J. (1993). Parts and wholes in face recognition. *Quarterly Journal of Experimental Psychology. Section A, Human Experimental Psychology, 46*(2), 225–245.

Tarr, M. J., & Gauthier, I. (2000). FFA: A flexible fusiform area for subordinate-level visual processing automatized by expertise. *Nature Neuroscience, 3*(8), 764–769.

Taylor, M. J., Batty, M., & Itier, R. J. (2004). The faces of development: A review of early face processing over childhood. *Journal of Cognitive Neuroscience, 16*(8), 1426–1442.

Tucker, D. M. (1993). Spatial sampling of head electrical fields: The geodesic sensor net. *Electroencephalography and Clinical Neurophysiology, 87*(3), 154–163.

Turkeltaub, P. E., Flowers, D. L., Verbalis, A., Miranda, M., Gareau, L., & Eden, G. F. (2004). The neural basis of hyperlexic reading: An fMRI case study. *Neuron, 41*(1), 11–25.

Webb, S. J., Dawson, G., Bernier, R., & Panagiotides, H. (2006). ERP evidence of atypical face processing in young children with autism. *Journal of Autism and Developmental Disorders, 36*(7), 881–890.

Webb, S. J., Jones, E. J., Merkle, K., Murias, M., Greenson, J., Richards, T., et al. (2010). Response to familiar faces, newly familiar faces, and novel faces as assessed by ERPs is intact in adults with autism spectrum disorders. *International Journal of Psychophysiology, 77*(2), 106–117.

Webb, S. J., Merkle, K., Murias, M., Richards, T., Aylward, E., & Dawson, G. (2009). ERP responses differentiate inverted but not upright face processing in adults with ASD. *Social Cognitive and Affective Neuroscience.* Advance online publication. doi: 10.1093/scan/nsp002

Wechsler, D. (1997). *Manual for the Wechsler Adult Intelligence Scale* (3rd ed.). San Antonio, TX: Psychological Corporation.

Wechsler, D. (2003). *The Wechsler Intelligence Scale for Children* (4th ed.). San Antonio, TX: Psychological Corporation.

Wolf, J. M., Tanaka, J. W., Klaiman, C., Cockburn, J., Herlihy, L., Brown, C., et al. (2008). Specific impairment of face processing abilities in children with autism spectrum disorder using the *Let's face it!* skills battery. *Autism Research, 1*(6), 329–340.

Wong, A. C., Gauthier, I., Woroch, B., DeBuse, C., & Curran, T. (2005). An early electrophysiological response associated with expertise in letter perception. *Cognitive, Affective, and Behavioral Neuroscience, 5*(3), 306–318.

Woodcock, R., McGrew, K., & Mather, N. (2001). *Woodcock–Johnson III Tests of Achievement.* Itasca, IL: Riverside Publishing.

Yin, R. (1970). Face recognition by brain-inujred patients: A dissociable ability. *Neuropsychologia, 8*, 395–402.

Zhou, Y., Dougherty, J. H., Jr., Hubner, K. F., Bai, B., Cannon, R. L., & Hutson, R. K. (2008). Abnormal connectivity in the posterior cingulate and hippocampus in early Alzheimer's disease and mild cognitive impairment. *Alzheimer's and Dementia, 4*(4), 265–270.

The specific impairment of fearful expression recognition and its atypical development in pervasive developmental disorder

Shota Uono[1], Wataru Sato[2], and Motomi Toichi[1]

[1]Graduate School of Medicine, Faculty of Human Health Science, Kyoto University, Kyoto, Japan
[2]The Hakubi Project, Primate Research Institute, Kyoto University, Inuyama, Aichi, Japan

Several studies have examined facial expression recognition in pervasive developmental disorder (PDD), including autism and Asperger's disorder, but the results have been inconsistent. We investigated the relationship between facial expression recognition and age, face recognition, and symptom severity. Subjects were 28 individuals with mild PDD subtypes and 28 age- and gender-matched controls. Among six emotions, fearful expression recognition was specifically impaired in PDD subjects. Age had positive effects on fearful expression recognition directly and indirectly via the development of face recognition in controls, but not in PDD subjects. Furthermore, fearful expression recognition was related to the severity of PDD symptoms. We conclude that individuals with PDD show an atypical development of facial expression recognition. Moreover, impaired fearful expression recognition is closely related to social dysfunction.

Keywords: Face recognition; Facial expression recognition; Fear; Pervasive developmental disorder; Social dysfunction.

Individuals with pervasive developmental disorder (PDD), including autism and Asperger's disorder, are characterized by a qualitative impairment of social interaction (American Psychiatric Association 2000). Kanner's original clinical study emphasized that individuals with autism have innately impaired affective contact with others (Kanner, 1943), and difficulty in the expression and perception of emotion is proposed to contribute to a failure to establish interpersonal relationships (Hobson, 1993). Considerable research has focused on the ability to recognize emotion from the facial expressions of others to elucidate the cause of this social dysfunction.

However, previous studies investigating emotion recognition in PDD have reported inconsistent findings. Several studies have demonstrated impaired facial expression recognition in PDD (Braverman, Fein, Lucci, & Waterhouse, 1989; Celani, Battacchi, & Arcidiacono, 1999), with others further suggesting that individuals with PDD are specifically impaired in recognizing fearful expressions (Ashwin, Chapman, Colle, & Baron-Cohen, 2006; Corden, Chilvers, & Skuse, 2008; Howard et al., 2000; Humphreys, Minshew, Leonard, & Behrmann, 2007; Pelphrey et al., 2002). However, some studies have reported that individuals with PDD showed no impairment in facial expression recognition (Adolphs, Sears, & Piven, 2001; Castelli, 2005; Grossman, Klin, Carter, & Volkmar, 2000).

These inconsistent findings regarding facial expression recognition in PDD may be due to a number of potential factors. First, the majority of the previous studies lacked a developmental perspective for facial expression recognition in individuals with PDD.

Correspondence should be addressed to: Shota Uono, Graduate School of Medicine, Faculty of Human Health Science, Kyoto University, Shogoin Kawahara-cho, Sakyo-ku, Kyoto 606-8507, Japan. E-mail: uonoshota1982@gmail.com

This study was supported by a Grant-in-Aid for JSPS Fellows, JSPS Funding Program for Next Generation World-Leading Researchers, and the Benesse Corporation. We would like to thank Yukari Ise, MA, for technical support. We are grateful to the volunteers who participated in the research and their parents.

The ability to recognize facial expressions improves with age during childhood and adolescence in typically developing individuals (for review, see Herba & Phillips, 2004), but little is known about the development of facial expression recognition in individuals with PDD. However, the review described above suggests atypical development of facial expression recognition in individuals with PDD (for a review, see also Harms, Martin, & Wallace, 2010). Recent studies with a large number of participants have shown deficits in the recognition of facial expressions, specifically fear, in adults (Ashwin et al., 2006; Corden et al., 2008; Humphreys, Minshew, et al., 2007), but not children, with PDD (Castelli, 2005; Grossman et al., 2000). Given that the ability to recognize faces improves with age in typically developing individuals, these data suggest that the ability to recognize facial expressions does not improve with age in individuals with PDD.

Second, previous studies did not examine the effects of the ability to perceive faces on facial expression recognition. Theoretical cognitive psychological studies have proposed that emotional facial recognition occurs through the basic visual processing of faces (e.g., Bruce & Young, 1986). This notion is supported by experimental studies showing that the face-recognition skill involving perceptual matching was positively correlated with the ability to recognize others' emotions (e.g., Bruce et al., 2000; Williams, Wishart, Pitcairn, & Willis, 2005). Neuropsychological studies have shown that individuals with prosopagnosia, who have difficulty in perceptually discriminating faces, have impaired emotion recognition (e.g., de Gelder, Pourtois, Vroomen, & Bachoud-Levi, 2000; Humphreys, Avidan, & Behrmann, 2007). These data suggest that individual differences in the basic ability to perceptually process faces may explain the inconsistent findings for impairment of emotion recognition in PDD.

In typically developing individuals, face-perception ability improves with age during childhood and adolescence (Carey, Diamond, & Woods, 1980; Mondloch, Geldart, Maurer, & Le Grand, 2003). This evidence suggests that the development of face perception leads to improved facial expression recognition in typically developing controls. Face perception has been shown to be impaired in children and adolescents with PDD (e.g., Boucher, Lewis, & Collis, 1998; Klin et al., 1999), suggesting that the atypical development of face-perception skills affects facial expression recognition in individuals with PDD. Previous studies have investigated the relationship between face perception and facial expression recognition in individuals with PDD (Hefter, Manoach, & Barton, 2005;

Riby, Doherty-Sneddon, & Bruce, 2008); however, the results are inconsistent. Hefter et al. demonstrated that participants with face-perception deficits recognize facial expressions as well as those with normal face-perception ability. On the other hand, Riby et al. showed that face-perception ability was positively correlated with facial expression recognition in individuals with PDD. However, these studies did not use all six basic facial expressions, and the chronological age of participants differed between studies. Thus, further studies are needed to clarify whether atypical development of face perception leads to deficits in the recognition of six basic facial expressions in individuals across a broader chronological age range.

Third, the degree of social dysfunction in individuals with PDD may relate to deficits in facial expression recognition. In normal participants, performance in face-perception tasks involving fearful faces correlates with higher social cognitive functions (e.g., theory of mind ability) (Corden, Critchley, Skuse, & Dolan, 2006; Marsh, Kozak, & Ambady, 2007). Although the relationship between emotion recognition and symptom severities in individuals with PDD has been investigated (e.g., Braverman et al., 1989; Corden et al., 2008; Tardif, Lainé, Rodriguez, & Gepner, 2007), little evidence exists of a relationship between fear recognition and symptom severity (cf. Humphreys, Minshew, et al., 2007). Thus, we tested whether the degree of impairment in facial expression recognition positively correlates with social dysfunction in individuals with PDD.

We investigated facial expression recognition deficits across development in individuals with high-functioning PDD and examined the recognition of facial expressions conveying the six basic emotions. We predicted that individuals with PDD would show impaired emotion recognition, particularly recognition of fearful expression. We also investigated the relationship between chronological age, face perception, and facial expression recognition. We tested the following model in typically developing controls and individuals with PDD by path analysis: (1) facial expression recognition and face perception improve with age; (2) the development of face perception leads to the improvement of facial expression recognition. Finally, we tested the relationship between impaired facial expression recognition and symptom severity in individuals with PDD. Based on the evidence described above, we predicted that recognition of fearful expressions would be negatively correlated with social dysfunction in individuals with PDD.

METHODS

Participants

The participants were 56 Japanese individuals, 28 with PDD, and 28 typically developing controls. Table 1 summarizes the participants' demographic characteristics. The two groups (PDD and control) were matched for chronological age—PDD group: $M \pm SD = 17.6 \pm 5.2$, range 9–30; control: $M \pm SD = 18.0 \pm 4.0$, range 9–28; independent t-test, $t(54) = 0.29$, $p > .1$—and gender—PDD group: five females and 23 males; control: four females and 24 males; Fisher's exact test, $p > .1$. Verbal and performance IQ in the PDD group was measured by the Japanese version of the Wechsler Adult Intelligence Scale (WAIS) and Wechsler Intelligence Scale for Children (WISC). All PDD participants had IQs within the normal range (full-scale IQ: $M = 103.3$, $SD = 13.4$; verbal IQ: $M = 105.2$, $SD = 14.7$; performance IQ: $M = 100.1$, $SD = 13.3$). All participants had normal or corrected-to-normal visual acuity.

Participants in the PDD group were diagnosed with either Asperger's disorder (12 males, three females) or pervasive developmental disorder not otherwise specified (PDD-NOS) (11 males, two females) at the time of the present study, using DSM-IV-TR (American Psychiatric Association, 2000). PDD-NOS includes heterogeneous subgroups of PDD with varying degrees of qualitative social impairment. The present study used subgroups that did not satisfy the criteria for Asperger's disorder because (1) they had similar impairment in qualitative social interaction without apparent restricted interests and stereotyped behaviors, or (2) their impairment in qualitative social interaction was milder than that observed in Asperger's disorder. Thus, our participants with PDD-NOS had milder pathologies than those with Asperger's disorder. The final diagnoses were made by a child psychiatrist (M.T.) based on the advice of clinical

psychologists, interviews with each subject, information from each subject's parents or teachers, and childhood clinical records (when available). Participants in the PDD group were outpatients who had been referred to Kyoto University Hospital or the Faculty of Human Health Science of Kyoto University Graduate School of Medicine because of their social maladaptation. The participants were referred from a variety of sources, and we found no systematic referral source bias between younger and older participants. They were all free of neurological or psychiatric problems other than PDD, and none were taking any medication. All participants aged 18 years and older and the parents of participants aged younger than 18 years provided written, informed consent to participate in this study, which was approved by the local ethics committee.

The level of symptom severity in individuals with PDD was assessed with the Childhood Autism Rating Scale (CARS) (Schopler, Reichler, & Renner, 1986) administered by a psychiatrist (M.T.). The CARS has been shown to be an effective tool for diagnosing autism in adolescents, adults, and children (Mesibov, Schopler, Schaffer, & Michal, 1989). The CARS includes 14 items assessing autism-related behavior and one item rating general impressions of autistic symptoms. Each item is rated on a scale of 1 to 4. A higher rating indicates more severe impairment. Total scores ranged from 15 to 60. The CARS scores (range 18–25.5) of all participants in the PDD group were below the cutoff score (27) for a diagnosis of autism (cf. Mesibov et al., 1989), indicating that symptom severity in the PDD group was milder than that of individuals with autism.

To investigate age differences in the PDD group according IQ and sex, we divided participants into younger ($n = 14$) and older ($n = 14$) groups. The results showed no significant differences in the distributions of IQ, $t(26) < 1.6$, $p > .1$, or of men and women (Fisher's exact test, $p > .1$) in the younger and older PDD groups.

One participant in the PDD group scored below the cutoff score for face perception, as described below. However, we found no difference between analyses that included and those that excluded this individual; thus, we included this participant in all analyses.

Stimuli and procedures

Expression recognition task

A total of 48 photographs of facial expressions depicting six basic emotions (anger, disgust, fear,

TABLE 1
Subject demographics in PDD and control (CON) groups

	PDD (n = 28)	CON (n = 28)
Male:female	23:5	24:4
Age range	9–30	9–28
Asperger:PDD-NOS	15: 3	–
	Mean (SD)	Mean (SD)
Age	17.6 (5.2)	18.0 (4.0)
Verbal IQ	105.5 (14.7)	–
Performance IQ	100.1 (13.3)	–
Full-scale IQ	103.3 (13.1)	–
CARS	21.3 (2.7)	–

happiness, sadness, and surprise) were used as stimuli. Half of these were pictures of Caucasian models, and the remaining half were pictures of Japanese models. These pictures were chosen from standardized photograph sets (Ekman & Friesen, 1976; Matsumoto & Ekman, 1988). A label-matching paradigm previously used by Sato et al. (2002) was employed to assess participants' recognition of emotional facial expressions. Pictures of people whose faces expressed various emotions were presented on the monitor one by one in a random order. Verbal labels identifying the six basic emotions were presented next to each photograph. Participants were asked to select the label that best described the emotion shown in each photograph. They were instructed to consider all six alternatives carefully before responding. No time limits were set, and no feedback was provided about performance. Participants saw each emotional expression eight times, resulting in a total of 48 trials for each participant. Prior to the experiment, we established that all participants understood the meaning of the emotional label and the task instructions, and participants were given two training trials to become familiar with the procedure. After ensuring that participants understood the task requirements, the experimental trials were initiated.

Face-perception task

The shortened version (13 items) of the Benton Facial Recognition Test (Benton, Sivan, Hamsher, Varney, & Spreen, 1994) was conducted. Performance on this test is based on perceptual factors and reflects basic visual face-processing mechanisms (e.g., Bentin, Deouell, & Soroker, 1999). Caucasian models were used for all of the face stimuli. Participants were required to match a target face with one picture or with up to three pictures of the same person (with different orientation and lighting) presented in a six-stimulus array of faces. No time limits were set, and no feedback was provided regarding performance.

Apparatus

The events were controlled by SuperLab Pro 2.0 (Cedrus, San Pedro, CA, USA) implemented on a Windows computer (HP xw4300 Workstation, Hewlett-Packard, Palo Alto, CA, USA). Stimuli were presented on a 19-inch CRT monitor (HM704UC, Iiyama, Tokyo, Japan; screen resolution 1024 × 768 pixels; refresh rate 100 Hz).

Data analysis

The t-tests, analyses of variance (ANOVAs) and the follow-up tests, and correlation analyses were conducted with SPSS 10.0J (SPSS, Tokyo, Japan). Randomization tests were conducted with programs developed by Edgington and Onghena (2007). Path analyses were conducted with XLSTAT-PLSPM (Addinsoft, New York, NY, USA).

Accuracy percentages for the expression recognition task were tested for difference from chance (i.e., 16.7%), using one-sample t-tests (two-tailed). The accuracy data were then subjected to a 2 (group) × 6 (facial emotion) repeated-measures ANOVA. Significant interactions were followed up by simple effects analyses (cf. Kirk, 1995). Additionally, we conducted randomization tests using the same designs to confirm the results without parametric assumptions (cf. Edgington & Onghena, 2007). Because a debate about testing interaction terms by randomization persists (cf. Anderson & ter Braak, 2003), we tested only main and simple main effects. We further tested the effects of stimulus-face ethnicity and PDD subtypes by t-tests. Correlations with IQ scores were calculated for the relationship between IQ and the impairment of facial expression recognition in the PDD group.

For the face-perception task, the total number of correct responses was calculated for each participant. The mean score difference between groups was analyzed by a t-test (two-tailed). We further tested the effects of PDD subtypes with t-tests. Correlations were calculated between face-perception performance and IQ scores.

To analyze the relationships between expression recognition, age, and face perception, Pearson's product-moment correlations between these variables were calculated for each group. We preliminarily confirmed that non-parametric correlation analysis (Spearman's rank coefficient correlation) produced identical results. Based on the results of the ANOVA for facial expression recognition, the results from the fearful expression task were used as a measure of facial expression recognition. Differences in the correlation coefficients between the PDD and control groups were also tested with the chi-square test.

Furthermore, path analyses were conducted for each group. Path analysis assesses the direct and indirect effects of explanatory variables on dependent variables; this cannot be accomplished with correlational analyses (Kothari, 1990). Based on our prediction of a relationship between chronological age, face perception, and facial expression recognition, our model hypothesized that chronological age would have direct and indirect effects on emotion recognition via the development of face perception. Path analyses were conducted with structural equation modeling (SEM), using partial least squares (PLS). Compared with the widely used covariance-based approach for SEM, the variance-based PLS approach is a powerful method for analyzing data from small samples (Chin & Newsted, 1999; Falk, & Miller, 1992). Analyses in unity mode were conducted in which latent variables associated with a single manifest variable were created (cf. Ziersch, 2005). Path coefficients were also tested for a difference from zero using bootstrap analyses (two-tailed) (Efron & Tibishirani, 1993). We analyzed whether the path coefficients of the control group were larger than those of the PDD group in each path, using randomization tests.

The CARS (Schopler et al., 1986) was used to assess the level of social dysfunction in individuals with PDD. Although the CARS factors are structured, the items included in the social functioning construct have been inconsistent among studies (Dilalla & Rogers, 1994; Magyar & Pandolfi, 2007; Stella, Mundy, & Tuchman, 1999). Therefore, we used the CARS items that were commonly classified as elements of the social functioning construct in all previous studies (cf. Magyar & Pandolfi, 2007). In the present study, we used the items "imitation," "nonverbal communication," "relationship to people," "verbal communication," and "visual response," and calculated the average score for these items. We calculated Pearson's correlation coefficients to investigate the relationship between impaired recognition of fearful expressions and symptom severity related to social domains.

The number of participants in the present study was relatively small for a correlational analysis; thus, we further analyzed whether outliers affected the results. In each correlational analysis, the Mahalanobis distance for each case was calculated to identify outliers (probability of group membership: $p < .05$). However, the results of the correlational analyses including outliers did not differ from those excluding them; thus, all participants were included in the analyses.

RESULTS

Expression-recognition task

One-sample t-tests showed that the accuracy percentage of expression recognition for each emotion category was greater than chance in both groups, $t(27) > 36.2, p < .001$.

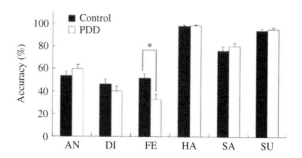

Figure 1. Mean (with *SE*) percentages of accurate facial expression recognition in typically developing controls and in individuals with PDD. Asterisk indicates significant difference between groups ($p < .05$). AN = anger; DI = disgust; FE = fear; HA = happiness; SA = sadness; SU = surprise.

The ANOVA for the accuracy percentages (Figure 1 and Table 2) revealed a significant interaction of group × facial emotion, $F(5, 270) = 4.16, p < .05$. A significant main effect of facial emotion, $F(5, 270) = 114.10, p < .001$, was also found. The main effect of group was not significant, $F(1, 54) = 1.14$. Followup analyses of the interaction revealed that the simple main effects of group, indicating less accurate recognition of emotional expressions in PDD subjects than in control group, were significant only for the fearful facial expressions, $F(1, 324) = 16.18, p < .005$.

Our preliminary analyses revealed that the normal distribution assumptions were not met for some conditions (Kolmogorov–Smirnov tests, $p < .05$ for the recognition of happy and surprised expressions in both control and PDD groups); thus, we reanalyzed the accuracy percentage data, using randomization tests. The results showed that, as in the case of the aforementioned parametric analyses, the main effect of facial emotion and the simple main effect of group in response to fearful expressions were significant ($p < .05$).

To investigate the possibility that impairment of fear recognition in the PDD group resulted from difficulty in recognizing cross-race faces (cf. Elfenbein & Ambady, 2002), we conducted a between-group comparison of fear recognition for each ethnicity. The results revealed that the recognition of fearful faces was less accurate in individuals with PDD than in typically developing controls irrespective of ethnicity— Caucasian: $t(54) = 2.06, p < .05$; Japanese: $t(54) = 3.15, p < .005$.

We further investigated whether PDD subtypes (Asperger's disorder and PDD-NOS) affected fear recognition differently. The results revealed that fear recognition was not significantly different between the PDD subtypes, $t(26) = 0.25, p > .1$. Recognition of

TABLE 2
Mean (with *SE*) scores of face perception task and mean (with *SE*) percentages of accurate emotion recognition

Group		Benton	Emotion recognition						
			AN	DI	FE	HA	SA	SU	ALL
Control	Mean	23.6	53.8	46.4	51.8	98.2	76.3	94.2	70.1
	(*SE*)	(1.5)	(4.3)	(4.3)	(3.7)	(0.9)	(3.7)	(1.7)	(1.5)
PDD	Mean	22.1	60.7	39.3	32.1	98.7	79.9	95.3	67.7
	(*SE*)	(2.3)	(4.1)	(4.6)	(5.0)	(0.8)	(3.2)	(1.9)	(1.6)

Notes: AN = anger; DI = disgust; FE = fear; HA = happy; SA = sad: SU = surprise; ALL = the mean of all conditions; PDD = pervasive developmental disorder

fearful faces was less accurate in both PDD subtype groups compared with typically developing controls—Asperger's disorder: $t(41) = 2.51, p < .05$; PDD-NOS: $t(39) = 3.08, p < .01$.

No significant correlations between the accuracy of fearful expression recognition and IQ scores were found in the PDD group ($r = -.07, .02,$ and $-.22$ for full-scale, verbal, and performance IQs, respectively; $ps > .1$). Even when the influence of chronological age was factored out, the results were identical ($r = -.07, .12,$ and $-.21$ for full-scale, verbal, and performance IQs, respectively; $ps > .1$).

Face-perception task

Benton Facial Recognition Task performance was less accurate in the PDD group than in the control group, $t(54) = 2.95, p < .05$ ($M \pm SE = 22.09 \pm 2.39$ and 23.57 ± 1.53 for PDD and control, respectively). The performance of all participants in both groups was above the cutoff score (18/27) (Benton et al., 1994) for impaired face perception, except for one participant in the PDD group.

We further analyzed whether PDD subtypes (Asperger's disorder and PDD-NOS) affected the performance of face perception. The results revealed that face perception was not significantly different between the PDD subtypes, $t(26) = 0.31, p > .1$. Both PDD groups showed less accurate face perception than typically developing controls—Asperger's disorder: $t(41) = 2.39, p < .05$; PDD-NOS: $t(39) = 2.82, p < .01$.

No significant correlations between face-perception scores and IQ scores were found in the PDD group ($r = -.10, -.02,$ and $-.17$ for full-scale, verbal, and performance IQs, respectively; $ps > .1$). Even when the influence of chronological age was factored out, the results were identical ($r = -.11, -.07,$ and $-.12,$ for full-scale, verbal, and performance IQs, respectively; $ps > .1$).

Relationships among fearful expression recognition, age, and face perception

Fearful expression recognition showed a significant positive correlation with chronological age in the control group, $r = .51, p < .01$, but not in the PDD group, $r = .07, p > .1$ (Figure 2A). The correlation between *face-perception* performance and chronological age showed a nonsignificant trend in both the control, $r = .37, p < .1$, and PDD groups, $r = .33, p < .1$ (Figure 2B).

The correlation between fearful expression recognition and face-perception performance was significant in the control group, $r = .53, p < .005$, but showed a nonsignificant trend in the PDD group, $r = .35, p < .1$ (Figure 2C).

Differences between the control and PDD group correlation coefficients were tested by the chi-square test. The results revealed that the between-group difference was marginally significant for the Age–Fear relationship ($\chi^2 = 3.03, p < .1$) but was not significant for the Age–Face ($\chi^2 = 0.03, p > .1$) and Face–Fear ($\chi^2 = 0.63, p > .1$) relationships.

Path analyses were conducted for each group to further examine the relationships among these variables. Based on the previous results in typically developing participants, the analyzed model assumed that age has positive effects on recognition of emotional expressions, both directly and indirectly via the development of face perception. The hypothesized model is presented in Figure 3.

For the control group, tests of path coefficients confirmed that all paths in the hypothesized model were significant (from age to expression recognition, from age to face perception, and from face perception to expression recognition; standardized coefficient = .37, .37, and .39, respectively; all $p < .05$).

The path coefficient tests revealed that no paths in the hypothesized model reached significance in the

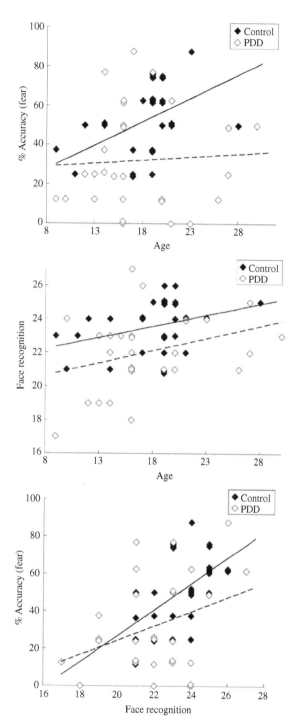

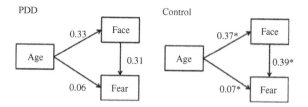

Figure 3. The hypothesized model for the development of fearful face recognition. Face perception and fearful expression recognition improve with age, and the development of face perception improves fearful face recognition. The hypothesized model shows better fits in typically developing controls (right panel), but not in individuals with PDD (left panel). Asterisks indicate significant path coefficients ($p < .05$).

PDD group (from age to expression recognition, from age to face perception, and from face perception to expression recognition; standardized coefficient = .06, .33, and .31, respectively; $ps > .1$).

We conducted group comparisons in each path, using randomization tests. The results revealed that the control group path coefficients were significantly larger than those of PDD group in the Age–Fear path ($p < .05$) but not in the Age–Face and Face–Fear paths ($ps > .1$).

The relationship between impaired fearful expression recognition and symptom severity

The average score of four items used as indices of social dysfunction ranged from 1.3 to 2.2. Correlation analyses revealed that fearful expression recognition in the PDD group was negatively and significantly correlated with social dysfunction, $r = -.51$, $p < .005$. Thus, individuals with PDD who showed worse recognition of fearful expressions had more severe symptoms in social domains (Figure 4). Even when the influence of chronological age, verbal IQ, and performance IQ was factored out, the correlation remained significant, $r = -.57$, $p < .005$.

DISCUSSION

The present study revealed that individuals with PDD were less accurate in recognizing fearful facial expressions than were typically developing controls. Consistent with our data, recent studies have shown impaired facial expression recognition, particularly of fearful faces in subjects with PDD (Ashwin et al., 2006; Corden et al., 2008; Humphreys, Minshew, et al.,

Figure 2. The relationships between chronological age, fearful expression recognition, and face perception. (a) The percentage of accurate fearful expression recognition is plotted against the chronological age of each participant. (b) The face-perception task score is plotted against the chronological age of each participant. (c) The percentage of accurate fearful expression recognition is plotted against the score in the face-perception task. Black and white diamonds represent each participant in the control and PDD group, respectively. Solid and broken lines represent linear regressions in the control and PDD group, respectively.

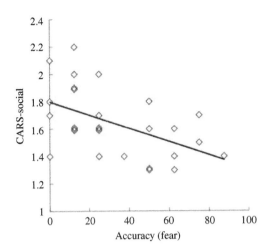

Figure 4. The relationship between the percentage of accurate fearful expression recognition and the degree of social dysfunction evaluated by the CARS. Severe social dysfunction predicts the poor recognition of fearful expressions in individuals with PDD.

2007). Although some reports have documented general impairment of facial expression recognition, the impairment of fear recognition was found irrespective of face ethnicity and PDD subtype in the present study. Our findings suggest that individuals with PDD have a greater tendency to show impaired fearful expression recognition compared to other facial expressions.

Our results also show that the ability to recognize fearful facial expressions improves with age in typically developing controls, but not in PDD subjects; hence, the impairment of fearful face recognition in the PDD group manifested in adult subjects. Consistent with these data, recent studies have shown impaired fearful face recognition in adults (Ashwin et al., 2006; Corden et al., 2008; Humphreys, Minshew, et al., 2007), but not in children with PDD (Castelli, 2005; Grossman et al., 2000). However, these findings do not imply normal emotion processing in children with PDD. For example, Dawson, Webb, Carver, Panagiotides, and McPartland (2004) demonstrated that children with PDD aged 3–5 years show atypical brain responses to fearful faces, suggesting that individuals with PDD have impaired fearful face processing during childhood. Developmental psychology studies have shown that the accurate recognition of fearful expressions emerges later than that for other emotions, except for disgust, even in typically developing children (Holder & Kirkpatrick, 1991; Vicari, Reilly, Pasqualetti, Vizzotto, & Caltagirone, 2000). Based on these findings, the paradigm used here, that is, matching facial photographs with the appropriate verbal label, may be less sensitive to group differences in fearful expression recognition in childhood.

The path analysis revealed that face-perception ability improves with age in controls, but not in PDD

subjects. Furthermore, typically developing controls, but not individuals with PDD, showed a significant positive relationship between face perception and fearful expression recognition. Hefter et al. (2005) showed that face-perception performance does not positively correlate with facial expression recognition in individuals with PDD, and previous studies have shown that face perception (for a review, see Maurer, Le Grand, & Mondloch, 2002) and facial expression recognition (Calder, Young, Keane, & Dean, 2000; Durand, Gallay, Seigneuric, Robichon, & Baudouin, 2007) rely on facial configuration processing in typically developing individuals. The detection of subtle changes in facial configuration (e.g., in the eye region) is required to discriminate between fearful and surprised faces (Ekman, 2003; Skuse, 2003). These findings suggest that the development of perceptual face processing facilitates fearful expression recognition in typically developing controls, but not in individuals with PDD.

The results revealed that ability to recognize fearful expressions did not improve with age in the PDD group independent of face-perception skill. Other components of facial expression processing may influence the performance of fear recognition in individuals with PDD. The finding of a group difference in the Age–Fear path, but not in the Age–Face and Face–Fear paths, suggests that deficits in emotion processing play an important role in impaired fearful expression recognition. Several researchers have proposed that emotional reactions in response to the facial expressions of others contribute to accurate facial expression recognition (e.g., Adolphs, 2002). Consistent with this theory, studies have suggested that a callous, unemotional trait (e.g., lack of empathy) specifically relates to impaired fearful face recognition (for a review, see Marsh & Blair, 2008). Minio-Paluello, Baron-Cohen, Avenanti, Walsh, and Aglioti (2009) found that individuals with PDD did not show empathetic bodily responses, and McIntosh, Reichmann-Decker, Winkielman, and Wilbarger (2006) reported that individuals with PDD did not exhibit spontaneous facial mimicry of other people's emotional expressions. Although individuals with PDD generally show high anxiety (e.g., Muris, Steerneman, Merckelbach, Holdrinet, & Meesters, 1998), an observational study suggested that their emotional response did not change in response to others' emotion (Corona, Dissanayake, Arbelle, Wellington, & Sigman, 1998). Taken together, these findings suggest that atypical emotional responses may explain poor fear recognition in individuals with PDD.

An alternative interpretation is that impaired face and emotion recognition is the result of reduced eye gaze fixation in individuals with PDD (for a review, see Senju & Johnson, 2009). For example, Corden et al.

(2008) demonstrated that reduced fixation on others' gaze predicted the degree of impairment in fear recognition shown by individuals with PDD. Similarly, our study found that the averaged CARS score involving attention to other individuals ("relationship to people" and "visual response") was negatively correlated with the impairment of fear recognition ($r = -.40$, $p < .05$). However, other studies have reported that individuals with PDD did not show gaze avoidance while observing others' emotional (Rutherford & Towns, 2008) and non-emotional faces (van der Geest, Kemner, Verbaten, & van Engeland, 2002). Bal et al. (2010) showed that children with PDD were not less accurate in recognizing fearful facial expressions, though they show less fixation on eye gaze when they observe fearful faces. We did not track the participants' eye movements and cannot comment on this issue. The gaze-avoidance hypothesis of emotion-recognition impairment is intriguing and warrants future investigation. The presence of confounding factors, such as chronological age and anxiety level, that may have contributed to inconsistent findings in previous studies make further studies necessary to clarify this issue.

Finally, we showed that impaired fearful expression recognition was related to social dysfunction in individuals with PDD. Consistent with this finding, previous studies have shown that accurate recognition of fearful faces is positively correlated with the theory of mind in typically developing individuals (Corden et al., 2006). Recognition of fearful faces is proposed to be important for the development of social cognitive function (such as theory of mind); fearful faces signal a threat and facilitate the interpretation of the other person's thoughts (Skuse, 2003). These findings suggest that fearful face recognition may contribute to the development of several social cognitive functions. Furthermore, our results suggest that previous inconsistent findings can be partially accounted for by the severity of social dysfunction in individuals with PDD.

A potential neural substrate for impaired recognition of fearful expressions in individuals with PDD is the amygdala. Previous neuropsychological studies have demonstrated that the amygdala plays an important role in fearful expression recognition (e.g., Sato et al., 2002). A recent functional magnetic resonance imaging (fMRI) study reported that the amygdala showed less activation to fearful faces in individuals with PDD (Ashwin, Baron-Cohen, Wheelwright, O'Riordan, & Bullmore, 2007). Furthermore, in line with the finding that fear recognition improved with age in controls, but not in individuals with PDD, structural MRI studies have found that the amygdala volume increased from childhood to adulthood

in normal controls, but not in individuals with PDD (Nacewicz et al., 2006; Schumann et al., 2004). These data suggest that abnormal amygdala development may contribute to impaired fearful expression recognition in individuals with PDD.

The fusiform gyrus, a region that shows face-specific responses (e.g., Kanwisher, McDermott, & Chun, 1997), is a possible candidate for the abnormal development of face-perception skills in PDD. Some neuroimaging studies have found less fusiform gyrus activation to face stimuli in adults and adolescents (11–14 years old) with PDD (Pierce, Müller, Ambrose, Allen, & Courchesne, 2001; Scherf, Luna, Minshew, & Behrmann, 2010; Schultz et al., 2000). Consistent with the finding that face-perception ability increases with age in normal subjects, studies have demonstrated that the fusiform gyrus shows increasing face-specific activity across development (Aylward et al., 2005; Golalrai et al., 2007). Furthermore, fMRI studies have suggested that the fusiform gyrus shows less activation to emotional facial expressions in individuals with PDD (Hall, Szechtman, & Nahmias, 2003; Wang, Dapretto, Hariri, Sigman, & Bookheimer, 2004). These results indicate that abnormal development of the functional integrity of the fusiform gyrus may play an important role in not only face perception but also facial expression recognition in individuals with PDD.

Some limitations of the present study should be noted. First, no time limits for stimulus presentation and subject responses were set. In the real world, rapid understanding of other people's emotions is critical for undertaking appropriate behaviors. More rapid stimulus presentation in the facial expression recognition task might result in impaired recognition of other emotional facial expressions in individuals with PDD. Second, face perception and facial expression recognition in younger participants should be investigated further, as social dysfunction in PDD appears in the first year of life (Osterling, Dawson, & Munson, 2002; Ozonoff et al., 2010).

Third, we used only Caucasian faces in the face-perception task. Studies have suggested that typically developing individuals have difficulty in recognizing the faces of other races (other-race effect) (Elfenbein & Ambady, 2002). The other-race effect may explain the impairment in face perception and emotion recognition and the lack of a relationship with facial expression recognition in PDD. However, our results are not consistent with a race effect because the individuals with PDD had difficulty in recognizing fear in other- and same-race faces. A preliminary study has suggested that individuals with PDD show a reduced other-race effect in a face-memory task (Sasson, 2007). Given that the other-race effect appears at 6 months of age (Kelly et al., 2007), it is intriguing, considering the diagnosis, to ask whether individuals with PDD show the other-race effect.

Fourth, IQ and CARS scores were assessed only in individuals with PDD. The present data suggest that the impairment in fear recognition among individuals with PDD cannot be explained by general intellectual abilities alone, because the mean IQ of individuals in the PDD group was in the normal range, and IQ was not correlated with fear recognition. Furthermore, no differences were found between typically developing individuals and those with PDD in terms of the ability to recognize facial emotions depicting anger, disgust, and fear; and we also found no differences between those with PDD and typically developing individuals in the ability to recognize anger and disgust (Figure 1 and Table 2). However, Harms et al. (2010) reviewed previous studies and indicated that the IQ profiles of PDD and control groups may affect performance on emotion-recognition tasks. Thus, further studies are needed to investigate the effect of IQ profiles on emotion recognition by individuals with PDD. The CARS is not an appropriate test for the assessment of social functioning in typically developing individuals, because even individuals with Asperger's disorder and PDD-NOS showed relatively low social dysfunction scores. Scales that could assess social function in both groups would be useful to investigate individual differences in face- and emotion-recognition tasks.

In summary, our results show that the recognition of fearful faces was specifically impaired in individuals with PDD compared to controls. Fearful expression recognition improved with age in the control group, but not in the PDD group. Furthermore, fearful expression recognition was also facilitated by face-perception ability in the control, but not in the PDD group. These results reveal atypical development of facial expression recognition in PDD. In addition, impaired recognition of fearful expressions in the PDD group was related to symptom severity, suggesting a close relationship between impaired fearful expression recognition and social dysfunction in the real world.

REFERENCES

Adolphs, R. (2002). Neural systems for recognizing emotion. *Current Opinion in Neurobiology, 12,* 169–177.

Adolphs, R., Sears, L., & Piven, J. (2001). Abnormal processing of social information from faces in autism. *Journal of Cognitive Neuroscience, 13,* 232–240.

American Psychiatric Association (2000). *Diagnostic and statistical manual of mental disorders* (DSM-IV-TR). Washington, DC: Author.

Anderson, M. J., & ter Braak, C. J. F. (2003). Permutation tests for multi-factorial analysis of variance. *Journal of Statistical Computation and Simulation, 73,* 85–113.

Ashwin, C., Baron-Cohen, S., Wheelwright, S., O'Riordan, M., & Bullmore, E.T. (2007). Differential activation of the amygdala and the 'social brain' during fearful face-processing in Asperger syndrome. *Neuropsychologia, 45,* 2–14.

Ashwin, C., Chapman, E., Colle, L., & Baron-Cohen, S. (2006). Impaired recognition of negative basic emotions in autism: A test of the amygdala theory. *Social Neuroscience, 1,* 349–363.

Aylward, E. H., Park, J. E., Field, K. M., Parsons, A. C., Richards, T. L., Cramer, S. C., et al. (2005). Brain activation during face perception: Evidence of a developmental change. *Journal of Cognitive Neuroscience, 17,* 308–319.

Bal, E., Harden, E., Lamb, D., Van Hecke, A. V., Denver, J. W., & Porges, S. W. (2010). Emotion recognition in children with autism spectrum disorders: Relations to eye gaze and autonomic state. *Journal of Autism and Developmental Disorders, 40,* 358–370.

Bentin, S., Deouell, L. Y., & Soroker, N. (1999). Selective visual streaming in face recognition: Evidence from developmental prosopagnosia. *Neuroreport, 10,* 823–827.

Benton, A. L., Sivan, A. B., Hamsher, K., Varney, N. R., & Spreen, O. (1994). *Contributions to neuropsychological assessment.* New York, NY: Oxford University Press.

Boucher, J., Lewis, V., & Collis, G. (1998). Familiar face and voice matching and recognition in children with autism. *Journal of Child Psychology and Psychiatry, 39,* 171–181.

Braverman, M., Fein, D., Lucci, D., & Waterhouse, L. (1989). Affect comprehension in children with pervasive developmental disorders. *Journal of Autism and Developmental Disorders, 19,* 301–316.

Bruce, V., Campbell, R. N., Doherty-Sneddon, G., Import, A. Langton, S., McAuley, S., et al. (2000). Testing face processing skills in children. *British Journal of Developmental Psychology, 18,* 319–333.

Bruce, V., & Young, A. (1986). Understanding face recognition. *British Journal of Psychology, 77,* 305–327.

Calder, A. J., Young, A. W., Keane, J., & Dean, M. (2000). Configural information in facial expression perception. *Journal of Experimental Psychology: Human Perception and Performance, 26,* 527–551.

Carey, S., Diamond, R., & Woods, B. (1980). The development of face recognition—a maturational component. *Developmental Psychology, 16,* 257–269.

Castelli, F. (2005). Understanding emotions from standardized facial expressions in autism and normal development. *Autism, 9,* 428–449.

Celani, G., Battacchi, M. W., & Arcidiacono, L. (1999). The understanding of the emotional meaning of facial expressions in people with autism. *Journal of Autism and Developmental Disorders, 29,* 57–66.

Chin, W. W., & Newsted, P. R. (1999). Structural equation modeling analysis with small samples using partial least squares. In R. E. Hoyle (Ed.), *Statistical strategies for small sample research* (pp. 307–341). Thousand Oaks, CA: Sage.

Corden, B., Chilvers, R., & Skuse, D. (2008). Avoidance of emotionally arousing stimuli predicts social-perceptual impairment in Asperger's syndrome. *Neuropsychologia, 46*, 137–147.

Corden, B., Critchley, H. D., Skuse, D., & Dolan, R. J. (2006). Fear recognition ability predicts differences in social cognitive and neural functioning in men. *Journal of Cognitive Neuroscience, 18*, 889–897.

Corona, R., Dissanayake, C., Arbelle, S., Wellington, P., & Sigman, M. (1998). Is affect aversive to young children with autism? Behavioral and cardiac responses to experimenter distress. *Child Development, 69*, 1494–1502.

Dawson, G., Webb, S. J., Carver, L., Panagiotides, H., & McPartland, J. (2004). Young children with autism show atypical brain responses to fearful versus neutral facial expressions of emotion. *Developmental Science, 7*, 340–359.

de Gelder, B., Pourtois, G., Vroomen, J., Bachoud-Levi, A. C. (2000). Covert processing of faces in prosopagnosia is restricted to facial expressions: Evidence from cross-modal bias. *Brain and Cognition, 44*, 425–444.

DiLalla, D. L., & Rogers, S. J. (1994). Domains of the Childhood Autism Rating Scale: Relevance for diagnosis and treatment. *Journal of Autism and Developmental Disorders, 24*, 115–128.

Durand, K., Gallay, M., Seigneuric, A., Robichon, F., & Baudouin, J. Y. (2007). The development of facial emotion recognition: The role of configural information. *Journal of Experimental Child Psychology, 97*, 14–27.

Edgington, O., & Onghena, P. (2007). *Randomization tests* (4th ed.). Boca Raton, FL: Chapman & Hall/CRC.

Efron, B., & Tibishirani, R. J. (1993). *An introduction to the bootstrap*. New York, NY: Chapman & Hall.

Ekman, P. (2003). *Emotions revealed*. New York, NY: Times Books.

Ekman, P., & Friesen, W. V. (1976). *Pictures of facial affect*. Palo Alto, CA: Consulting Psychologists Press.

Elfenbein, H. A., & Ambady, N. (2002). On the universality and cultural specificity of emotion recognition: A meta-analysis. *Psychological Bulletin, 128*, 203–235.

Falk, R. F., & Miller, N. B. (1992). *A primer for soft modeling*. Akron, OH: University of Akron Press.

Golarai, G., Ghahremani, D. G., Whitfield-Gabrieli, S., Reiss, A., Eberhardt, J. L., Gabrieli, J. D., et al. (2007). Differential development of high-level visual cortex correlates with category-specific recognition memory. *Nature Neuroscience, 10*, 512–522.

Grossman, J. B., Klin, A., Carter, A. S., & Volkmar, F. R. (2000). Verbal bias in recognition of facial emotions in children with Asperger syndrome. *Journal of Child Psychology and Psychiatry, 41*, 369–379.

Hall, G. B., Szechtman, H., & Nahmias, C. (2003). Enhanced salience and emotion recognition in autism: A PET study. *American Journal of Psychiatry, 160*, 1439–1441.

Harms, M. B., Martin, A., & Wallace, G. L. (2010). Facial emotion recognition in autism spectrum disorders: A review of behavioral and neuroimaging studies. *Neuropsychology Review, 20*, 290–322.

Hefter, R. L., Manoach, D. S., & Barton, J. J. (2005). Perception of facial expression and facial identity in subjects with social developmental disorders. *Neurology, 65*, 1620–1625.

Herba, C., & Phillips, M. (2004). Annotation: Development of facial expression recognition from childhood to adolescence: Behavioural and neurological perspectives.

Journal of Child Psychology and Psychiatry, 45, 1185–1198.

Hobson, R. P. (1993). *Autism and the development of mind*. Hove, UK: Lawrence Erlbaum Associates Ltd.

Holder, H. B., & Kirkpatrick, S. W. (1991). Interpretation of emotion from facial expressions in children with and without learning disabilities. *Journal of Learning Disabilities, 24*, 170–177.

Howard, M. A., Cowell, P. E., Boucher, J., Broks, P., Mayes, A., Farrant, A., et al. (2000). Convergent neuroanatomical and behavioural evidence of an amygdala hypothesis of autism. *Neuroreport, 11*, 2931–2935.

Humphreys, K., Avidan, G., & Behrmann, M. (2007). A detailed investigation of facial expression processing in congenital prosopagnosia as compared to acquired prosopagnosia. *Experimental Brain Research, 176*, 356–373.

Humphreys, K., Minshew, N., Leonard, G. L., & Behrmann, M. (2007). A fine-grained analysis of facial expression processing in high-functioning adults with autism. *Neuropsychologia, 45*, 685–695.

Kanner, L. (1943). Autistic disturbances of affective contact. *Nervous Child, 2*, 217–250.

Kanwisher, N., McDermott, J., & Chun, M. M. (1997). The fusiform face area: A module in human extrastriate cortex specialized for face perception. *Journal of Neuroscience, 17*, 4302–4311.

Kelly, D. J., Quinn, P. C., Slater, A. M., Lee, K., Ge, L., & Pascalis, O. (2007). The other-race effect develops during infancy: Evidence of perceptual narrowing. *Psychological Science, 18*, 1084–1089.

Kirk, R. E. (1995). *Experimental design: Procedures for the behavioral sciences*. Pacific Grove, CA: Brooks-Cole.

Klin, A., Sparrow, S. S., de Bildt, A., Cicchetti, D. V., Cohen, D. J., & Volkmar, F. R. (1999). A normed study of face recognition in autism and related disorders. *Journal of Autism and Developmental Disorders, 29*, 499–508.

Kothari, C. R. (1990). *Research methodology: Methods and techniques* (2nd ed.). New Delhi India: Wiley Eastern.

Magyar, C. I., & Pandolfi, V. (2007). Factor structure evaluation of the Childhood Autism Rating Scale. *Journal of Autism and Developmental Disorders, 37*, 1787–1794.

Marsh, A. A., & Blair, R. J. (2008). Deficits in facial affect recognition among antisocial populations: A meta-analysis. *Neuroscience and Biobehavioral Reviews, 32*, 454–465.

Marsh, A. A., Kozak, M. N., & Ambady, N. (2007). Accurate identification of fear facial expressions predicts prosocial behavior. *Emotion, 7*, 239–251.

Matsumoto, D., & Ekman, P. (1988). *Japanese and Caucasian facial expressions of emotion*. San Francisco, CA: Intercultural and Emotion Research Laboratory, Department of Psychology, San Francisco State University.

Maurer, D., Le Grand, R., & Mondloch, C. J. (2002). The many faces of configural processing. *Trends in Cognitive Sciences, 6*, 255–260.

McIntosh, D. N., Reichmann-Decker, A., Winkielman, P., & Wilbarger, J. L. (2006). When the social mirror breaks: Deficits in automatic, but not voluntary, mimicry of emotional facial expressions in autism. *Developmental Science, 9*, 295–302.

Mesibov, G. B., Schopler, E., Schaffer, B., & Michal, N. (1989). Use of the Childhood Autism Rating Scale with autistic adolescents and adults. *Journal of the American Academy of Childhood and Adolescent Psychiatry, 28,* 538–541.

Minio-Paluello, I., Baron-Cohen, S., Avenanti, A., Walsh, V., & Aglioti, S. M. (2009). Absence of embodied empathy during pain observation in Asperger syndrome. *Biological Psychiatry, 65,* 55–62.

Mondloch, C. J., Geldart, S., Maurer, D., & Le Grand, R. (2003). Developmental changes in face processing skills. *Journal of Experimental Child Psychology, 86,* 67–84.

Muris, P., Steerneman, P., Merckelbach, H., Holdrinet, I., & Meesters, C. (1998). Comorbid anxiety symptoms in children with pervasive developmental disorders. *Journal of Anxiety Disorders, 12,* 387–393.

Nacewicz, B. M., Dalton, K. M., Johnstone, T., Long, M. T., McAuliff, E. M., Oakes, T. R., et al. (2006). Amygdala volume and nonverbal social impairment in adolescent and adult males with autism. *Archives of General Psychiatry, 63,* 1417–1428.

Osterling, J. A., Dawson, G., & Munson, J. A. (2002). Early recognition of 1-year-old infants with autism spectrum disorder versus mental retardation. *Development and Psychopathology, 14,* 239–251.

Ozonoff, S., Iosif, A. M., Baguio, F., Cook, I. C., Hill, M. M., Hutman, T., et al. (2010). A prospective study of the emergence of early behavioral signs of autism. *Journal of the American Academy of Child and Adolescent Psychiatry, 49,* 256–266.

Pelphrey, K. A., Sasson, N. J., Reznick, J. S., Paul, G., Goldman, B. D., & Piven, J. (2002). Visual scanning of faces in autism. *Journal of Autism and Developmental Disorders, 32,* 249–261.

Pierce, K., Müller, R. A., Ambrose, J., Allen, G., & Courchesne, E. (2001). Face processing occurs outside the fusiform 'face area' in autism: Evidence from functional MRI. *Brain, 124,* 2059–2073.

Riby, D. M., Doherty-Sneddon, G., & Bruce, V. (2008). Exploring face perception in disorders of development: Evidence from Williams syndrome and autism. *Journal of Neuropsychology, 2,* 47–64.

Rutherford, M. D., & Towns, A. M. (2008). Scan path differences and similarities during emotion perception in those with and without autism spectrum disorders. *Journal of Autism and Developmental Disorders, 38,* 1371–1381.

Sasson, N. (2007). *Face processing and the own-race bias: A study of typical development and autism.* Saarbrücken, Germany: VDM Verlag Dr. Müller.

Sato, W., Kubota, Y., Okada, T., Murai, T., Yoshikawa, S., & Sengoku, A. (2002). Seeing happy emotion in fearful and angry faces: Qualitative analysis of the facial expression recognition in a bilateral amygdala damaged patient. *Cortex, 38,* 727–742.

Scherf, K. S., Luna, B., Minshew, N., & Behrmann, M. (2010). Location, location, location: Alterations in the functional topography of face- but not object- or

place-related cortex in adolescents with autism. *Frontiers in Human Neuroscience, 4,* 2–16.

Schopler, E, Reichler, R. J., & Renner, B. R. (1986). *The Childhood Autism Rating Scale (CARS) for diagnostic screening and classification of autism.* New York, NY: Irvington.

Schultz, R. T., Gauthier, I., Klin, A., Fulbright, R. K., Anderson, A. W., Volkmar, F., et al. (2000). Abnormal ventral temporal cortical activity during face discrimination among individuals with autism and Asperger syndrome. *Archives of General Psychiatry, 57,* 331–340.

Schumann, C. M., Hamstra, J., Goodlin-Jones, B. L., Lotspeich, L. J., Kwon, H., Buonocore, M. H., et al. (2004). The amygdala is enlarged in children but not adolescents with autism; the hippocampus is enlarged at all ages. *Journal of Neuroscience, 24,* 6392–6401.

Senju, A., & Johnson, M.H. (2009). Atypical eye contact in autism: Models, mechanisms and development. *Neuroscience and Behavioral Reviews, 33,* 1204–1214.

Skuse, D. (2003). Fear recognition and the neural basis of social cognition. *Child and Adolescent Mental Health, 8,* 50–60.

Stella, J., Mundy, P., & Tuchman, R. (1999). Social and nonsocial factors in the Childhood Autism Rating Scale. *Journal of Autism and Developmental Disorders, 29,* 307–317.

Tardif, C., Lainé, F., Rodriguez, M., & Gepner, B. (2007). Slowing down presentation of facial movements and vocal sounds enhances facial expression recognition and induces facial-vocal imitation in children with autism. *Journal of Autism and Developmental Disorders, 37,* 1469–1484.

van der Geest, J. N., Kemner, C., Verbaten, M. N., & van Engeland, H. (2002). Gaze behavior of children with pervasive developmental disorder toward human faces: A fixation time study. *Journal of Child Psychology and Psychiatry, 43,* 669–678.

Vicari, J., Reilly, J. S., Pasqualetti, P., Vizzotto, A., & Caltagirone, C. (2000). Recognition of facial expressions of emotions in school-age children: The intersection of perceptual and semantic categories. *Acta Paediatrica, 89,* 836–845.

Wang, A. T., Dapretto, M., Hariri, A. R., Sigman, M., & Bookheimer, S. Y. (2004). Neural correlates of facial affect processing in children and adolescents with autism spectrum disorder. *Journal of the American Academy of Child and Adolescent Psychiatry, 43,* 481–490.

Williams, K. R., Wishart, J. G., Pitcairn, T. K., & Willis, D. S., (2005). Emotion recognition by children with Down syndrome: Investigation of specific impairments and error patterns. *American Journal on Mental Retardation, 110,* 378–392.

Ziersch, A. M. (2005). Health implications of access to social capital: Findings from an Australian study. *Social Science & Medicine, 61,* 2119–2131.

Cortical deficits of emotional face processing in adults with ADHD: Its relation to social cognition and executive function

Agustin Ibáñez[1,2,3#], Agustin Petroni[2,4#], Hugo Urquina[1], Fernando Torrente[1], Teresa Torralva[1], Esteban Hurtado[3,5], Raphael Guex[1,8], Alejandro Blenkmann[2,6,7], Leandro Beltrachini[2,7], Carlos Muravchik[7], Sandra Baez[1], Marcelo Cetkovich[1], Mariano Sigman[2,4], Alicia Lischinsky[1], and Facundo Manes[1]

[1]Institute of Cognitive Neurology (INECO) and Institute of Neuroscience, Favaloro University, Buenos Aires, Argentina

[2]National Scientific and Technical Research Council (CONICET), Buenos Aires, Argentina

[3]Laboratory of Cognitive Neuroscience, Universidad Diego Portales, Santiago, Chile

[4]Integrative Neuroscience Laboratory, Physics Department, University of Buenos Aires, Buenos Aires, Argentina

[5]School of Psychology, Pontificia Universidad Católica de Chile, Santiago, Chile

[6]Institute of Cellular Biology and Neuroscience "Prof E. De Robertis" (IBCN), School of Medicine, University of Buenos Aires - CONICET, Buenos Aires, Argentina

[7]Laboratory of Industrial Electronics, Control and Instrumentation (LEICI), National University of La Plata, La Plata, Argentina

[8]Laboratory for Behavioral Neurology and Imaging of Cognition, University of Geneva, Geneva, Switzerland

Although it has been shown that adults with attention-deficit hyperactivity disorder (ADHD) have impaired social cognition, no previous study has reported the brain correlates of face valence processing. This study looked for behavioral, neuropsychological, and electrophysiological markers of emotion processing for faces (N170) in adult ADHD compared to controls matched by age, gender, educational level, and handedness. We designed an event-related potential (ERP) study based on a dual valence task (DVT), in which faces and words were presented to test the effects of stimulus type (faces, words, or face-word stimuli) and valence (positive versus negative). Individual signatures of cognitive functioning in participants with ADHD and controls were assessed with a comprehensive neuropsychological evaluation, including executive functioning (EF) and theory of mind (ToM). Compared to controls, the adult ADHD group showed deficits in N170 emotion modulation for facial stimuli. These N170 impairments were observed in the absence of any deficit in facial structural processing, suggesting a specific ADHD impairment in early facial emotion modulation. The cortical current density mapping of N170 yielded a main neural source of N170 at posterior section of fusiform gyrus (maximum at left hemisphere for words and right hemisphere for faces and simultaneous stimuli). Neural generators of N170 (fusiform gyrus) were reduced in ADHD. In those patients, N170 emotion processing was associated with performance on an emotional inference ToM task, and N170 from simultaneous stimuli was associated with EF, especially working memory. This is the first report to reveal an adult ADHD-specific impairment in the cortical modulation of emotion for faces and an association between N170 cortical measures and ToM and EF.

Correspondence should be addressed to: Agustín Ibáñez, Laboratory of Experimental Psychology and Neuroscience, Institute of Cognitive Neurology (INECO) and CONICET, Castex 3293 (1425), Buenos Aires, Argentina. E-mail: aibanez@neurologiacognitiva.org

#These authors contributed equally to this work.

This research was partially supported by grants from CONICET, FINECO, and Diego Portales University.

http://dx.doi.org/10.1080/17470919.2011.620769

Keywords: Adult ADHD; N170; Valence; Face; Word.

Attention-deficit hyperactivity disorder (ADHD) is a neuropsychiatric condition with onset in childhood that extends over adolescent and adult life with a considerable symptomatic burden and functional impairment (Malloy-Diniz, Fuentes, Leite, Correa, & Bechara, 2007). Its medical profile includes problems of self-regulation and self-motivation, distractibility, procrastination, and prioritization (Barkley, 2001, 2010; Safren, 2006). A recent meta-analysis suggested that the prevalence of adult ADHD is currently underestimated (Simon, Czobor, Balint, Meszaros, & Bitter, 2009). It has been shown that adults with ADHD have impaired social cognition (Uekermann et al., 2010). However, no studies have focused on the brain correlates of the adult ADHD deficits in emotion processing. The aim of this study is to identify cortical markers of emotion processing in adult ADHD and explore their relation to individual neuropsychological profiles.

ADHD in childhood is more related to hyperactivity and impulsiveness, whereas in adulthood it presents a different profile, with fewer externalizing symptoms and a higher rate of psychiatric comorbidity (Klassen, Katzman, & Chokka, 2010). Nevertheless, deficits in executive functions (e.g., the capacity for formulating goals, planning, and execute plans) (Lezak, 1982) have been consistently demonstrated in adults with ADHD (Adler, 2010). Some studies have shown deficits in adults with ADHD in domains related to executive functioning, such as working memory (Torralva et al., 2010), phonologic fluency (Schecklmann et al., 2009), and inhibitory control (Rapport, Van Voorhis, Tzelepis, & Friedman, 2001; Wodushek & Neumann, 2001).

Despite the fact that deficits in social cognition are an evident clinical phenomenon in ADHD, very little research has been developed in this area (Uekermann et al., 2010). In ADHD children, a few reports have suggested various deficits in domains such as facial affect recognition (Pelc, Kornreich, Foisy, & Dan, 2006; Sinzig, Morsch, & Lehmkuhl, 2008), theory of mind (ToM) (Buitelaar, van der Wees, Swaab-Barneveld, & van der Gaag, 1999; Sodian, Hulsken, & Thoermer, 2003; but for different results see Charman et al., 2001), social skills (King et al., 2009; Matthys, Cuperus, & Van, 1999), and empathy (Braaten & Rosen, 2000; Dyck, Ferguson, & Shochet, 2001). In adults with ADHD, there are even fewer studies that have reported deficits in

domains related to facial emotion processing (Marsh & Williams, 2006) and prosody perception (Shapiro, Gordon, Hack, & Killackey, 1993). Facial emotion processing seems to be the social cognition process that is most affected in ADHD adults (Marsh & Williams, 2006). In general terms, these social cognition impairments are consistent with frontostriatal dysfunction in ADHD (Uekermann et al., 2010), suggesting the central nature of social dysfunction in this disorder (Hoza, Waschbusch, Pelham, Molina, & Milich, 2000; Maedgen & Carlson, 2000). Despite the classical association of frontostriatal deficits and executive functions, a link between social cognition deficits, frontostriatal network, and ADHD has been highlighted (Sonuga-Barke, 2003; Uekermann et al., 2010).

Emotional inference of facial clues is one of the most important steps in the development of complex social cognition behaviors (Grossmann, 2010). Faces are multidimensional stimuli directly related to important social incentives (Ohman & Mineka, 2001). Moreover, the central role of eyes and gaze in social cognition has been acknowledged (Itier & Batty, 2009). Facial emotional expression gives an automatic and fast shortcut to alarm signs, mentalizing, and intersubjective communication. People with low social competence are impaired in recognizing emotions from facial expressions (Edwards, Manstead, & MacDonald, 1984; Feldman, Philippot, & Custrini, 1991; Philippot & Feldman, 1990). Thus, the ability to identify emotions from faces in ADHD participants can be easily assessed by the presentation of faces.

An approach which combines measures from neuropsychological and neurophysiological markers represents a reference standard in order to understand abnormal cognitive processing in neuropsychiatry and individual differences. This study seeks to identify possible behavioral, neuropsychological, and electrophysiological markers of abnormal emotion processing for faces in adult ADHD compared with controls matched by age, gender, educational level, and handedness.

Event-related potentials (ERPs) provide excellent temporal resolution of cognitive brain processing. The N170 is a cortical marker specifically linked to facial processing, with neural generators in the fusiform gyrus and superior temporal sulcus (Deffke et al., 2007; Sadeh, Zhdanov, Podlipsky, Hendler, & Yovel, 2008). The N170 represents an early cortical response specialized for facial processing compared with objects or words (Proverbio, Riva, Martín & Zani, 2010; Rossion, Joyce, Cottrell, & Tarr, 2003). The N170 can be modulated by emotion processing (Ibáñez et al., 2010d; Righart & de Gelder, 2008). Thus, this component represents an ideal brain marker to assess possible cortical markers of emotion processing for faces in ADHD.

To our knowledge, only a single study has previously assessed facial processing in ADHD indexed by the N170, which was done with adolescents. Williams et al. (2008) reported an abnormal emotion-related N170, suggesting that the structural facial processing stage is affected in adolescents with ADHD. However, these results must be taken with caution because participants with ADHD also had comorbid depression and anxiety. For adults with ADHD, even though evidence of deficits in the processing of emotion has been reported (Herrmann et al., 2009), no N170 valence effects elicited by facial processing have been previously assessed.

We designed an ERP study based on a dual valence task (DVT) (Ibáñez et al., 2011a; Petroni et al., in press), in which faces and words were presented to test the effects of stimulus type (ST) (faces, words, or face-word stimuli), valence (positive vs. negative), and compatibility (compatible vs. incompatible word and face valence combinations). Adult participants with ADHD and controls classified stimuli according to its emotional valence (positive or negative).

In order to identify individual signatures of cognitive functioning in participants with ADHD and controls, a comprehensive neuropsychological assessment was carried out: general neuropsychology, executive functioning, and ToM. Because one plausible hypothesis is the assumption that emotion facial processing would be related to more basic (executive functions) as well as more complex (ToM) deficits in adult ADHD, we combined, in a multivariate analysis, the ERP measures with the neuropsychological assessment.

It has previously been proposed that processing of facial emotion is intertwined with complex social skills (Grossmann, 2010; Itier & Batty, 2009) and executive functioning (Pessoa, 2009). Both executive and social cognition deficits in ADHD have been reported elsewhere (Kain & Perner, 2003; Uekermann et al., 2010). This relationship between facial processing and other social or executive functions can be explained by the existence of a relatively higher-order cognitive process, in which the structural representation of the face is associated with semantic and cognitive information (Balconi & Lucchiari, 2005; Bruce & Young, 1986). Consequently, we predict that the adult participants with ADHD will show abnormal cortical processing

of facial valence indexed by the N170. Moreover, these cortical deficits will be associated with basic tasks of ToM related to emotional inference. We also address two additional hypotheses: (1) Cortical deficits of emotional processing in participants with ADHD will be independent of encoding of structural facial processing (since emotional processing impairments in ADHD seem to be not exclusive for face processing); and (2) valence deficits will be related to executive functioning in participants with ADHD.

MATERIALS AND METHODS

Participants

Ten adult participants with ADHD, one female, ($M = 33.1$ years of age, $SD = 3.6$), three left-handed, completed the dual valence task. Ten healthy controls, matched for gender, age ($M = 33.0$, $SD = 3.8$ years old), handedness, and years of education were recruited (see below and Table 1). Participants with ADHD and controls received a thorough neuropsychological battery that comprised measures of general neuropsychology, executive functioning, and social cognition. A questionnaire was given to healthy participants to rule out hearing, visual, psychiatric, or neurological deficits. All participants gave signed, informed consent in agreement with the Helsinki Declaration. All experimental procedures were approved by the ethics committee of the Institute of Cognitive Neurology, Buenos Aires, Argentina.

Participant criteria and recruitment process

All participants with ADHD fulfilled DSM-IV criteria for ADHD. Diagnosis was made by three experts (A.L., F.T., and F.M.). Participants were recruited from among the patients of the Institute of Cognitive Neurology (INECO, Buenos Aires, Argentina), from the adult ADHD clinic. From the initial set, eight patients presented with ADHD combined type (ADHD/C), and two patients presented with predominantly inattentive type (ADHD/I). All patients were taking methylphenidate medication, which was suspended on the day of ERP recordings. ADHD diagnosis based on the DSM-IV criteria was assessed with the following protocol for adults:

1. ADHD Rating Scale for Adults (Barkley & Murphy, 1998), in patient and informant versions. It identifies current and retrospective childhood symptoms corresponding to DSM-IV characterization of ADHD (see supplementary material for details).
2. Depression Inventory II (BDI-II) (Beck, Steer, Ball, & Ranieri, 1996) and the Young Mania Rating Scale (YMRS) (Young, Biggs, Ziegler, & Meyer, 1978), to assess depression and mania, respectively. Scales were administered to patients and controls.
3. Neuropsychological assessment (see next section for details).

Neuropsychological assessment

Both participants with ADHD and controls received a comprehensive neuropsychological battery that lasted approximately 120 min. It included general neuropsychology, executive functioning, and social cognition.

General neuropsychology

we used general neuropsychology to evaluate participants' basic cognitive functioning (see supplementary material for details). Memory was evaluated by the Rey Verbal Learning Test (RVLT), which comprises verbal learning, immediate and delayed recall, and a distractor list. Attention and concentration were assessed by the Trail Making Test A (TMT-A) (Partington, 1949). Phonological and semantic fluency were assessed by the Controlled Oral Word Association Test (COWAT) (Benton et al., 1994). An arithmetic test, Wechsler Adult Intelligence Scale III (WAIS III) (Wechsler, 1997), was also included.

Executive functioning

Several tests were compiled to evaluate executive functioning. The INECO Frontal Screening (Torralva, Roca, Gleichgerrcht, López, & Manes, 2009) was used to assess frontal lobe function indexed by several subtasks: Motor Programming, Conflicting Instructions, Verbal Inhibitory Control, Abstraction, Backward Digit Span, Spatial Working Memory, and Go/No Go. Backward Digit Span and TMT-B (Partington, 1949) were used to assess attentional flexibility, attentional speed, and sequencing and planning skills. Numbers Key and Searching Symbols (Wechsler, 1997) were used to evaluate visual perception and organization, visual scanning, and the efficient production of multiple motor responses. Ordering Letters and Numbers (Letters & Numbers hereafter) was used to assess mental manipulation and working memory (Wechsler, 1997). Finally, a working memory index was derived from performance on the Digit Span, Arithmetic, and Letter–Number Sequencing subtests (Hill et al., 2010).

Social cognition

Only one social cognition test was included: the Reading Mind in the Eyes Test (RMET) (Baron-Cohen et al., 2001), which assesses individual differences in the ability to infer the affective mental states of other humans. All neuropsychological performance data are shown in Table 1.

Procedure

Dual valence task

In a two-alternative, forced-choice task, participants classified words or faces displayed on a computer screen according to their valence, into one of two categories (positive or negative), as quickly as possible. When responses were not correct, an X appeared briefly in the center of the screen. The task comprised two blocks of 320 trials.

Trial structure. A trial (Figure 1) started with a fixation cross for 1000 ms. Then a stimulus was presented for 100 ms, followed by a fixation cross until participants responded. If the response was incorrect, a red cross was presented as feedback for 100 ms, and the trial ended. Otherwise, the trial ended without feedback. After responses or feedback, an ISI of 1000 ms was added (not shown in Figure 1).

Simultaneous stimuli block. In this block, participants were exposed in each trial to a face in the center of the screen and a word 4° below, simultaneously for 100 ms. Participants had to respond as to whether the face was angry or happy, ignoring the word presented below. Congruent trials presented a face and a word of the same valence, and incongruent trials presented stimuli with opposite valences (e.g., an angry face with a pleasant word).

Single stimulus block. In a second, counterbalanced block, participants were exposed on each trial to a face or a word in the center of the screen, and responded as to whether the stimulus was angry or happy in the case of faces, or pleasant or unpleasant, in the case of

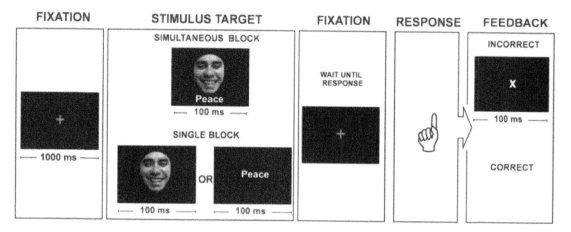

Figure 1. Experimental design. The trial started with a fixation cross, followed by a target stimulus (face or word) or two simultaneous stimuli (face and word), depending on which of the two main blocks was being performed. After the target stimulus, a fixation cross appeared until the response. If the response was incorrect, a red cross appeared and the trial ended. Otherwise, the trial ended without feedback. After responses or feedback, an ISI of 1000 ms was added (not shown).

words. Single-stimulus block trials were presented one by one with strict alternation between words and faces.

Two response keys were used. Each block was separated into two subblocks of 160 trials, in which the response keys were inverted explicitly. Each subblock was preceded by written instructions with the correct correspondence between stimulus category and response key and six trials of practice. This procedure was taken from a previous two-choice task (Hurtado, Gonzalez, Haye, Manes, & Ibáñez, 2009; Ibáñez et al., 2010d, 2011a).

Stimulus construction and validation

Facial pictures were taken from a data set used in previous studies (Hurtado, Gonzalez, Haye, Manes, & Ibáñez, 2009; Ibáñez et al., 2010d, 2011a, 2011b) (see supplementary data for stimuli validation). A set of 10 happy and 10 angry pictures controlled for intensity, brightness, color, and contrast was included. Each of the 10 actors was present in one happy and one angry stimulus. Thirty-three pleasant and 32 unpleasant words controlled for arousal, content, length, and frequency were also selected from a previous study (Ibáñez, López, & Cornejo, 2006; see supplementary data for validation details).

ERP recordings

EEG signals were sampled at 500 Hz from a Biosemi 128-channel Active Two system. Data were band-pass filtered (0.1 to 100 Hz) while recording and (0.3 to 30 Hz) off-line to remove unwanted frequency components. During recording, the reference was set by default to link mastoids and re-referenced off-line to average electrodes. Two bipolar derivations monitored vertical and horizontal ocular movements (EOG). EEG data were segmented from 200 ms before to 800 ms after the stimulus onset. All segments with eye-movement contamination were removed from further analysis by an automatic (Gratton, Coles, and Donchin method for removing eye-blink artifacts) visual procedure. Artifact-free segments were averaged to obtain ERPs.

Source localization

Distributed source models (8000 dipoles) of the N170 component for each condition were estimated by the standardized, low-resolution, brain electromagnetic tomography algorithm (sLORETA) (Pascual-Marqui, 2002). These locations were derived by performing a location-wise inverse weighting of data, with a minimum norm least-squares analysis of their estimated variances, leading to a smooth solution. An average head model built from a sample of 152 MRIs provided by the International Consortium of Brain Mapping (ICBM) was used (Mazziotta et al., 2001). To make a more realistic model, we considered white matter anisotropy by using a diffusion tensor atlas of 81 healthy participants (Mori et al., 2008) coregistered with the ICBM model. The forward problem was solved by a finite element method (Zhang et al., 2004).

Possible solutions were constrained for location to the cortical surface but were not constrained for orientation (Valdez-Hernandez et al., 2009). This head model is useful for source localization when individual MRI data are not available. Given that temporal differences occur between participants and STs, the local minimum within the N170 window was considered for each of them. Average signal for the N170 representative electrodes A9, A10, A11, A12 (left) and B6, B7, B8, B9 (right) was obtained, within a 150–210-ms time window for faces and simultaneous stimuli, and within a 160–230-ms time window for word stimuli for each subject. N170 peak amplitude was found as the local minimum of this average. Potentials of all channels at local minimum were extracted for each participant and condition. Standardized current density power was obtained for each condition and subject. Finally, the average of this source images for faces, simultaneous, and words stimuli were obtained for the each group.

Data analysis

Off-line processing and analysis of EEG data were performed by Matlab software, EEGLAB toolbox, and T-BESP software (http://neuro.udp.cl/software). To analyze scalp topography of the ERP components, we used regions of interest (ROIs), as recommended for dense arrays (e.g., Aravena et al., 2010; Ibáñez et al., 2010c; San Martín, Manes, Hurtado, Isla, & Ibáñez, 2010), since it improves statistical power. ROIs were chosen after visual inspection of each component. Each N170 ROI (left and right) included four adjacent electrodes around T8 and T7 (Rossion & Jacques, 2008): the N170 ROIs were A9, A10, A11, and A12 for the left and B6, B7, B8, and B9 for the right hemisphere (see Figure 2, for the channel location selection). For ERP analysis, the 160–210-ms time window for N170 was visually selected for mean amplitude analysis.

Accuracy and N170 mean amplitudes were averaged for faces, words, and simultaneous stimuli separately and analyzed by repeated-measures ANOVA with ST (faces, words, simultaneous) and Valence (positive vs. negative) as within-subject factors. In the simultaneous stimuli condition, the factor Congruency was considered. Congruency had two categories: congruent (a positive face plus a positive word or a negative face plus a negative word) and incongruent (a negative face plus a positive word or a positive face plus a negative word). Finally, only for ERP data, the factor Hemisphere (left and right locations) was considered. A between-subject factor Group (ADHD vs. controls) was included.

In order to perform multivariate comparisons between ERPs and neuropsychology, we calculated global scores for ERPs as follows:

1. Stimulus discrimination (face-minus word) and stimulus interference (face minus simultaneous-stimuli) scores were calculated for N170 mean amplitude.
2. Valence discrimination scores were calculated by subtracting positive from negative stimuli, for N170 mean amplitude (for faces, words, and simultaneous stimuli).
3. Congruency. For simultaneous stimuli, the difference between congruent (e.g., positive face and positive word) and incongruent (i.e., negative face and positive word) was calculated for N170 mean amplitude.

To test whether ERP measures of ST and valence were associated with individual cognitive profiles, global scores were correlated with all neuropsychological tests (general, social, and executive neuropsychology), using Spearman's rank correlations corrected for multiple comparisons (false discovery rate (FDR) correction, which controls the fraction of rejections that are false positives).

RESULTS

Demographic and clinical assessment

Table 1 shows the overall results from the demographic, clinical, and neuropsychological assessments.

Demographic data

No differences regarding age, $F(1, 18) = 0.001$, $p = .96$; gender, $\chi^2(1) = 0.000, = 1$; educational level, $F(1, 18) = 2.240, p = .15$; or handedness, $\chi^2(1) = 0.260, p = .60$, were observed between groups.

Clinical evaluation

ADHD participants showed significantly higher scores on behavioral measures of ADHD symptoms than did control subjects (Barkley ADHD Rating Scale for Adults). There was an expected significant between-group difference between the ADHD-RS-Inattention scale, $F(1, 18) = 13.598, p < .005$, and the ADHD-RS-Hyperactivity/impulsivity subscale, $F(1, 18) = 5.66, p = .02$, indicating that ADHD participants had significantly higher scores for inattention and impulsivity than did control subjects. A difference between groups for BDI-II scores, $F(1, 18) = 6.438, p = .02$, was observed, indicating

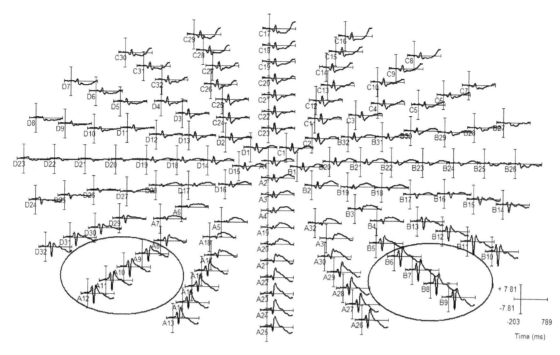

Figure 2. Channel locations and selected electrodes. Figure shows the overall ERP response to DVT, and the ellipses contain selected electrodes for left (A8 to A12) and right N170 (B6 to B9).

high levels of depression in the ADHD group. No differences between groups were observed for the Young scale, $F(1, 18) = 0.545, p = .47$.

Neuropsychological assessment

General neuropsychology

No group differences regarding memory were observed for the RVLT total score, $F(1, 18) = 1.108$, $p = .30$, and the delayed, $F(1, 18) = 2.568, p = .13$. However, the RVLT recognition revealed a deficit in the ADHD group, $F(1, 18) = 9.184, p < .01$. No differences were observed in attention and concentration assessed with the TMT-A, $F(1, 18) = 1.049, p = .32$. No group differences were found on the arithmetic evaluation: WAIS-III, $F(1, 18) = 0.253, p = .62$. The phonological fluency task, $F(1, 18) = 7.862, p < .05$, yielded lower scores in the ADHD group.

Executive functioning

The global score on the IFS showed a trend toward lower performance for the ADHD group compared with controls, $F(1, 18) = 3.79, p = .07$. On closer examination, only the IFS subtasks of Abstraction Capacity, $F(1, 18) = 4.47, p = .04$, and Spatial

Working Memory, $F(1, 18) = 6.37, p = .02$, yielded lower scores in the ADHD group. As regards the other measures of executive functioning, attention deficits in the ADHD group were revealed, as measured by digit repetition, $F(1, 18) = 34.184, p < .001$. No differences were observed on the Working Memory Index, $F(1, 18) = 2.66, p = .11$. In contrast, no deficits in attentional flexibility, attentional speed, or sequencing were observed in the ADHD group as measured by the TMT-B, $F(1, 18) = 0.016, p = .90$; Backward Digit Span, $F(1, 18) = 2.070, p = .17$; and Letters and Numbers task, $F(1, 18) = 0.76, p = .39$.

Social cognition

When we compared percentage accuracy on the RMET, a small deficit was indicated in the ADHD group, shown by a trend, $F(1, 18) = 3.48, p = .08$, suggesting that patients had a subtle deficit in the emotional inference process.

Dual valence paradigm

Behavioral results

Both groups performed the task with an accuracy of 88% or more (see Table 2 for mean fractions and *SD*).

TABLE 1

Results are shown as mean (*SD*), and statistical comparison test results are shown in the right-hand column. Statistical comparison test result *p* values are shown when significance was achieved. In all other cases, *ns* is used to indicate a "nonsignificant" difference

		ADHD (n = 10)	*Control* (n = 10)	p
Demographics	Age (years)	33.1 (3.42)	33.3 (3.64)	*ns*
	Gender (M:F)	1:9	1:9	*ns*
	Education (years)	15.9 (0.87)	17.8 (0.89)	*ns*
Clinical profile	Handedness (L:R)	3:7	2:8	*ns*
	Barkley Inattention	12.30 (2.60)	1.80 (1.14)	.005
	Hyperactivity	9.20 (2.21)	2.70 (1.60)	.02
	BDI-II	17.90 (4.29)	5,50 (2.32)	.02
	YMRS	1.10 (0.64)	0.50 (0.50)	*ns*
General neuropsychology	TMT-A	32.20 (4.41)	38.60 (4.48)	*ns*
	Phonological fluency	17.30 (1.41)	22.90 (1.82)	.05
	RVLT	47.10 (3.80)	52.20 (2.33)	*ns*
	DL	7.90 (1.07)	6.60 (0.60)	*ns*
	Delayed	10.80 (0.92)	12.90 (0.99)	*ns*
	Recognition	12.80 (0.54)	14.80 (0.46)	.01
	Arithmetic (WAIS-III)	14.20 (1.26)	15.10 (0.96)	*ns*
Executive functions	WMI	100.60 (4.38)	110.02 (3.91)	*ns*
	Digit repetition	11.80 (1.32)	19.50 (0.93)	.001
	Digits backward	4.44 (0.34)	5.10 (0.34)	*ns*
	TMT-B	71.00 (6.77)	72.20 (4.23)	*ns*
	Letters and Numbers	11.20 (0.80)	12.20 (0.80)	*ns*
	IFS total score	25.30 (0.85)	27.90 (056)	.045
Social cognition	RMET	71.32 (3.02)	79.21 (2.43)	.08

Notes: BDI-II: Beck Depression Inventory-II; YMRS: Young Mania Rating Scale; RVLT: Rey Auditory Verbal Learning Task; DL: Distractor List; TMT: Trail Making Test; IFS: INECO Frontal Screening; WMI: Working Memory Index.

TABLE 2

Performance on the DVT for patients and controls (fractions). The signs + and – depict emotional valences. The double signs in the last four columns indicate valence for faces and words, respectively

	Face +	*Face –*	*Word +*	*Word –*	*Sim ++*	*Sim +–*	*Sim – –*	*Sim –+*
ADHD mean	0.92	0.93	0.90	0.90	0.93	0.91	0.90	0.90
ADHD *SD*	0.09	0.05	0.06	0.10	0.07	0.05	0.05	0.05
Controls mean	0.92	0.90	0.91	0.9	0.92	0.89	0.89	0.88
Controls *SD*	0.07	0.07	0.06	0.07	0.05	0.08	0.07	0.07

The performance on the overall task was very similar for both groups: 91%, $SD = 0.02$, for the ADHD group and 90%, $SD = 0.02$, for the control group, $F(1, 18) = 0.04$, $p = .89$. For faces, word, and simultaneous stimuli, no differences were found for ST, $F(2, 36) = 1.8$, $p = .17$, or group differences, $F(1, 18) = 0.08$, $p = .77$. For faces, no main effects of valence, $F(1, 18) = 0.07$, $p = .79$, or group, $F(1, 18) = 0.95$, $p = .34$, were found. For words, no main effects for valence, $F(1, 18) = 0.26$, $p = .61$, or group, $F(1, 18) = 0.003$, $p = .95$, were found. Finally, for simultaneous stimuli, an effect of valence was observed—accuracy:

positive > negative, $F(2, 18) = 4.10$, $p = .03$—but no group effect, $F(2, 17) = 0.55$, $p = .58$. No congruency effects, $F(2, 18) = 0.06$, $p = .94$, or group differences were observed, $F(2, 18) = 0.55$, $p = .58$, for simultaneous stimuli comparing the congruency of face and word valence. There were no any interactions between any of the previously mentioned factors. Table 2 shows the descriptive statistics.

Summarizing the behavioral results, accuracy was high across all conditions for both groups, and, despite the small mean differences between conditions, no significant effects were observed for ST in either group. For valence effects, only a small difference was obtained in simultaneous stimuli for both ADHD and control groups (performance was better for positive than for negative valence). No other effects yielded significant differences.

ERPs

ST effects

In a comparison of the N170 amplitudes elicited by faces, words, and simultaneous stimuli, a main effect of ST was obtained, $F(2, 36) = 4.40$, $p < .01$, mainly caused by an amplitude enlargement of the N170 for faces. The ST effect was more accentuated over the right hemisphere, as evidenced by ST x Hemisphere interaction, $F(2, 36) = 8.71$, $p < .001$. We performed a post-hoc analysis of this interaction (Tukey HSD test, $MS = 1.44$, $df = 36$) and found that faces elicited enhanced N170 amplitudes compared with simultaneous stimuli ($p < .001$) and words ($p < .0005$) in the right hemisphere. Although face stimuli presented right > left amplitude differences, this effect was not significant ($p = .20$). In the left hemisphere, words showed a trend toward enhanced amplitude compared to the right hemisphere ($p = .057$). No N170 amplitude differences were observed for words compared with faces ($p = .99$) on the left side. However, significant word N170 amplitude enlargement was obtained compared with simultaneous stimuli ($p < .005$) in the left hemisphere. In summary, faces elicited an enhanced amplitude over the right hemisphere (compared with words or simultaneous stimuli), and words elicited an enhanced amplitude over the left hemisphere (compared with simultaneous stimuli). No group differences or other factor interactions were observed for ST modulation. Both groups presented N170 modulation of faces to the right and words to the left. Figure 3A shows the ERPs for ST modulation and Figure 3B the scalp topography for both groups.

Valence effects

Faces. No main effects of valence, $F(1, 18) = 1.07$, $p = .31$; group, $F(1, 18) = 0.05$, $p = .81$; or hemisphere, $F(1, 18) = 2.04$, $p = .17$, were observed. A significant interaction between valence x group, $F(1, 18) = 6.49$, $p = .02$, and a strong interaction between valence x group x hemisphere, $F(1, 18) = 18.32$, $p < .0005$, were found, evidencing different cortical patterns of emotional processing between groups

in the right hemisphere. Post-hoc comparisons performed on this last interaction (Tukey HSD test, $MS = 59.05$, $df = 18.07$) showed that N170 of controls distinguished facial valence in the right hemisphere, but ADHD participants lacked N170 valence modulation. Increased N170 amplitude for positive faces compared with negative ones in the right hemisphere yielded significant effects ($p < .0005$) in control participants. In contrast to controls, no effects of valence were observed in the left ($p = .31$) or right hemispheres ($p = .87$) for ADHD patients. Moreover, when we compared the specific valences between groups in the right hemisphere, no differences were observed with negative stimuli ($p = .98$). Nevertheless, a strong effect indicated that the ADHD group had a significantly reduced amplitude for positive stimuli, compared with controls ($p < .01$). No other relevant pairwise comparisons were significant. Figure 3 and Table 3 show the N170 effects on valence for controls and patients.

Words. For word stimuli, N170 was not modulated by the valence, $F(1, 18) = 0.03$, $p = .84$. The interaction between word valence x hemisphere was not significant, $F(1, 18) = 0.32$, $p = .57$. No group effects or interactions between group and other factors were observed. Means are shown in Table 3. In brief, word valence was not discriminated by N170 in either controls or patients in either hemisphere.

Simultaneous stimuli. Similar results were observed for simultaneous stimuli as were reported for facial valence modulation. An interaction between valence x group x hemisphere, $F(1, 18) = 5.63$, $p < .05$, suggested that controls still presented valence effects of simultaneous stimuli in the right hemisphere. Post-hoc comparisons performed on this last interaction (Tukey HSD test, $MS = 30.15$, $df = 18.26$) showed that controls presented facial valence modulation in the right hemisphere ($p < .05$), but participants with ADHD lacked an N170 valence modulation in the right hemisphere ($p = .89$). No other pairwise comparisons yielded significant effects.

We performed additional analyses to investigate valence effects in relation to congruency between faces and words in the simultaneous stimuli. No effects of valence congruency were observed in the N170 window, in either hemisphere or group, nor was there any interaction. Table 3 shows the means and SDs for these conditions.

Source activity

Figure 4A shows the distributed activation evoked by the ST conditions (face, words, and simultaneous) in both, controls and ADHD participants. The

TABLE 3
N170 amplitude values in response to stimulus type, valence, and congruency factors

	ADHD mean (SD)		Controls mean (SD)	
	Left	Right	Left	Right
a. Stimulus type effects				
Faces	−2.49 (1.64)	−2.59 (1.88)	−2.29 (1.64)	−3.95 (1.88)
Words	−1.77 (1.28)	−0.68 (1.15)	−2.91 (1.28)	−1.77 (1.15)
Simultaneous	−0.51 (1.58)	−0.88 (1.95)	−0.94 (1.58)	−2.07 (1.95)
b. Face valence effects				
Positive	−2.70 (1.63)	−2.81 (1.88)	−2.51 (1.62)	−4.94 (1.73)
Negative	−2.37 (1.69)	−2.47 (1.92)	−2.06 (1.65)	−2.85 (1.91)
c. Word valence effects				
Positive	−1.56 (1.39)	−0.76 (1.15)	−2.96 (1.39)	−1.76 (1.21)
Negative	−1.98 (1.18)	−0.60 (1.21)	−2.86 (1.11)	−1.77 (1.24)
d. Simultaneous valence effects				
Positive	−0.74 (1.57)	−1.14 (1.97)	−1.05 (1.12)	−2.48 (1.97)
Negative	−0.28 (1.72)	−0.73 (1.95)	−0.82 (1.62)	−1.57 (1.92)
e. Congruency effects				
Congruent	−0.37 (1.23)	−0.82 (1.90)	−0.96 (1.56)	−2.01 (1.82)
Incongruent	−0.64 (1.87)	−0.95 (2.02)	−0.91 (1.61)	−2.13 (2.02)

source of N170 neural activity was observed at different posterior portions of the fusiform gyrus (FG): left hemisphere for words, peak at −30, −81, and −20 for controls, and −25, 87, and −21 for patients; right hemispheres for faces, peak at 40, 67, and −12 for controls, and 25, −86, and −18 for patients; and simultaneous stimuli, peak at 26, −76, and −16 for controls and 20, −86, and −20 for patients. Table 4 shows the results of the estimation of cortical sources for N170. Standardized current density power was higher in the case of controls against patients, as consistent with greater amplitude of sources for controls. Controls presented decreasing FG activation from face to simultaneous and word stimuli. Consistent with the ERP results, the patient group presented a reduced activation of fusiform gyrus and N170 peak compared to controls. Figure 4B and 4C shows the average intensity of the source peak and the FG for N170 window in all conditions and groups.

Multivariate analysis

ERP correlations with general neuropsychology

ADHD patients. Phonological fluency was correlated with N170 stimulus discrimination (face–word; $r = .52$, $p = .03$) and face valence (positive–negative, $r = .64$, $p = .02$).

Controls. The TMT-A was positively correlated with N170 stimulus discrimination (face–word; $r = .31$, $p = .04$).

Executive functioning

ADHD patients. Working memory (Backward Digit Span) correlated with face valence (positive–negative, $r = .58$, $p < .04$). The N170 of valence discrimination for simultaneous stimuli (positive–negative) correlated with Digit Repetition, ($r = .71$, $p = 0.001$), Letters and Numbers (WAIS-III) ($r = .68$, $p = .02$), and the Working Memory Index ($r = .54$, $p < .02$).

Controls. The TMT-B correlated positively with the N170 valence discrimination for simultaneous stimuli ($r = .589$, $p < .05$) ($r = .64$, $p = .02$). The N170 valence discrimination for simultaneous stimuli correlated with the Working Memory Index ($r = .68$, $p < .01$).

Theory of mind

ADHD patients. RMET scores correlated significantly with N170 face valence discrimination (happy–angry; $r = .51$, $p = .03$).

Controls. RMET scores correlated significantly with N170 face valence discrimination (happy–angry; $r = .32$, $p = .04$).

DISCUSSION

The primary goal of this report was to investigate cortical markers of facial and semantic emotion processing in adults with ADHD and controls matched for gender,

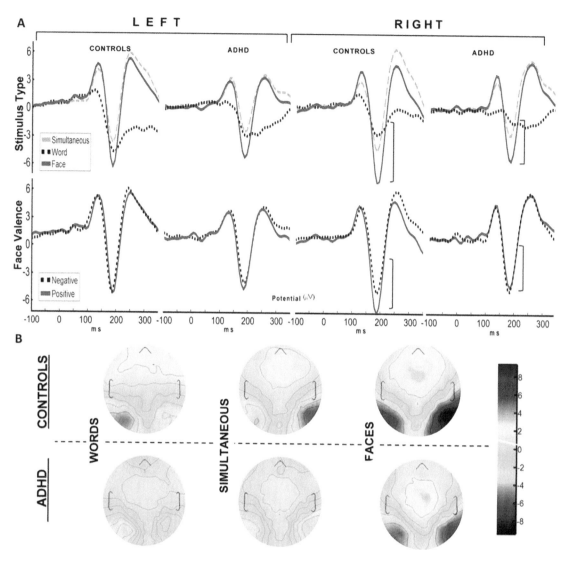

Figure 3. Main ERP results. (A) N170 component for both hemispheres and groups. First row: stimulus type effects. Second row: Face Valence effects. (B) Topographical maps of N170 window for stimulus type effects (faces, words, and simultaneous stimuli) in both ADHD and control groups.

handedness, educational level, and age. The secondary goal was to assess the individual variability of cognitive processing that related to the cortical markers of the DVT in both groups. Although both groups showed high accuracy on the DVT, there were important between-group differences in cortical processing. Compared with controls, the adult ADHD group showed deficits in N170 emotion modulation for facial stimuli. Those N170 impairments were observed despite there being no deficit in processing of facial structure, suggesting an ADHD-specific impairment in early facial emotion modulation. The two groups showed slightly dissimilar cognitive profiles associated with N170 processing. Notably, in ADHD participants, N170 emotion processing was associated with performance on an emotional inference ToM task,

and N170 for simultaneous stimuli was associated with executive functioning, especially working memory. In summary, this is the first report to reveal an adult ADHD-specific impairment in the cortical modulation of face valence (independent of facial processing per se) and an association of cortical measures with emotional ToM and executive functioning.

Behavioral performance (DVT)

Accuracy was high in both groups, evidencing an adequate comprehension and execution of the task. This high accuracy means that we can be confident that our findings for the ADHD group cannot be explained as reflecting inattention or distractibility.

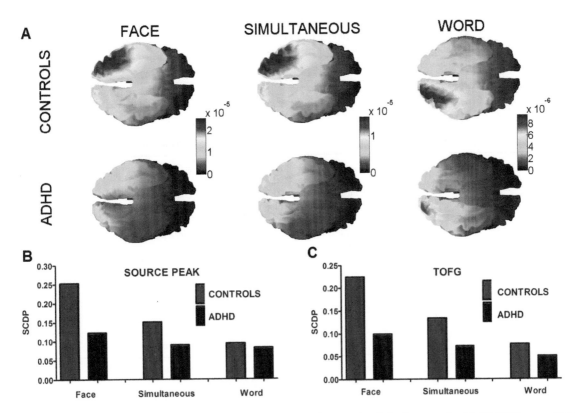

Figure 4. Cortical standardized current density power mapping of N170 (face, word, and simultaneous stimuli). (A) N170 source imaging estimation for controls (above) and patients (below) with ADHD. (B) Average values of estimated standardized current density power at maximum peaks of activation for each condition at N170 window. (C) Average values of N170 estimated standardized current density power at temporo-occipital fusiform gyrus (TOFG).

Moreover, the demonstration of cortical differences in the absence of behavioral divergences highlights the power of using ERPs as a tool for measuring the subclinical brain processes that are related to cognition (e.g., Gray, Ambady, Lowenthal, & Deldin, 2004; Guerra et al., 2009; Ibáñez et al., 2006; Ibáñez, San Martín, Hurtado, & López, 2008a; Ibáñez, San Martín, Hurtado, & López, 2008b; Ibáñez et al., 2010a, 2010b, 2010c; Kotchoubey et al., 2006), and pathophysiology in ADHD (e.g., Herrmann et al., 2009; López et al., 2006).

N170 results for ST, valence, and simultaneous stimuli processing

Our results replicate a previous DVT study carried out with healthy volunteers with no disorder (Ibáñez et al., 2011a; Petroni et al., in press) and other studies about the modulation of N170 amplitude via ST modulation (face > word; Rossion et al., 2003) and facial valence modulation (positive > negative: Ibáñez et al., 2011; Schacht & Sommer, 2009). It replicates the finding that there are no effects of word valence (Schacht & Sommer, 2009, but see Ibáñez et al., 2010d) and no

effects of congruency between the valence of faces and words (Krombholz, Schaefer, & Boucsein, 2007). In summary, the N170 component seems to be part of early facial structural processing that is sensitive to the specific valence of faces and less responsive to other more complex processes related to compatibility or arousal.

For the ADHD group, cortical deficits in emotion modulation for faces were observed. Moreover, on closer examination, a deficit in face valence modulation followed reduced N170 amplitude for positive stimuli in the right hemisphere. This main ERP result suggests a specific impairment of right hemisphere early processing of emotional faces, triggered by positive emotions. This finding is consistent with recent results in other domains. Firstly, ADHD seems to involve a deficit in positive valence picture processing at middle latency (EPN) (Herrmann et al., 2009) and abnormal affective processing of positive stimuli (Conzelmann et al., 2009). Recently, it has been proposed that, in ADHD, a possible reduction in amygdala activity (see Plessen et al., 2006) in response to positive stimuli may lead to reduced activation of the reward system and in turn to impaired processing of positive emotional stimuli (Herrmann et al., 2009;

TABLE 4
Estimation of N170 neural generators

				Faces		
				Source peak		Right TOFG
	MNI coordinates			Anatomical description:	SCDP	SCDP
	X	Y	Z	(HOCSA) Harvard-Oxford Cortical Structural Atlas)	Mean	Mean
Control	40	−67	−12	33% Occipital Fusiform Gyrus, 21% Lateral Occipital Cortex, inferior division, 1% Temporal Occipital Fusiform Cortex, 1% Inferior Temporal Gyrus, temporo-occipital part	0.253	0.224
ADHD	25	−86	−18	42% Occipital Fusiform Gyrus, 8% Lateral Occipital Cortex, inferior division, 6% Occipital Pole, 1% Lingual Gyrus	0.123	0.099

				Simultaneous		
				Source peak		Right TOFG
	MNI coordinates			Anatomical Description	SCDP	SCDP
	X	Y	Z	(Harvard-Oxford Cortical Structural Atlas)	Mean	Mean
Control	26	−76	−16	68% Occipital Fusiform Gyrus, 4% Lingual Gyrus, 1% Lateral Occipital Cortex, inferior división	0.151	0.133
ADHD	20	−86	−20	10% Occipital Fusiform Gyrus, 5% Lateral Occipital Cortex, inferior division, 3% Occipital Pole, 1% Lingual Gyrus	0.090	0.072

				Words		
				Source peak		Left TOFG
	MNI coordinates			Anatomical description:	SCDP	SCDP
	X	Y	Z	(Harvard-Oxford Cortical Structural Atlas)	Mean	Mean
Control	−30	−81	−20	38% Occipital Fusiform Gyrus, 11% Lateral Occipital Cortex, inferior division, 1% Lingual Gyrus	0.094	0.076
ADHD	−25	−87	−21	27% Occipital Fusiform Gyrus, 13% Lateral Occipital Cortex, inferior division, 2% Lingual Gyrus, 1% Occipital Pole	0.082	0.050

Notes: TOFG: temporo-occipital fusiform gyrus; SCDP: standardized current density power.

Herrmann, Biehl, Jacob, & Deckert, 2010). This positive bias may be a specific deficit in ADHD, opposed to other emotional impairments present in comorbid disorders (such as the negative bias reported in depression and mania; e.g., Lennox, Jacob, Calder, Lupson, & Bullmore, 2004). Secondly, ADHD appears to involve predominantly right hemispheric dysfunction (for a review, see Barr, 2001; see also Booth et al., 2005). Thirdly, impaired emotional facial processing is the most consistently reported form of social cognitive impairment in ADHD (Uekermann et al., 2010). In children and adolescents with ADHD, abnormal N170 facial processing has been reported (Williams et al., 2008) as well as abnormal activity of frontal

and posterior cingulated cortex activated by emotional expressions, indexed by fMRI (Marsh & Blair, 2008).

Consistent with previous reports, the main source of the N170 was estimated as being located in the right temporo-occipital fusiform gyrus (TOFG) for faces (Rossion et al., 2002, 2003; Sadeh et al., 2008) and in the left TOFG for words (Maillard et al., 2010; Rossion et al., 2003). Theoretical models of emotion face perception (Vuilleumier & Pourtois, 2007) propose a parallel and interactive system indexing object recognition (e.g., triggered by the FG) and emotional discrimination (e.g., triggered by the amygdala). The amygdala mediates emotional processing and valence and is involved in the processing of facial affect. The more basic and structural face integration process seems to be preserved in patients, yet more subtle processes, such as the emotional processing of the face, seem to be affected at early stages of processing (e.g., triggered by reduced connectivity between amygdala and FG). This speculative remark is consistent with recent reports of ADHD abnormal amygdala activation to emotional stimuli (e.g., Brotman et al., 2010; Herrmann et al., 2010).

Our result confirms previous reports of abnormal facial processing in ADHD (children and adolescents), and suggests that in adult ADHD the early stage of cortical face valence processing is affected. In addition, this impairment is related to emotional inference of mental states and executive functioning. Summarizing our data and related reports, we may propose that early right hemisphere dysfunction of processing positive facial expressions in adult ADHD is a neurocognitive marker of basic social cognition deficits, calling for further examination.

ADHD cognitive profile and its association with ERP processing

Our patients evidenced mild to moderate levels of depressive symptoms as revealed by scores on the BDI-II, as is usual in this clinical population and congruent with previous reports from our team (Torralva et al., 2010; Torrente et al., 2010) and other studies (e.g., LeBlanc & Morin, 2004). Notwithstanding, in the current study, no associations between depression and ERP processing were found, suggesting a relative independence of those domains. Further research is called for in this area.

Participants with ADHD presented deficits in recall performance on the RAVLT, as well as some executive impairment, which is not a new issue (e.g., Torralva et al., 2010). In addition, we found a subtle deficit in ToM indexed by the RMRT, and this task correlated with cortical deficits in face valence. At the same time, in the ADHD patients, the ERPs for simultaneous stimuli valence discrimination were associated with higher levels of executive functioning and working memory. Both executive and ToM deficits in ADHD have been reported elsewhere, and these deficits are often both associated with the disorder (Kain & Perner, 2003; Uekermann et al., 2010).

The finding of a combined executive and social impairment is consistent with current neural models of cognition (Pessoa, 2009) and particularly with dysfunction of frontostriatal structures in ADHD (for reviews, see Bush, Valera, & Seidman, 2005; Marsh & Williams, 2006; Uekermann et al., 2010). Thus, a subtle frontostriatal deficit in ADHD could be the neural signature of the combined profile of executive and social cognitive deficits that was reported in this study.

Recent reports on individual differences and N170 processing have stressed the importance of combining a multilevel analysis of cortical measures of facial processing with neuropsychological assessment (Herzmann, Kunina, Sommer, & Wilhelm, 2010; Ibáñez, Haye, González, Hurtado, & Henríquez, 2009; Marsh & Williams, 2006; Petroni et al., in press). Combining neuropsychology with brain function measures of emotion processing may lead to improved clinical assessment of emotional disturbances in ADHD (Williams, 2008). Our results highlight this by demonstrating an ADHD-specific pattern of association between executive and social deficits on ERP and neuropsychology measures; this supports a frontostriatal model of ADHD related to emotional and cognitive functioning. In addition, between-group individual differences suggested a different cognitive profile for people with ADHD, probably reflecting the use of different cognitive strategies in ADHD and controls (see Durston et al., 2003).

Limitations and future studies

Our results suggest an abnormal brain processing of emotional facial stimuli in adult ADHD, consistent with a broad body of research about frontostriatal dysfunction. This preliminary report should be recreated in the near future considering several possible improvements. First, a larger sample including groups of different ADHD subtypes and gender differences is required. The sample size of this study was small, since we included only ADHD patients with no comorbidity. Our results are restricted to males with ADHD. All patients were under medication, and stimulants may have permanent effects on brain function. Assessing other types of emotion (e.g., six basic emotions) and comparing those effects in drug-naive participants (in order to avoid the possible long-term effects of medication) would be additional steps. Finally, it would be relevant to compare the present result with other disorders that occur comorbidly with ADHD, such as bipolar disorder and schizophrenia (Barr, 2001; Ibáñez et al., 2011c; Lus & Mukaddes, 2009; Peralta et al., 2010).

CONCLUSION

In this report, we identified brain markers of impaired facial emotion modulation in participants with ADHD. Those deficits were related to subtle differences in ToM and executive functioning, supporting the frontostriatal dysfunction hypothesis of ADHD. By a multilevel approach, we highlighted the advantage of combing neuropsychological assessment with brain measures from translational neuroscience.

Although a broad range of studies have assessed social cognitive impairments in ADHD, as well as their relation to executive function, the clinical and everyday impact of those deficits is still unclear (Marsh & Williams, 2006; Nijmeijer et al., 2008). Further research on social cognitive deficits would help us to understand various problems that tend to occur in the clinical profile of children and adults with ADHD, such as having fewer friends and difficulty in keeping friends (Nijmeijer et al., 2008), an increased risk of mood and anxiety disorders or antisocial personality disorder (see Nijmeijer et al., 2008), and higher rates of marriage difficulties (Biederman et al., 1993; Murphy & Barkley, 1996). Social cognitive deficits could account at least partially for troubles in school life and employment, together with basic cognitive deficits, especially in relation to discipline and acceptance of norms, both of which have been described as problematic areas of functioning for people with ADHD (Biederman et al., 1993; Murphy & Barkley, 1996). Unfortunately, current social skills training programs for children and adults with ADHD (Hesslinger et al., 2002; Pelham & De Jong, 1992; Safren et al., 2005) do not include facial affect perception. Thus, basic social cognitive deficits (facial emotion processing) should be expanded not only as a research area in ADHD but also as an important topic for therapy in future.

REFERENCES

Adler, L. A. (2010). Monitoring adults with ADHD: A focus on executive and behavioral function. *Journal of Clinical Psychiatry, 71*, e18.

Aravena, P., Hurtado, E., Riveros, R., Cardona, F., Manes, F., & Ibáñez, A. (2010). Applauding with closed hands: Neural signature of action sentence compatibility effects, *PLoS ONE, 5*(7), e11751. doi: 10.1371/journal.pone.0011751

Balconi, M., & Lucchiari, C. (2005). Event-related potentials related to normal and morphed emotional faces. *Journal of Psychology, 139*(2), 176–192.

Barkley, R. (2001). The executive functions and self-regulation: An evolutionary neuropsychological perspective. *Neuropsychological Reviews, 11*, 1–29.

Barkley, R. (2010). Differential diagnosis of adults with ADHD: The role of executive function and self-regulation. *Journal of Clinical Psychiatry, 71*, 1–17.

Barkley, R. A., & Murphy, K. R. (1998). Attention-deficit hyperactivity disorder: A clinical workbook (2nd edn.). New York, NY: Guilford Press.

Baron-Cohen, S., Wheelwright, S., Hill, J., Raste, Y., & Plumb, I. (2001). The "Reading the Mind in the Eyes" Test, Revised Version: A study with normal adults, and adults with Asperger syndrome or high-functioning autism. *Journal of Child Psychology and Psychiatry, 42*(2), 241–251.

Barr, W. (2001). Schizophrenia and attention-deficit disorder: Two complex disorders of attention. *Annals of the New York Academy of Sciences, 931*, 239–250.

Beck, A. T., Steer, R. A., Ball, R., & Ranieri, W. (1996). Comparison of Beck Depression Inventories -IA and -II in psychiatric outpatients. *Journal of Personality Assessment, 67*, 588–597.

Benton, A. L., Hamsher, K., & Sivan, A. B. (1994). *Multilingual Aphasia Examination* (3rd edn.). Iowa City, IA: AJA Associates.

Biederman, J., Faraone, S. V., Spencer, T., Wilens, T., Norman, D., Lapey, K. A., et al. (1993). Patterns of psychiatric comorbidity, cognition, and psychosocial functioning in adults with attention-deficit hyperactivity disorder. *American Journal of Psychiatry, 150*, 1792–1798.

Booth, J. R., Burman, D. D., Meyer, J. R., Lei, Z., Trommer, B. L., Davenport, N. D., et al. (2005). Larger deficits in brain networks for response inhibition than for visual selective attention in attention-deficit hyperactivity disorder (ADHD). *Journal of Child Psychology and Psychiatry, 46*(1), 94–111.

Braaten, E. B., & Rosen, L. A. (2000). Self-regulation of affect in attention deficit-hyperactivity disorder (ADHD) and non-ADHD boys: Differences in empathic responding. *Journal of Consulting and Clinical Psychology, 68*, 313–321.

Brotman, M. A., Rich, B. A., Guyer, A. E., Lunsford, J. R., Horsey, S. E., Reising, M. M., et al. (2010). Amygdala activation during emotion processing of neutral faces in children with severe mood dysregulation versus ADHD or bipolar disorder. *American Journal of Psychiatry, 167*(1), 61–69.

Bruce, V., & Young, A. (1986). Understanding face recognition. *British Journal of Psychology, 77*, 305–327.

Buitelaar, J. K., van der Wees, M., Swaab-Barneveld, H., & van der Gaag, R. J. (1999). Theory of mind and emotion-recognition functioning in autistic spectrum disorders and in psychiatric control and normal children. *Development and Psychopathology, 11*, 39–58.

Bush, G., Valera, E. M., & Seidman, L. J. (2005). Functional neuroimaging of attention-deficit/hyperactivity disorder: A review and suggested future directions. *Biological Psychiatry, 57*(11), 1273–1284.

Charman, T., Baron-Cohen, I., Baird, G., Cox, A., Wheelwright, S., & Swettenham, J. (2001). Commentary: The Modified Checklist for Autism in Toddlers. *Journal of Autism and Developmental Disorders, 31*, 145–148.

Conzelmann, A., Mucha, R. F., Jacob, C. P., Weyers, P., Romanos, J., Gerdes, A. B., et al. (2009). Abnormal affective responsiveness in attention-deficit/hyperactivity disorder: Subtype differences. *Biological Psychiatry, 65*(7), 578–585.

Deffke, I., Sander, T., Heidenreich, J., Sommer, W., Curio, G., & Trahms, L. (2007). MEG/EEG sources of the 170-ms response to faces are co-localized in the fusiform gyrus. *NeuroImage, 35*, 1495–1501.

Durston, S., Tottenham, N. T., Thomas, K. M., Davidson, M. C., Eigsti, I. M., Yang, Y., et al. (2003). Differential patterns of striatal activation in young children with and without ADHD. *Biological Psychiatry, 53*(10), 871–878.

Dyck, M. J., Ferguson, K., & Shochet, I. M. (2001). Do autism spectrum disorders differ from each other and from non-spectrum disorders on emotion recognition tests? *European Child & Adolescent Psychiatry*, *10*, 105–116.

Edwards, R., Manstead, A. S., & MacDonald, C. J. (1984). The relationship between children's sociometric status and ability to recognize facial expressions of emotion. *European Journal of Social Psychology*, *14*, 235–238.

Feldman, R. S., Philippot, J. R., & Custrini, R. J. (1991). Social competence and nonverbal behaviour. In R. S. Feldman & B. Rime (Eds.), *Fundamentals of nonverbal behavior* (pp. 329–350). New York, NY: Cambridge University Press.

Gray, H. M., Ambady, N., Lowenthal, W. T., & Deldin, P. (2004). P300 as an index of attention to self-relevant stimuli. *Journal of Experimental Social Psychology*, *40*, 216–224.

Grossmann, T. (2010). The development of emotion perception in face and voice during infancy. *Restorative Neurology and Neuroscience*, *28*, 219–236.

Guerra, S., Ibáñez, A., Martín, M., Bobes, M. A., Reyes, A., Mendoza, R., et al. (2009). N400 deficits from semantic matching of pictures in probands and first degrees relatives from multiplex schizophrenia families. *Brain and Cognition*, *70*(2), 221–230.

Herrmann, M. J., Biehl, S. C., Jacob, C., & Deckert, J. (2010). Neurobiological and psychophysiological correlates of emotional dysregulation in ADHD patients. *Attention-Deficit Hyperactivity Disorders*, *2*(4), 233–239.

Herrmann, M. J., Schreppel, T., Biehl, S. C., Jacob, C., Heine, M., Boreatti-Hümmer, A., et al. (2009). Emotional deficits in adult ADHD patients: An ERP study. *Social Cognitive and Affective Neuroscience*, *4*, 340–345.

Herzmann, G., Kunina, O., Sommer, W., & Wilhelm, O. (2010). Individual differences in face cognition: Brain–behavior relationships. *Journal of Cognitive Neuroscience*, *22*(3), 571–589.

Hesslinger, B., Tebartz van Elst, L., Nyberg, E., Dykierek, P., Richter, H., & Berner, M. (2002). Psychotherapy of attention-deficit hyperactivity disorder in adults – a pilot study using a structured skills training program. *European Archives of Psychiatry and Clinical Neuroscience*, *252*, 177–184.

Hill, B. D., Elliott, E. M., Shelton, J. T., Pella, R. D., O'Jile, J. R., & Gouvier, W. D. (2010). Can we improve the clinical assessment of working memory? An evaluation of the Wechsler Adult Intelligence Scale-Third Edition using a working memory criterion construct. *Journal of Clinical and Experimental Neuropsychology*, *32*(3), 315–323.

Hoza, B., Waschbusch, D. A., Pelham, W. E., Molina, B. S., & Milich, R. (2000). Attention-deficit/hyperactivity disordered and control boys' responses to social success and failure. *Child Development*, *71*, 432–446.

Hurtado, E., Gonzalez, R., Haye, A., Manes, F., & Ibáñez, A. (2009). Contextual blending of ingroup/outgroup face stimuli and word valence: LPP modulation and convergence of measures. *BMC Neuroscience*, *10*, 69.

Ibáñez, A., Gleichgerrcht, E., Hurtado, E., González, R., Haye, A., & Manes, F. (2010d). Neural markers of early contextual blending: N170 modulation of ingroup/outgroup relative position and associated valence. *Frontiers in Human Neuroscience*, *4*, 188. doi: 10.3389/fnhum.2010. 00188

Ibáñez, A., Haye, A., González, R., Hurtado, E., & Henríquez, R. (2009). Multi-level analysis of cultural phenomena: The role of ERP approach to prejudice. *Journal for Theory in Social Behavior*, *39*, 81–110.

Ibáñez, A., Hurtado, E., Lobos, A., Trujillo, N., Escobar, J., Baez, S., et al. (in press).Subliminal presentation of other faces (but not own face) primes behavioral and evoked cortical processing of empathy for pain. *Brain Research*. doi: 10.1016/j.brainres.2011.05.014

Ibáñez, A., Hurtado, E., Riveros, R., Urquina, H., Cardona, J. F., Petroni, A., et al. (2011a). Facial and semantic emotional interference: A pilot study on the behavioral and cortical responses to the dual valence association task. *Behavioral and Brain Functions*, *7*, 8.

Ibáñez, A., López, V., Cornejo, C. (2006). ERPs and contextual semantic discrimination: Evidence of degrees of congruency in wakefulness and sleep. *Brain and Language*, *98*(3), 264–275.

Ibáñez, A., Manes, F., Escobar, J., Trujillo, N., Andreucci, P., & Hurtado, E. (2010b). Gesture influences the processing of figurative language in non-native speakers. *Neuroscience Letters*, *471*, 48–52.

Ibáñez, A., Riveros, R., Aravena, P., Vergara, V., Cardona, J. F., García, L., et al. (2010a). When context is hard to integrate: Cortical measures of congruency in schizophrenics and healthy relatives from multiplex families. *Schizophrenia Research*. doi: 10.1016/j.schres.2010.04.008

Ibáñez, A., Riveros, R., Hurtado, E., Gleichgerrcht, E., Urquina, H., Herrera, E., Amoruso, L., Martin-Reyes, M., & Manes, F. (2011c). The face and its emotion: Cortical Deficits in Structural Processing and Early Emotional Discrimination in Schizophrenic and Relatives. *Psychiatry Research*, doi: 10.1016/j.psychres.2011.07.027

Ibáñez, A., San Martín, R., Hurtado, E., & López, V. (2008a). Methodological considerations related to sleep paradigm using event related potentials. *Biological Research*, *41*, 271–275.

Ibáñez, A., San Martín, R., Hurtado, E., & López, V. (2008b). ERP studies of cognitive processing during sleep. *International Journal of Psychology*, *44*(4), 290–304. doi: 10.1080/00207590802194234

Ibáñez, A., Toro, P., Cornejo, C., Urquina, H., Manes, F., Weisbrod, M., et al. (2010c).High contextual sensitivity of metaphorical expressions and gesture blending: A video ERP design. *Psychiatry Research, Neuroimaging*, 10.1016/j.pscychresns.

Itier, R. J., & Batty, M. (2009). Neural bases of eye and gaze processing: The core of social cognition. *Neuroscience & Biobehavioral Reviews*, *33*, 843–863.

Kain, W., & Perner, J., (2003). Do children with ADHD not need their frontal lobes for theory of mind? A review of brain imaging and neuropsychological studies. In M. Brune, H. Ribbert, & W. Schiefenho (Eds.), *The social brain: Evolution and pathology* (pp. 197–230). Chichester, UK: Wiley.

King, S., Waschbusch, D. A., Pelham, W. E., Jr., Frankland, B. W., Andrade, B. F., & Jacques, S. (2009). Social information processing in elementary-school aged children with ADHD: Medication effects and comparisons with typical children. *Journal of Abnormal Child Psychology*, *37*, 579–589.

Klassen, L. J., Katzman, M. A., & Chokka, P. (2010). Adult ADHD and its comorbidities, with a focus on bipolar disorder. *Journal of Affective Disorders, 124*, 1–8.

Kotchoubey, B., Jetter, U., Lang, S., Semmler, A.,Mezger, G.,Schmalohr, D., et al. (2006). Evidence of cortical learning in vegetative state. *Journal of Neurology, 53*(10), 1374–1376.

Krombholz, A., Schaefer, F., & Boucsein, W. (2007). Modification of N170 by different emotional expression of schematic faces. *Biological Psychology, 76*(3), 156–162.

LeBlanc, N., & Morin, D. (2004). Depressive symptoms and associated factors in children with attention-deficit hyperactivity disorder. *Journal of Child and Adolescent Psychiatric Nursing, 17*, 49–55.

Lennox, B. R., Jacob, R., Calder, A. J., Lupson, V., & Bullmore, E. T. (2004). Behavioural and neurocognitive responses to sad facial affect are attenuated in patients with mania. *Psychological Medicine, 34*(5), 795–802.

Lezak, M. (1982). The problem of assessing executive functions. *International Journal of Psychology, 17*, 281–297.

López, V., López-Calderón, J., Ortega, R., Kreither, J., Carrasco, X., Rothhammer, P., et al. (2006). Attention-deficit hyperactivity disorder involves differential cortical processing in a visual spatial attention paradigm. *Clinical Neurophysiology, 117*(11), 2540–2548.

Lus, G., & Mukaddes, N. M. (2009). Co-morbidity of bipolar disorder in children and adolescents with attention-deficit/hyperactivity disorder (ADHD) in an outpatient Turkish sample. *World Journal of Biological Psychiatry, 10*(4 Pt 2), 488–494.

Maedgen, J. W., & Carlson, C. L. (2000). Social functioning and emotional regulation in the attention-deficit hyperactivity disorder subtypes. *Journal of Clinical Child Psychology, 29*, 30–42.

Maillard, L., Barbeau, E. J., Baumann, C., Koessler, L., Benar, C., Chauvel, P., et al. (2010). From perception to recognition memory: Time course and lateralization of neural substrates of word and abstract picture processing. *Journal of Cognitive Neuroscience, 23*(4), 782–800.

Malloy-Diniz, L., Fuentes, D., Leite, W. B., Correa, H., & Bechara, A. (2007). Impulsive behavior in adults with attention-deficit/hyperactivity disorder: Characterization of attentional, motor and cognitive impulsiveness. *Journal of the International Neuropsychological Society, 13*, 693–698.

Marsh, A. A., & Blair, R. J. (2008). Deficits in facial affect recognition among antisocial populations: A meta-analysis. *Neuroscience & Biobehavioral Reviews, 32*, 454–465.

Marsh, P. J., & Williams, L. M. (2006). ADHD and schizophrenia phenomenology: Visual scanpaths to emotional faces as a potential psychophysiological marker? *Neuroscience & Biobehavioral Reviews, 30*, 651–665.

Matthys, W., Cuperus, J. M., & Van, E. H. (1999). Deficient social problem-solving in boys with ODD/CD, with ADHD, and with both disorders. *Journal of the American Academy of Child and Adolescent Psychiatry, 38*, 311–321.

Mazziotta, J., Toga, A., Evans, A., Fox, P., Lancaster, J., Zilles, K., Woods, R., Paus, T., et al. (2001). A probabilistic atlas and reference system for the human brain: International Consortium for Brain Mapping (ICBM).

Philosophical Transactions of the Royal Society B, 356, 1293–1322.

Mori, S., Oishi, K., Jiang, H., Jiang, L., Li, X., Akhter, K., Hua, K., Faria, A.V., Mahmood, A., Woods, R., Toga, A.W., Pike, G.B., Neto, P.R., Evans, A., Zhang, J., Huang, H., Miller, M.I., Van, Z.P., Mazziotta, J. (2008). Stereotaxic white matter atlas based on diffusion tensor imaging in an ICBM template. *Neuroimage, 40*, 570–582.

Murphy, K., & Barkley, R. (1996). Attention-deficit hyperactivity disorder adults: Comorbidities and adaptive impairments. *Comprehensive Psychiatry, 37*(6), 393–401.

Nijmeijer, J. S., Minderaa, R. B., Buitelaar, J. K., Mulligan, A., Hartman, C. A., & Hoekstra, P. J. (2008). Attention-deficit/hyperactivity disorder and social dysfunctioning. *Clinical Psychology Review, 28*(4), 692–708.

Ohman, A., & Mineka, S. (2001). Fears, phobias, and preparedness: Toward an evolved module of fear and fear learning. *Psychological Review, 108*, 483–522.

Partington, J. (1949). Detailed instructions for administering Partington's pathways test. *Psychological Service Center Journal, 1*, 46–48.

Pascual-Marqui, R. D. (2002). Standardized low-resolution brain electromagnetic tomography (sLORETA): Technical details. *Methods & Findings in Experimental & Clinical Pharmacology, 24*, 5–12.

Pelc, K., Kornreich, C., Foisy, M. L., & Dan, B. (2006). Recognition of emotional facial expressions in attention-deficit hyperactivity disorder. *Pediatric Neurology, 35*, 93–97.

Pelham, T. L., & DeJong, A. R. (1992). Nationwide practices for screening and reporting prenatal cocaine abuse: A survey of teaching programs. *Child Abuse & Neglect, 16*(5), 763–770.

Peralta, V., de Jalón, E. G., Campos, M. S., Basterra, V., Sanchez-Torres, A., & Cuesta, M. J. (2011). Risk factors, pre-morbid functioning and episode correlates of neurological soft signs in drug-naive patients with schizophrenia-spectrum disorders. *Psychological Medicine, 41*, 1279–1289.

Pessoa, L. (2009). How do emotion and motivation direct executive control? *Trends in Cognitive Sciences, 13*(4), 160–166.

Petroni, A., Urquina, H., Guex, R., Hurtado, E., Manes, F., Sigman, M., et al. (in press). Early cortical measures of valence, stimulus type discrimination and interference: Association to executive function and social cognition.

Philippot, P., & Feldman, R. S. (1990). Age and social competence in preschoolers' decoding of facial expression. *British Journal of Social Psychology, 29*(Pt 1), 43–54.

Plessen, K. J., Bansal, R., Zhu, H., Whiteman, R., Amat, J., & Quackenbush, G. A. (2006). Hippocampus and amygdala morphology in attention-deficit/hyperactivity disorder. *Archives of General Psychiatry, 63*, 795–807.

Proverbio, A. M., Riva, F., Martín, E., & Zani, A. (2010). Face coding is bilateral in the female brain. *PLoS ONE, 5*, e11242.

Rapport, L., Van Voorhis, A., Tzelepis, A., & Friedman, S. (2001). Executive functioning in adult attention-deficit hyperactivity disorder. *Clinical Neuropsychology, 15*, 479–491.

Righart, R., & de Gelder, B. (2008). Rapid influence of emotional scenes on encoding of facial expressions: An ERP study. *Social Cognitive and Affective Neuroscience, 3*, 270–278.

Rossion, B., Gauthier, I., Goffaux, V., Tarr, M. J., & Crommelinck, M. (2002). Expertise training with novel objects leads to left-lateralized face-like electrophysiological responses. *Psychological Science*, *13*(3), 250–257.

Rossion, B., & Jacques, C. (2008). Does physical interstimulus variance account for early electrophysiological face-sensitive responses in the human brain? Ten lessons on the N170. *NeuroImage*, *39*(4), 1959–1979.

Rossion, B., Joyce, C. A., Cottrell, G. W., & Tarr, M. J. (2003). Early lateralization and orientation tuning for face, word, and object processing in the visual cortex. *NeuroImage*, *20*(3), 1609–1624.

Sadeh, B., Zhdanov, A., Podlipsky, I., Hendler, T., & Yovel, G. (2008). The validity of the face-selective ERP N170 component during simultaneous recording with functional MRI. *NeuroImage*, *42*(2), 778–786.

Safren, S. (2006). Cognitive-behavioral approaches to ADHD treatment in adulthood. *Journal of Clinical Psychiatry*, *67*, 46–50.

Safren, S. A., Otto, M. W., Sprich, S., Winett, C. L., Wilens, T. E., & Biederman, J. (2005). Cognitive-behavioral therapy for ADHD in medication-treated adults with continued symptoms. *Behaviour Research and Therapy*, *43*, 831–842.

San Martín, R., Manes, F., Hurtado, E., Isla, P., & Ibáñez, A. (2010). Size and probability of rewards modulate the feedback error-related negativity associated with wins but not losses in a monetarily rewarded gambling task. *NeuroImage*, *51*, 1194–1204.

Schacht, A., & Sommer, W. (2009). Emotions in word and face processing: Early and late cortical responses. *Brain and Cognition*, *69*(3), 538–550.

Schecklmann, M., Ehlis, A., Plichta, M., Romanos, J., Heine, M., Boreatti-Hummer, A., et al. (2009). Diminished prefrontal oxygenation with normal and above-average verbal fluency performance in adult ADHD. *Journal of Psychiatric Research*, *43*, 98–106.

Shapiro, L. P., Gordon, B., Hack, N., & Killackey, J. (1993). Verb-argument structure processing in complex sentences in Broca's and Wernicke's aphasia. *Brain and Language*, *45*, 423–447.

Simon, V., Czobor, P., Balint, S., Meszaros, A., & Bitter, I. (2009). Prevalence and correlates of adult attention-deficit hyperactivity disorder: Meta-analysis. *British Journal of Psychiatry*, *194*, 204–211.

Sinzig, J., Morsch, D., & Lehmkuhl, G. (2008). Do hyperactivity, impulsivity and inattention have an impact on the ability of facial affect recognition in children with autism and ADHD? *European Child & Adolescent Psychiatry*, *17*, 63–72.

Sodian, B., Hulsken, C., & Thoermer, C. (2003). The self and action in theory of mind research. *Consciousness and Cognition*, *12*, 777–782.

Sonuga-Barke, E. J. (2003). The dual pathway model of AD/HD: An elaboration of neuro-developmental characteristics. *Neuroscience and Biobehavioral Reviews*, *27*(7), 593–604.

Torralva, T., Gleichgerrcht, E., Torrente, F., Roca, M., Strejilevich, S. A., Cetkovich, M., et al. (2011). Neuropsychological functioning in adult bipolar disorder and ADHD patients: A comparative study. *Psychiatry Research*, *186*, 261–266.

Torralva, T., Roca, M., Gleichgerrcht, E., López, P., & Manes, F. (2009). INECO Frontal Screening (IFS): A brief, sensitive, and specific tool to assess executive functions in dementia. *Journal of the International Neuropsychological Society*, *15*, 777–786.

Torrente, F., Lischinsky, A., Torralva, T., López, P., Roca, M., & Manes, F. (2011). Not always hyperactive? Elevated apathy scores in adolescents and adults with ADHD. *Journal of Attention Disorder*, *15*, 545–556.

Uekermann, J., Kraemer, M., Abdel-Hamid, M., Schimmelmann, B. G., Hebebrand, J., & Daum, I. (2010). Social cognition in attention-deficit hyperactivity disorder (ADHD). *Neuroscience & Biobehavioral Reviews*, *34*, 734–743.

Valdes-Hernandez, P. A., Von, E.N., Ojeda-Gonzalez, A., Kochen, S., Aleman-Gomez, Y., Muravchik, C., Valdes-Sosa, P. A. (2009). Approximate average head models for EEG source imaging. *Journal of Neuroscience Methods*, *185*, 125–132.

Vuilleumier, P., & Pourtois, G. (2007). Distributed and interactive brain mechanisms during emotion face perception: Evidence from functional neuroimaging. *Neuropsychologia*, *45*(1), 174–194.

Wechsler, D. (1997). *Wechsler Adult Intelligence Scale-III: Administration and scoring manual*. San Antonio, TX: Harcourt Assessment.

Wheeler, J., & Carlson, C. L. (1994). The social functioning of children with ADD with hyperactivity and ADD without hyperactivity: A comparison of their peer relations and social deficits. *Journal of Emotional and Behavioral Disorders*, *2*, 2–12.

Williams, J. (2008). Working toward a neurobiological account of ADHD: Commentary on Gail Tripp and Jeff Wickens, dopamine transfer deficit. *Journal of Child Psychology and Psychiatry*, *49*(7), 705–711.

Williams, L. M., Hermens, D. F., Palmer, D., Kohn, M., Clarke, S., Keage, H., et al. (2008). Misinterpreting emotional expressions in attention-deficit/hyperactivity disorder: Evidence for a neural marker and stimulant effects. *Biological Psychiatry*, *63*(10), 917–926.

Wodushek, T., & Neumann, C. (2001). Inhibitory capacity in adults with symptoms of attention-deficit/hyperactivity disorder (ADHD). *Archives of Clinical Neuropsychology*, *18*(3), 317–330.

Young, R. C., Biggs, J. T., Ziegler, V. E., & Meyer, D. A. (1978). A rating scale for mania: Reliability, validity and sensitivity. *British Journal of Psychiatry*, *133*, 429–435.

Zhang, L., Gerstenberger, A., Wang, X., Liu, W.K. 2004. Immersed finite element method. *Computer Methods in Applied Mechanics and Engineering*, *193*, 2051–2067.

Neural correlates of social approach and withdrawal in patients with major depression

Birgit Derntl[1,2], Eva-Maria Seidel[1,2], Simon B. Eickhoff[2,3], Thilo Kellermann[2,3], Ruben C. Gur[4], Frank Schneider[2,3], and Ute Habel[2,3]

[1]Institute of Clinical, Biological, and Differential Psychology, Faculty of Psychology, University of Vienna, Vienna, Austria

[2]Department of Psychiatry, Psychotherapy, and Psychosomatics, RWTH Aachen University, Aachen, Germany

[3]JARA-BRAIN Jülich Aachen Research Alliance, Translational Brain Medicine, Germany

[4]Neuropsychiatry Division, Department of Psychiatry, University of Pennsylvania, Philadelphia, PA, USA

Successful human interaction is based on correct recognition, interpretation, and appropriate reaction to facial affect. In depression, social skill deficits are among the most restraining symptoms leading to social withdrawal, thereby aggravating social isolation and depressive affect. Dysfunctional approach and withdrawal tendencies to emotional stimuli have been documented, but the investigation of their neural underpinnings has received limited attention. We performed an fMRI study including 15 depressive patients and 15 matched, healthy controls. All subjects performed two tasks, an implicit joystick task as well as an explicit rating task, both using happy, neutral, and angry facial expressions. Behavioral data analysis indicated a significant group effect, with depressed patients showing more withdrawal than controls. Analysis of the functional data revealed significant group effects for both tasks. Among other regions, we observed significant group differences in amygdala activation, with patients showing less response particularly during approach to happy faces. Additionally, significant correlations of amygdala activation with psychopathology emerged, suggesting that more pronounced symptoms are accompanied by stronger decreases of amygdala activation. Hence, our results demonstrate that depressed patients show dysfunctional social approach and withdrawal behavior, which in turn may aggravate the disorder by negative social interactions contributing to isolation and reinforcing cognitive biases.

Keywords: Depression; Approach; Withdrawal; Emotion; Amygdala; fMRI.

Emotional facial expressions are salient cues in social life, and accurately recognizing, interpreting, and responding to them is essential for successful social interaction. While evidence has accumulated on impaired facial emotion recognition and its dysfunctional neural underpinnings in patients suffering from depression (e.g., Dannlowski et al., 2007; Douglas & Porter, 2010; Suslow et al., 2010; for review, see Leppänen, 2006), deficits in behavioral tendencies prompted by an emotional expression and the corresponding dysfunctional neural correlates have received limited attention in this clinical population. Regarding behavioral tendencies, Gray's theory (Gray, 1982) of a behavioral approach, BAS, and a behavioral inhibition

Correspondence should be addressed to: Birgit Derntl, Institute for Clinical, Biological, and Differential Psychology, Faculty of Psychology, University of Vienna, Liebiggasse 5, 1010 Vienna, Austria. E-mail: birgit.derntl@univie.ac.at

This study was supported by the Medical Faculty of the RWTH Aachen University (START 609811 to B.D.) and the DFG (IRTG 1328). S.B.E. was supported by the Human Brain Project (R01-MH074457-01A1) and the Helmholtz Initiative on Systems Biology. Work at the University of Pennsylvania was supported by Grant MH-60722 from the National Institutes of Health.

system, BIS has been examined most extensively, supposing two antipodal motivational systems: one appetitive (approach) and one aversive (withdrawal), both forming the basis of human behavior (cf. Elliot & Covington, 2001; Puca, Rinkenauer, & Breidenstein, 2006). Though several aspects of this theory are still under debate (for overview, see Corr, 2008), in its original version Gray (1972) postulated that the BIS is associated with punishment and frustrative non-reward and thus consequently with avoidance behavior. However, the BAS is assumed to be related to reward and relief from punishment regulating appetitive motivation and approach behavior. It is widely acknowledged that the two systems do not function independently but instead influence and consequently inhibit each other (Gray, 1990; Gray & McNaughton, 2000). Due to this interaction, Gray (1994) speculated that depression is based on a combination of an elevated BIS and a decreased BAS sensitivity, with the BIS inhibiting the BAS, prompting a reduction in positive affect and appetitive motivation (Davidson, 1992), cognitive misinterpretations of social signals, and disturbances in social interaction (for review, see Segrin, 2000), eventually leading to social withdrawal and isolation. In this regard, several studies demonstrated a direct association between severity of depressive symptoms and self-reported BAS sensitivity (Campbell-Sills, Liverant, & Brown, 2004; Kasch, Rottenberg, Arnow, & Gotlib, 2002). Moreover, supportive results were reported by Tse and Bond (2004), who demonstrated that depressed patients tend to interpret social information in a negative way, feeling rejected by others and, thus, avoiding social interaction. Additionally, results from two prospective studies showed that self-reported BAS sensitivity could predict the clinical course of depression (Kasch et al., 2002; McFarland, Shankman, Tenke, Bruder, & Klein, 2006).

In a recent study from our lab (Seidel et al., 2010a), we investigated approach and withdrawal tendencies in depressed patients applying two different tasks: (1) an implicit joystick task, where participants were asked to pull or push the lever toward pictures of facial emotional expressions depending on the color of the frame encircling the expression; and (2) an explicit rating task, where subjects were instructed to indicate whether they would move toward or away from a person showing the presented emotional expression. Direct comparison of implicit and explicit results enabled analysis of more automatic vs. more conscious responses to facial expressions of emotion. Although patients correctly recognized the emotional expressions, they reacted differently to these social cues: especially female patients displayed

stronger withdrawal tendencies in the explicit condition, which were less pronounced in the implicit condition. Thus, we speculated that while automatic behavioral responses are still intact, conscious ratings seem to be negatively influenced by cognitive biases, fitting Beck's cognitive negativity model of depression (Beck, 2008).

Regarding the neural correlates, Gray (1982) assumed that a network comprising the septohippocampal system constitutes the BIS. Using fMRI, Reuter and colleagues (2004) demonstrated elevated activation of the amygdala, insula, and temporofrontal regions during viewing of disgust-eliciting stimuli (BIS condition) in healthy controls. Thus, they reported activation of several regions known to be involved in various emotional competencies, such as emotion recognition (e.g., Derntl et al., 2009a, 2009b, 2009c; Fitzgerald, Angstadt, Jelsone, Nathan, & Phan, 2006; Habel et al., 2007; Moser et al., 2007; for review, see Fusar-Poli et al., 2009) or empathy (e.g., Derntl et al., 2010; Lamm, Batson, & Decety, 2007; Schulte-Rüther, Markowitsch, Shah, Fink, & Piefke, 2008; for review, see Lamm & Singer, 2009).

The neural network of the BAS is assumed to comprise the basal ganglia and the prefrontal cortex (Pickering & Gray, 1999); this is supported by results from electrophysiological studies demonstrating a relationship between asymmetrical frontal activity and self-reported BAS sensitivity (e.g., Coan & Allen, 2003; Harmon-Jones & Allen, 1997). Thus, these data support Davidson's hypothesis that activity of the left prefrontal cortex is associated with the BAS whereas activity of the right prefrontal cortex is related to the BIS. Moreover, Davidson (1998) assumed that decreased left frontal activity reflects a reduced BAS constituting a vulnerability marker for depression. However, there are no neuroimaging data explicitly prompting the BIS/BAS on behavioral tendencies in social situations in depressed patients.

Since social withdrawal is one of the core symptoms in depression that affects multiple psychosocial domains, it is mandatory to further elucidate the neural underpinnings of the behavioral approach and withdrawal tendencies toward facial emotional expressions. Therefore, this study aims to investigate the neural correlates of social approach and withdrawal in depressed patients.

On the basis of our previous results (Seidel et al., 2010a), we hypothesized stronger withdrawal behavior in depressed patients as measured with self-report questionnaire data. Regarding the neural correlates, we expected stronger activation of patients during avoidance irrespective of emotion (in both tasks), particularly in brain regions involved in the BIS network;

that is, the right prefrontal cortex and temporofrontal regions. For the explicit task, we hypothesized less amygdala activation for positive stimuli as shown in previous studies (e.g., Suslow et al., 2010).

METHODS

Sample

Fifteen depressed patients (9 women, mean age = 34.1 years, SD = 12.0) fulfilling DSM-IV (APA, 1994) criteria for major depression and 15 healthy controls (9 women, mean age = 32.9 years, SD = 10.9) matched for gender, age, and education participated. All subjects were native German speakers and right-handed as assessed by the Edinburgh Handedness Inventory (Oldfield, 1971). The study was approved by the local institutional review board and conducted according to the Declaration of Helsinki (1964). Written informed consent was obtained and all subjects were paid for participation (€30).

Depressed patients were recruited from the inpatient units of the Department of Psychiatry, Psychotherapy, and Psychosomatics, University Hospital Aachen. They had had no substance abuse for the last 6 months and no comorbid psychiatric (Axis I or II) or neurological disorder, as assessed by the German version of the Structured Clinical Interview for DSM-IV (SCID) (Wittchen, Zaudig, & Fydrich, 1997). None of these patients experienced psychotic symptoms during the current or previous episodes. Severity of affective symptoms was assessed with the German versions of the Beck Depression Inventory (BDI) (mean = 25.6, SD = 6.1; Beck et al., 1961), and the 17-item version of the Hamilton Depression Rating (HAMD) Scale (mean = 19.9, SD = 7.3; Hamilton, 1960). The mean age of onset was 29.47 years (SD = 10.8), with mean illness duration of 4.67 years (SD = 7.1). All but two of the depressed patients were taking antidepressant medication at the time of testing (Selective reuptake inhibitors, SSRI, n = 2; selective noradrenalin reuptake inhibitors, SNRI, n = 5; SSRI + SNRI, n = 4; SNRI + quetiapine, n = 2).

The nonpsychiatric control group consisted of 15 healthy adults (9 women) with no history of psychiatric (including substance abuse) or neurological illness. Subjects with such disorders or their first-degree relatives were also excluded.

All participants additionally completed a neuropsychological test battery assessing crystallized verbal intelligence (Mehrfachwortwahltest – Version B: MWT-B; Lehrl, 1996), executive functions (Trail Making Test, Parts A and B: TMT–A/–B; Reitan, 1958), and working memory (digit span, Wechsler

Adult Intelligence Scale: WAIS–III; Von Aster, Neubauer, & Horn, 2006). Moreover, questionnaire data from the German version of the BIS/BAS scale by Carver and White (1994) and the Action Regulating Emotion Systems (ARES) scales (Hartig & Moosbrugger, 2003) were obtained.

Patients and controls differed significantly in their crystallized intelligence (MWT-B: t = 3.293, p = .003), with patients showing lower scores, whereas performance in the other neurocognitive tasks did not differ significantly (TMT-A: t = 1.274, p = .213; TMT-B: t = −0.013, p = .989; digit span forward: t = .555, p = .584; digit span backward: t = −0.425, p = .675). Regarding group comparison of BIS/BAS scores, we observed significantly higher BAS scores in controls (t = 5.713, p < .001) and significantly higher BIS scores in patients (t = 10.241, p < .001). Demographic and neuropsychological characteristics are shown in Table 1.

Moreover, we explored the ability to recognize facial expressions of emotions in patients and controls by applying an emotion-identification task: Vienna Emotion Recognition Tasks (VERT-K; Derntl, Kryspin-Exner, Fernbach, Moser, & Habel, 2008). It consists of 36 colored photographs of facial expressions of five basic emotions (happiness, sadness, anger, fear, and disgust) as well as neutral expressions out of the same stimulus set (Gur et al. 2002). The stimulus material is balanced for valence and gender. Only evoked expressions were shown in randomized order. The instruction was to recognize the emotion depicted as soon and as accurately as possible and to choose one out of six possible emotion categories. The faces remained on the screen until the participant selected a label.

Functional tasks

Similar versions of the functional tasks have been validated in a recent behavioral study in patients with major depression and are described in more detail there (Seidel et al., 2010a). Briefly, we applied two tasks tapping the implicit and explicit behavioral tendencies separately.

Implicit joystick task

Experimental paradigms of studies investigating behavioral tendencies (i.e., approach and avoidance) are based on the findings of Cacioppo, Priester, and Berntson (1993). It has been shown that pushing a lever is faster than pulling in response to aversive stimuli and pulling is faster than pushing in response to appetitive stimuli (Chen & Bargh, 1999; Duckworth,

TABLE 1
Overview on demographic characteristics, neuropsychological performance (raw scores), and self-report questionnaire data of patients and controls. Moreover, for patients, mean values of the clinical rating scales are listed

	Patients (n = 15)	*Controls* (n = 15)	t *values*	p *values*
Gender (M:F)	6:9	6:9	–	–
Age (range)	34.1 (11.95)	32.9 (10.93)	0.303	.764
Years of education	16.1 (3.72)	17.0 (4.00)	0.614	.544
Verbal IQ	106.9 (9.58)	119.9 (12.84)	3.293	.003
TMT-A (seconds)	19.2 (3.76)	20.6 (6.54)	1.274	.213
TMT-B (seconds)	40.1 (13.54)	38.3 (11.28)	−0.013	.989
Digit span (raw score)	15.6 (4.24)	15.6 (4.52)	0.009	.993
BIS	33.4 (4.38)	19.3 (2.94)	10.24	<.001
BAS	26.0 (4.47)	33.8 (2.73)	5.713	<.001
BDI	25.6 (6.05)			
HAMD	19.9 (7.30)			
GAF	49.0 (9.49)			
Mean age of onset	29.5 (10.81)			
Mean illness duration	4.7 (7.12)			

Notes: Standard deviations appear in parentheses. TMT = Trail Making Test, BIS = Behavioral Inhibition Scale, BAS = Behavioral Activation Scale, BDI = Beck Depression Inventory, HAMD = Hamilton Depression Rating Scale.

Bargh, Garcia, & Chaiken, 2002; Neumann & Strack, 2000). On the basis of these findings, implicit tendencies were measured with an MR-compatible joystick with y-axis limitation (Mag Design and Engineering, Sunnyvale, CA, USA). Sixty pictures of facial expressions for each condition (happy, angry, and neutral) displayed by 60 different Caucasian actors (balanced for gender) were presented. These pictures were taken from the same standardized stimulus set that has been frequently applied in behavioral and neuroimaging studies (for development, see Gur et al., 2002). All stimuli depicted evoked facial expressions. The faces were presented twice, once within a blue and once within a yellow frame. Subjects were instructed to pull the lever when the stimulus appeared with a blue frame, yielding approaching behavior, or to push the lever when the stimulus was framed with a yellow line, yielding avoidance behavior, irrespective of facial expression. The randomly presented faces remained on the screen for a maximum of 3 s (or until a response was given) followed by a randomized, variable inter-stimulus interval (ISI), ranging from 1900 to 5100 ms in steps of 400 ms (during which subjects viewed a fixation cross). Six runs, containing 60 pictures each, were separated by a short break where the scan was not interrupted but the participant could take a break for 23 s indicated by the word "break" on the screen, and then for 4 s the word "attention" appeared to remind the subjects to prepare for the task again. To keep data comparable to previous studies (e.g., Marsh, Ambady, & Kleck, 2005; Seidel et al., 2010a, 2010b), we defined reaction time (RT) as the time from stimulus onset to when the lever reached its maximal point.

The difference in RTs for pushing vs. pulling revealed the dominant behavioral tendency (i.e., approach or avoidance). To familiarize participants with the task, a short practice run was conducted inside the scanner, using an asterisk stimulus within blue and yellow frames. Figure 1a illustrates the implicit task.

Explicit rating task

In the explicit rating task, we presented 90 evoked expressions (30 per anger, happiness, and neutral, balanced for gender and only Caucasian actors) of 30 poses randomly taken from the joystick paradigm stimulus set. Here, participants were asked to imagine standing face to face with the person and to indicate whether they would approach, avoid, or show no tendency at all by pressing the corresponding button out of the three possibilities ("+" standing for approach, "−" indicating avoidance, and "0" meaning no tendency). Stimuli were presented maximally for 5 s with a randomized, variable ISI ranging from 1900 to 5100 ms in steps of 400 ms (during which subjects viewed a fixation cross). Manual responses triggered immediate progression to the next ISI.

Participants were told not to refer their rating to the attractiveness or trustworthiness of the person but only to the emotional expression. The aim of this rating scale was to measure the conscious behavioral tendency compared to the automatic motor RT (joystick task). We have already validated this task in a sample of depressed patients (Seidel et al., 2010a) and healthy controls (Seidel et al., 2010b). Figure 1b illustrates the explicit rating task.

(a)

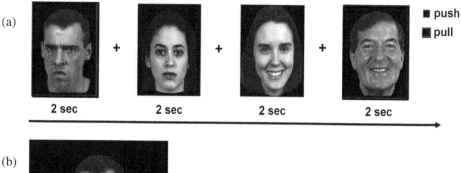

(b)

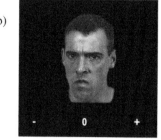

Imagine you are standing face to face to this person.
Would you go towards or away from him/her?

Figure 1. Illustration of the implicit (a) and the explicit task (b). In the implicit task, participants had to push the joystick lever when the face was surrounded by a yellow frame, whereas a blue frame meant pull the lever. In the explicit task, participants were instructed to decide whether they wanted to move toward or away from the person displayed by pressing the corresponding button.

All stimuli were presented with goggles (VisuaStimDigital, Resonance Technology, Inc., Los Angeles, CA, USA). The presentation of images, recording of responses, and acquisition of scanner triggers were achieved with the Presentation software package (Neurobehavioral Systems, Inc., Albany, CA, USA).

Behavioral data analysis

Statistical analyses of behavioral data were performed with SPSS (Statistical Package for the Social Sciences, Version 15.0, SPSS, Inc., Somer, NY, USA). Due to the significant difference in verbal IQ, we included MWT-B scores as a covariate in the behavioral data analysis.

For the analysis of RTs (in milliseconds) of the motor responses in the *implicit joystick task*, we computed a repeated measures ANOVA with expression (anger, happiness, and neutral) and direction (approach and avoid) as within-subject factors, and diagnosis as between-subject factor.

The analysis of group differences in the *explicit rating task* was similarly performed with a 2 (diagnosis) × 3 (expression), repeated-measures ANOVA on the average of participants' ratings. Correlations (partial correlations controlling for verbal IQ [VIQ]) were computed between clinical characteristics (BDI, HAM-D, and GAF), questionnaire data (BIS and BAS), and reactions in the implicit or explicit task.

We also applied a 2 (diagnosis) × 6 (facial expression), repeated-measures ANOVA on the performance in the *emotion recognition task* (percent correct) and added MWT-B scores as a covariate.

fMRI acquisition parameters and data processing

Data acquisition

Functional MR images were acquired on a 3T Siemens MRI whole-body scanner (SIEMENS Trio) at the Department of Psychiatry, Psychotherapy and Psychosomatics, RWTH Aachen University. We used a standard head coil and foam paddings to reduce head motion. Functional imaging was performed with a gradient echo EPI sequence with the following BOLD imaging parameters: TR = 2200 ms, TE = 30 ms, FoV = 200 mm, 36 slices, slice thickness = 3.1 mm, in-plane resolution = 3.1 × 3.1 mm, flip angle = 90°, and distance factor = 15%.

The measurement time of the joystick task was about 30 min, and the rating task took about 10 min. Additionally, a high-resolution structural image (3-D Magnetization Prepared Rapid Gradient Echo: MP-RAGE) was acquired at the end of the measurement with the following parameters: TR = 1900 ms; TE = 2.52 ms; TI = 900 ms; flip angle = 9°; 256 matrix; FoV = 250 mm; 176 slices per slab. The time needed was 4 min.

Data preprocessing

Five dummy scans before the beginning of the experiment were discarded to allow for magnetic saturation. Functional data processing was performed with Statistical Parametric Mapping (SPM5) software (Wellcome Department of Imaging Neuroscience, London, UK) implemented in Matlab (Mathworks, Inc., Sherborn, MA, USA). Functional images were realigned to correct for head movement between scans by an affine registration (Ashburner & Friston, 2003). Each subject's T1 scans were coregistered to the mean image of the realigned functional images. The mean functional image was subsequently normalized to the Montreal Neurological Institute (MNI) single-subject template (Collins, Neelin, Peters, & Evans, 1994; Evans et al., 1992), using linear proportions and a nonlinear sampling as derived from a segmentation algorithm (Ashburner & Friston, 2005). Normalization parameters were then applied to the functional images and coregistered to the T1 image. Images were resampled at a $1.5 \times 1.5 \times 1.5$ mm voxel size and spatially smoothed, using an 8-mm full-width-at-half-maximum Gaussian kernel.

For this event-related design, each of the six experimental conditions in the implicit task (anger approach, anger avoidance, happy approach, happy avoidance, neutral approach, and neutral avoidance) and three conditions in the explicit task (angry, happy, and neutral faces) were modeled with a separate regressor convolved with the canonical hemodynamic response function and its first-order temporal derivative.

Statistical analysis was performed at the individual and group level. Since we were specifically interested in group differences in behavioral tendencies and their neural correlates, we explored neural activation with specific t-contrasts highlighting the significant group differences. Concerning the *implicit task*, we applied several t-contrasts, directly comparing approach and avoidance behavior as well as specific interesting behavioral tendencies that were hypothesis driven, such as approaching happy and avoiding angry faces. For the *explicit task*, we directly compared neural activation during processing of happy and angry faces (vs. neutral faces) of patients and controls by applying independent-sample t-tests.

Region of interest

We performed a ROI analysis for the amygdala region with the aim of maximizing the sensitivity to group as well as hemispheric lateralization differences in the amygdala. Furthermore, we aimed to determine its exact role in approach and avoidance behavior. The amygdala was chosen for several reasons: it plays a major role in emotion processing and several studies have revealed dysfunctional amygdala activation in depressed patients during processing of facial expressions of emotions (e.g., Dannlowski et al., 2007; Suslow et al., 2010). Values for amygdala ROIs were extracted with the probabilistic cytoarchitectonic maps (Amunts et al., 2005), as available in the Anatomy Toolbox in SPM5 (Eickhoff et al., 2005; Eickhoff, Heim, Zilles, & Amunts, 2006). Mean parameter estimates were extracted for left and right amygdala ROI in each condition, and Levene tests for homogeneity of variances indicated homoscedasticity for all parameter estimates of all tasks (*explicit*: happy left: $p = 0.623$, happy right: $p = .779$; anger left: $p = .990$, anger right: $p = .934$; neutral left: $p = .920$, neutral right: $p = .911$—*implicit*: happy approach left: $p = .939$, happy approach right: $p = .1.000$; anger approach left: $p = .702$, anger approach right: $p = .960$; neutral approach left: $p = .912$, neutral approach right: $p = .961$; happy avoid left: $p = .216$, happy avoid right: $p = .955$; anger avoid left: $p = .805$, anger avoid right: $p = .986$; neutral avoid left: $p = .901$, neutral avoid right: $p = .689$).

A three-way ANOVA was applied with diagnosis as between-subject factor as well as condition and laterality as repeated factors. To control for the significant group difference in verbal IQ, we included MWT-B scores as a covariate in the analysis. Greenhouse-Geisser corrected p values are presented.

Corollary analyses

Correlation analyses were performed for each task between performance (RT—implicit task—and scores for approach/avoidance—explicit task) and amygdala activation (mean parameter estimates taken from the ROI analysis) for the whole group. Moreover, amygdala activation was also correlated with performance in the self-report BIS/BAS scores. Additionally, for depressed patients, we also analyzed any association between clinical characteristics (BDI, HAM-D, and GAF) and amygdala parameter estimates.

To account for multiple comparisons, we applied a combined height and extent threshold technique based on Monte-Carlo simulations, using AlphaSim (Cox, 1996). According to 1000 simulations based on a height threshold of $p < .001$ (uncorrected) and the spatial properties of the residual image, an extent threshold of 55 contiguous voxels suffices to comply with a family wise error of $p < .05$. This correction for thresholding will be referred to as "height and extent corrected threshold" (HET), and group results as well as direct comparisons between patients and controls are depicted at this threshold.

RESULTS

Behavioral data

Implicit task

Applying the repeated-measures ANOVA on the RT data revealed no significant emotion effect, $F(2, 56) = 0.603$, $p = .551$; no significant diagnosis effect, $F(1, 28) = 1.390$, $p = .248$; and no significant emotion-by-diagnosis interaction, $F(2, 56) = 0.025$, $p = .976$. Moreover, we observed neither a significant direction effect, $F(1, 28) = 1.001$, $p = .327$, nor a significant effect of VIQ, $F(1, 28) = 0.013$, $p = .909$, and no other significant interaction emerged: emotion-by-VIQ: $F(2, 56) = 0.693$, $p = .505$; direction-by-VIQ: $F(1, 28) = 0.541$, $p = .469$; direction-by-diagnosis: $F(1, 28) = 0.002$, $p = .962$; emotion-by-direction: $F(2, 56) = 0.393$, $p = .677$; emotion-by-direction-by-VIQ: $F(2, 56) = 1.901$, $p = .162$; emotion-by-direction-by-diagnosis: $F(2, 56) = 0.833$, $p = .440$.

Explicit task

Analysis of the rating data, including VIQ as a covariate, showed a significant emotion effect, $F(2, 56) = 11.495$, $p < .001$, $\eta_p^2 = .315$, with highest values and thus strongest approach toward happy faces followed by neutral expressions. Angry faces elicited least approach but strongest avoidance. Furthermore, we observed a significant diagnosis effect, $F(1, 28) = 8.895$, $p = .006$, $\eta_p^2 = .262$, with stronger avoidance in patients, and no significant emotion-by-diagnosis interaction, $F(2, 56) = 1.443$, $p = .246$. VIQ showed no significant impact, $F(1, 28) = 0.081$, $p = .779$, and no significant emotion-by-VIQ interaction emerged, $F(2, 56) = 3.514$, $p = .073$. For illustration of the behavioral performance in the explicit task, see Figure 2.

Corollary analysis

Using partial correlations to control for VIQ differences, we analyzed whether behavioral performance showed any association with BIS/BAS scores and observed a significant positive correlation between BAS scores and happy ratings ($r = .586$, $p = .002$)

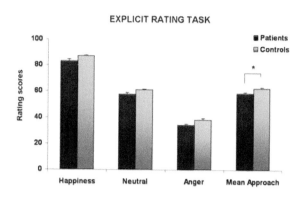

Figure 2. Results of the explicit rating task demonstrating approaching behavior toward happy, neutral, and angry facial expressions and a mean approach score. Data analysis revealed a significant group difference ($p = .005$), indicating stronger withdrawal in the patient group across all emotions.

as well as a significant negative correlation between BIS scores and happy ratings ($r = -.502$, $p = .009$) across the whole group. Correlations with neutral and angry ratings or with the implicit task did not reach significance (all $p > .09$).

Further analysis between symptom severity (BDI and HAMD scores) and RTs in the implicit task revealed no significant association (all $p > .12$). However, correlation analysis for the explicit rating task and psychopathological parameters revealed significant associations between the HAMD scores and the rating results for happy faces ($r = -.602$, $p = .017$) and for angry faces ($r = .725$, $p = .002$). They indicate that patients with more severe symptoms show less approach toward happy faces and more avoidance toward faces expressing anger.

Emotion recognition

To clarify whether patients correctly classified the emotional facial expressions presented in the implicit and explicit task, serving as a necessary prerequisite to adequately fulfill the implicit and explicit task, we conducted a repeated-measures ANOVA, including VIQ as a covariate, which revealed a significant emotion effect, $F(5, 140) = 2.975$, $p = .030$, $\eta_p^2 = .115$, with highest accuracy for happy and lowest for disgusted faces, a significant emotion-by-diagnosis interaction, $F(5, 140) = 2.653$, $p = .042$, $\eta_p^2 = .103$, but no significant effect of diagnosis, $F(1, 28) = 1.764$, $p = .197$, nor VIQ, $F(1, 28) = 2.365$, $p = .138$, or emotion-by-VIQ interaction, $F(5, 140) = 2.245$, $p = .080$. Post-hoc analysis disentangling the significant emotion-by-diagnosis interaction revealed only a significant difference for fear recognition ($p = .001$)

with better accuracy in the patient group. All other comparisons remained not significant (all $p > .298$).

Functional data

Implicit task

The t-contrast comparing patients and controls for *approach* reactions showed more activation in the left calcarine gyrus (BA 17) and the left postcentral gyrus (BA 1). The reverse t-contrast showed that approach in controls (compared to patients) was associated more strongly with neural activation in the cerebellum bilaterally (vermis), the right calcarine gyrus (BA 17), the left middle temporal gyrus, the left superior frontal gyrus, the right temporoparietal junction, the hippocampus bilaterally, and the left precuneus. Directly comparing patients and controls for *avoidance* reactions revealed stronger neural activation in the left calcarine gyrus, left postcentral gyrus, and left paracentral lobule in the patient group. In controls (compared to patients), avoidance reactions yielded stronger neural responses of the cerebellum bilaterally, the right calcarine gyrus, the right middle frontal gyrus, the right fusiform gyrus, the right lingual gyrus, and the left superior frontal gyrus (for a detailed list, see Table 2).

Since one major interest of our study was to analyze group differences in approaching happy and avoiding angry faces (i.e., the behavioral tendencies typically prompted by these emotions), we applied the respective t-contrasts. For *approaching happy faces*, the t-contrast comparing patients and controls demonstrated that patients exhibited stronger activation of the left calcarine gyrus (BA 17), the left postcentral gyrus (BA 1), and the left inferior occipital gyrus. The reverse contrast showed that controls recruited the right cerebellum, the left superior frontal gyrus (BA 6), the right fusiform gyrus, and the right superior parietal lobule more strongly (for detailed information, see Table 3).

For *avoiding angry faces*, patients (compared to controls) showed stronger activation of the left calcarine gyrus (BA 17) and the left postcentral gyrus (BA 1). Controls (compared to patients) showed elevated activation in the cerebellum bilaterally, the right calcarine gyrus (BA 17), the left supplementary motor area, the left dorsal anterior cingulate, the right fusiform gyrus, the right middle and the left superior frontal gyrus, the right superior parietal lobule, and the right insula (for more details, see Table 3).

Directly comparing *approach and avoidance toward happy faces* in patients vs. controls revealed significantly stronger activation of the right orbitofrontal gyrus and the right supramarginal gyrus in patients, while controls demonstrated elevated activation of the posterior cingulate bilaterally only (see Table 3 for detailed information).

Additionally, we analyzed *avoidance vs. approach of angry faces*, again directly comparing patients and controls. While patients recruited the left caudate and the right cerebellum, controls showed stronger activation of the inferior frontal gyrus bilaterally and the right orbitofrontal gyrus (see Table 3 for detailed information; see Figure 3 for illustration of group-specific neural activation during the implicit task).

Explicit task

Analysis of functional data of the explicit task by means of t-contrasts directly comparing patients and controls for each condition revealed significant group differences for angry and happy expressions (vs. neutral, $t = 3.19$, $p < .05$ HET corr.). During perception of *angry faces* and imagination of approaching or avoiding these faces, patients (compared to controls) showed stronger activation in the left precuneus, the right calcarine gyrus (BA 17), and the left posterior cingulate cortex. Controls (compared to patients) demonstrated elevated responses of a whole network of regions including the cerebellum bilaterally, the fusiform gyrus bilaterally, the left superior occipital gyrus, the right middle frontal gyrus, and the right cuneus.

Comparing patients and controls for the happy condition (i.e., processing of *happy faces* and imaging moving toward or away from the face), we observed stronger activation of the right calcarine gyrus and the left inferior occipital gyrus in patients. Controls (compared to patients) showed a much more widespread network including the left lingual gyrus, the cerebellum bilaterally, and the left amygdala. Figure 4 depicts this significant difference in amygdala activation. For a detailed list of activated regions, see Table 4.

ROI analysis

Implicit task

Applying a repeated-measures ANOVA, including VIQ as a covariate, revealed only a trend for a diagnosis effect, $F(1, 28) = 3.735$, $p = .065$, $\eta_p^2 = .130$, indicating stronger activation in the control group, while no other significant main effect emerged– emotion: $F(2, 56) = 1.555$, $p = .221$; laterality: $F(1, 28) = 0.210$, $p = .651$; direction: $F(1, 28) = 0.290$, $p = .595$; VIQ: $F(1, 28) = 1.495$, $p = .233$) or interaction (all $p > .271$).

TABLE 2
Significant group effects of approach and avoidance (threshold: $t > 3.14$, $p < .05$ HET corr.) are given, and regions are listed with MNI coordinates, cluster size (κ), and t values

Contrast	Cluster	MNI			t value	κ
		x	y	z		
Approach						
Patients > controls	L. calcarine gyrus	0	−89	−6	6.01	520
	L. postcentral gyrus	−44	−33	63	4.16	226
Controls > patients	R. cerebellum (vermis)	3	−59	−12	5.31	1264
	R. calcarine gyrus (BA 17)	12	−89	8	4.16	264
	L. middle temporal gyrus	−50	−56	2	3.98	262
	L. superior frontal gyrus	−29	−8	65	5.25	250
	R. temporoparietal junction	30	−39	39	4.71	233
	L. Hippocampus	−24	−27	−2	4.13	143
	R. middle frontal gyrus	36	2	59	4.09	138
	R. fusiform gyrus	24	−42	−20	4.05	119
	L. cerebellum	−23	−77	−26	4.10	106
	R. superior parietal lobe	5	−83	47	4.40	101
	L. supramarginal gyrus	−57	−29	41	4.09	97
	L. Precuneus	−9	−75	56	3.62	68
	R. Hippocampus	30	−8	−38	3.71	65
	R. lingual gyrus	27	−89	−15	4.04	58
	R. precentral gyrus (BA 44)	62	11	21	4.49	57
Avoidance						
Patients > controls	L. calcarine gyrus (BA 17)	0	−87	−5	6.39	587
	L. postcentral gyrus (BA 1)	−44	−32	62	4.68	353
	L. paracentral lobule (BA 4)	−5	−26	54	4.72	167
Controls > patients	R. cerebellum (vermis)	2	−60	−14	4.66	631
	R. calcarine gyrus (BA 17)	12	−89	8	4.68	337
	L. cerebellum	−23	−77	−26	4.02	103
	R. middle frontal gyrus	38	2	59	4.17	97
	R. fusiform gyrus	36	−45	−9	3.87	90
	R. lingual gyrus (BA 18)	27	−89	−15	4.42	87
	L. superior frontal gyrus	−29	−8	65	4.10	66

Explicit task

The repeated-measures ANOVA, including VIQ as a covariate, revealed a significant effect of diagnosis, $F(1, 28) = 4.312$, $p = .043$, $\eta_p^2 = .122$, with stronger amygdala activation in controls, but neither a significant emotion effect, $F(2, 56) = 0.435$, $p = .650$, nor laterality effect, $F(1, 28) = 0.641$, $p = .431$, nor VIQ effect, $F(1, 28) = 0.422$, $p = .522$. Moreover, no interaction reached significance (all $p > .224$).

Corollary analyses

Implicit task

Since we observed no significant laterality effect, we used the mean amygdala parameter estimates to investigate an association between BIS scores and avoidance of angry expressions, and between BAS scores and approach of happy and neutral faces. We observed a significant positive association between BAS scores and amygdala activation during approach of happy faces ($r = .392$, $p = .024$). However, no other correlation reached significance (all $p > .170$). Exploring the association of psychopathology (BDI, HAMD, and GAF) with amygdala activation in the implicit task also revealed no significant correlation (all $p > .268$).

Explicit task

Correlating mean amygdala parameter estimates with BIS/BAS values revealed a significant correlation between amygdala activation during perception of angry faces and BIS scores ($r = −.533$, $p = .050$). However, BAS values did not show any significant association with processing of happy or neutral faces in the amygdala (all $p > .287$).

Analysis of associations between psychopathological rating scales and amygdala activation revealed

TABLE 3
Significant group effects of approaching happy and avoiding angry faces (threshold: $t > 3.14$, $p < .05$ HET corr.) are given, and regions are listed with MNI coordinates, cluster size (κ), and t values

Contrast	Cluster	MNI x	MNI y	MNI z	t value	κ
Approach happy						
Patients > controls	L. calcarine gyrus	0	−89	−6	5.51	437
	L. postcentral gyrus (BA 1)	−45	−32	62	4.55	409
	L. inferior occipital gyrus	−51	−77	−9	4.29	73
Controls > patients	R. cerebellum	3	−59	−14	4.74	326
	L. superior frontal gyrus	−29	−8	65	4.31	111
	R. fusiform gyrus	32	−80	−3	4.02	61
	R. superior parietal lobe	29	−39	38	3.99	56
Avoid anger						
Patients > controls	L. calcarine gyrus	0	−87	−5	5.38	364
	L. postcentral gyrus (BA 1)	−44	−32	60	3.96	163
Controls > patients	R. cerebellum (vermis)	3	−60	−14	5.09	1126
	R. calcarine gyrus (BA 17)	12	−89	8	4.87	436
	L. cerebellum	−24	−77	−26	3.81	178
	L. SMA	0	9	56	3.81	178
	L. dorsal ACC	−8	17	33	3.72	153
	R. fusiform gyrus	36	−45	−8	4.51	153
	L. middle frontal gyrus	35	0	59	4.22	142
	L. superior frontal gyrus	−29	−8	65	4.22	108
	R. superior parietal lobe	5	−84	45	4.27	65
	R. insula	38	0	−5	3.77	61
Approach happy vs. avoid happy						
Patients > controls	R. orbitofrontal gyrus	30	39	3	4.28	66
	R. supramarginal gyrus	50	−48	30	3.77	57
Controls > patients	L. posterior cingulate cortex	−21	−51	18	4.46	238
	R. posterior cingulate cortex	11	−24	38	3.78	72
Avoid anger vs. approach anger						
Patients > controls	L. caudate nucleus	−12	5	17	4.22	64
	R. cerebellum	6	−45	−18	3.67	62
Controls > patients	R. orbitofrontal gyrus	14	26	−17	4.18	67
	L. inferior frontal gyrus	−36	−5	27	4.10	87
	R. inferior frontal gyrus	38	−6	23	4.09	95

Note: SMA = Supplementary Motor Area; ACC = Anterior Cingulate Cortex.

significant negative correlations between BDI values and amygdala activation for happy faces ($r = -.600$, $p = .018$). Stronger depressive symptomatology was associated with less amygdala involvement. Amygdala activation during perception of angry faces showed a significant positive association with global assessment of functioning scores ($r = 0.574$, $p = .025$); hence, a higher level of functioning indicated a stronger amygdala response to angry faces. HAMD scores did not show any significant association with amygdala response (all $p > .396$). Figure 5 illustrates the significant correlations between amygdala activation and clinical parameters.

DISCUSSION

The aim of the present fMRI study was to examine the neural underpinnings of social approach and withdrawal in patients suffering from depression. Therefore, we explored behavioral approach and avoidance tendencies in response to evoked expressions of happy, neutral, and angry facial expressions. Moreover, in contrast to previous studies (e.g., Hasler, Allen, Sbarra, Bootzin, & Bernert, 2010; Kasch et al., 2002; McFarland et al., 2006), we directly compared rather automatic with more conscious behavioral reactions to these salient emotional cues by relying on an implicit as well as an explicit task.

In comparison to age-, gender-, and education-matched controls and despite unimpaired emotion recognition ability, we observed significant avoidance behavior in the patient sample that was accompanied by dysfunctional neural activation including the amygdala region.

Amygdala and social withdrawal

In the present study, we observed significantly less amygdala activation in patients compared to controls

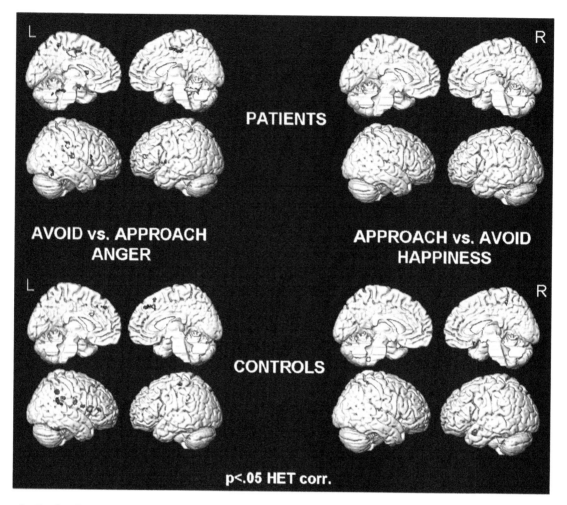

Figure 3.   Results of the implicit task, revealing widespread activation in patients (top) and controls (bottom) for avoid versus approach angry faces (left). On the right side, the direct comparison of approach versus avoid happy faces is depicted with less activation in patients (top) than controls (bottom).

during approach and withdrawal in the explicit task, whereas in the implicit task, only a trend in this direction emerged. Across all participants, correlation analysis revealed significant negative associations of the self-report BIS data with amygdala activation during processing of angry faces, indicating lower amygdala activation with higher BIS scores. In patients, this finding further extended to symptom severity and global functioning. BDI scores correlated negatively with amygdala activation during processing of happy expressions, indicating that the more depressed patients feel, the less amygdala participation they reveal during expected explicit approach toward positive faces. This specific problem of depressed patients is also reflected in significant correlations between amygdala activation and BDI scores, indicating that the more depressed patients feel, the more they tend to avoid, and this will also be reflected in less amygdala activation. Correspondingly, the higher their level

of functioning, the more amygdala activation was demonstrated during explicit avoidance and approach of angry faces. Thus, the more amygdala activation was demonstrated, the better patients coped with their life and everyday needs.

Dysfunctional activation of the amygdala, a central structure in the limbic emotion network, has been observed in several neuroimaging studies addressing emotion processing in patients with major depression. Mostly, previous studies reported a hyperactivation of the amygdala when patients were confronted with mood-congruent facial expressions, such as facial expressions depicting sadness (e.g., Phillips, Drevets, Rauch, & Lane, 2003; Whalen, Shin, Somerville, McLean, & Kim, 2002). Recently, Suslow and colleagues (2010) applied a backward-masking design using happy and sad faces and neutral expressions as masks, and they observed this mood-congruency effect—elevated amygdala

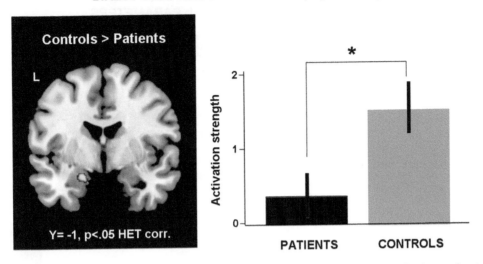

Figure 4. Results from the *t*-contrast directly comparing neural activation during processing of happy faces of patients and controls. Controls showed significantly stronger activation of the left amygdala (x, y, z: -18, -2, -2; $\kappa = 69$, $t = 4.06$, $p < .05$ HET corr.), which is also apparent in the parameter estimates marked with an asterisk.

TABLE 4

Significant group effects for the explicit task are given ($t > 3.19$, $p < .05$ HET corr.), and regions are listed with MNI coordinates, cluster size (κ), and *t* values

Contrast	Cluster	MNI			t *value*	κ
		x	*y*	*z*		
HAPPY						
Patients > controls	R. calcarine gyrus (BA 17)	14	-98	-6	3.98	132
	L. inferior occipital gyrus (BA 19)	-50	-80	-9	3.97	60
Controls > patients	L. lingual gyrus (BA 18)	-3	-65	0	4.30	130
	L. cerebellum	-5	-63	-20	3.80	119
	L. cerebellum	-8	-47	-11	4.14	85
	R. cerebellum	18	-44	-48	4.04	71
	L. amygdala	-18	-2	-20	4.06	69
	R. cerebellum	47	-48	-29	3.96	68
Anger						
Patients > controls	–	–	–	–	–	–
Controls > patients	L. cerebellum	-8	-77	-32	4.92	868
	L. fusiform gyrus	-39	-57	-21	4.53	239
	L. cerebellum	-2	-65	-2	5.56	206
	R. fusiform gyrus	45	-50	-29	4.30	176
	L. superior occipital gyrus	-15	-93	35	4.55	130
	L. cerebellum	-20	-69	-47	3.99	96
	R. inferior frontal gyrus	50	30	14	3.71	95
	R. cerebellum	38	-75	-24	4.21	88
	L. cerebellum	-3	-60	-21	4.03	68
	L. cerebellum	-8	-47	-11	3.97	56

activation during processing of sad faces in patients—but also decreased activation during processing of happy faces. Similarly, for approaching happy faces in the explicit task, we also observed decreased amygdala activation of the patient sample. Suslow et al. (2010) assume that the significantly lower amygdala response to positive emotional faces might

indicate less engagement in the encoding of positively valenced stimuli, eventually leading to disturbed relationships in the sense of less attunement and mutual involvement (Bouhuys, Geerts, Mersch, & Jenner, 1996; Surguladze et al., 2004). Hence, in addition to a stronger avoidance response to angry faces, depressed patients also withdraw from happy

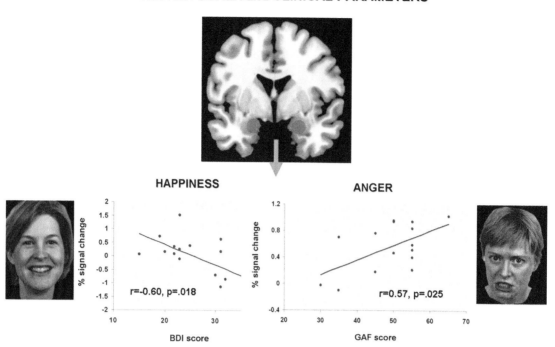

Figure 5. Results from correlation analyses showing a significant negative association between amygdala activation during processing of happy faces and BDI scores (left), revealing that the more severe the symptoms (higher BDI scores), the less amygdala activation was observable. Moreover, a significant positive correlation between amygdala activation during processing of angry faces and global assessment of functioning (GAF) scores (right) emerged, indicating that the higher the GAF scores (the better patients cope with everyday living), the higher the amygdala activation.

faces. Thus, patients also avoid persons with a positive, smiling expression, which in most cultures (or at least the Western culture of all these patients) is considered the nonverbal sign of an invitation to join someone, to take a step closer, and, thus, it is a prompt followed by most people in everyday life. Hence, our results demonstrate that patients show an inadequate and abnormal behavioral tendency in a positive socio-emotional context.

Neural correlates of social withdrawal

Patients and controls performed equally well in the implicit task; however, group differences were detected for the neural correlates of approach and avoidance behavior.

Direct comparison of implicit avoidance and approach of angry faces in patients vs. controls revealed stronger activation of the left caudate and the right cerebellum in patients.

The left caudate nucleus is known to be involved in reward processing (Balleine, Delgado, & Hikosako, 2007) as well as feedback processing (Graybiel, 2005;

Packard & Knowlton, 2002) and emotion regulation (e.g., Beer & Lombardo, 2007). Concerning emotion processing and depression, our results support findings from Scheuerecker et al. (2010), who reported stronger response of the left caudate nucleus in depressed patients than controls during an emotion-matching task. Interestingly, increased activation of the caudate region has also been shown during emotional stress situations in cocaine-dependent patients (Li & Sinha, 2008). Probably, our task instructing patients to avoid and approach other people, or at least simulate it with the joystick, put them in a strong emotional stress situation reflected in the increased caudate activation. However, as mentioned by Beer and Lombardo (2007), the putative role of the caudate in regulating emotions and stress responses is still speculative and needs further examination.

Explicit evaluation of angry faces elicited stronger activation of the left precuneus, the right calcarine gyrus, and the posterior cingulate. In their meta-analysis, Cavanna and Trimble (2006) demonstrated that activation of the anterior precuneus has repeatedly been observed during mental and motor imagery, but evidence has accumulated that this region also plays a particular role in social cognition, self-agency, and self-processing (e.g., Kjaer, Nowak, & Lou, 2002; Koenigsberg et al., 2009; Lou et al., 2004; Vogeley & Fink, 2003). Moreover, den Ouden, Frith, and Blakemore (2005) speculate that the precuneus together with the posterior cingulate is specifically involved in processing of intentions related to the self. Thus, patients seem to be specifically aware of the intention to withdraw from angry faces, which was the major tendency in patients (and most controls).

Regarding social withdrawal in controls (compared to patients), we observed significantly stronger activation of the right orbitofrontal gyrus and the inferior frontal gyrus bilaterally. Activity in the inferior frontal gyrus has been repeatedly observed during various emotional processes, such as passive viewing of faces (Dapretto et al., 2006), and emotion recognition and evaluation (Carr et al., 2003; Seitz et al., 2008); only recently has it been reported to play a major role in emotional perspective taking (Schulte-Rüther, Markowitsch, Fink, & Piefke, 2007; Schulte-Rüther et al., 2008; Derntl et al., 2010). Schulte-Rüther and colleagues (2007) assumed that this activation might mirror the degree of interpersonal emotional involvement. This suggests that controls can strongly participate and respond emotionally when confronted with a salient stimulus communicating the request to go away (cf. Horstmann, 2003), thereby facilitating avoidance behavior (cf. Marsh, et al., 2005).

Neural correlates of social approach

Analyzing approach versus avoidance of happy faces revealed that patients (compared to controls) relied on activation of the right orbitofrontal gyrus and the right supramarginal gyrus. Only recently, Hsu, Langenecker, Kennedy, Zubieta, and Heitzeg (2010) observed significant positive correlations between self-report scores on recent negative life stressors and activation of the orbitofrontal gyrus bilaterally during processing of negative words in depressed patients. Furthermore, Surguladze and colleagues (2010) observed stronger activation of this region in depressed patients during processing of facial expressions of strong disgust. According to Kringelbach and Rolls (2004), the orbitofrontal gyri are specifically engaged in representing the emotional impact of anticipated negative outcomes. Moreover, this region has also been found to be ineffective in downregulating amygdala activation during effortful reappraisal of negative stimuli (Johnstone, van Reekum, Urry, Kalin, & Davidson, 2007). Hence, approaching someone, even when the person is smiling, seems to be associated with anticipated negative outcome and inefficient emotion regulation in patients suffering from depression.

Moreover, patients (compared to controls) recruited the right supramarginal gyrus more strongly during approach of happy faces. Activation of this region has consistently been observed during memory retrieval (e.g., Naghavi & Nyberg, 2005; Wagner, Shannon, Kahn, & Buckner, 2005), and has been shown to mediate attention toward stimuli that are potentially important for the individual (e.g., Downar, Crawley, Mikulis, & Davis, 2002). Consequently, Ciarimelli, Grady, and Moscovitch (2008) demonstrated elevated activation of the inferior parietal lobe, including the supramarginal gyrus, when individuals subjectively felt as if they were reliving their memories and were confident about their memories, and when these memories were strong. Depressed patients have to endure many negative experiences in social interaction, particularly regarding approach behavior; thus, they may anticipate exclusion and memorize prior experiences of social rejection, which might be reflected in the neural activation.

In controls (compared to patients), the posterior cingulate bilaterally was associated with approach toward happy faces. Notably, patients recruited this region more strongly when they processed angry faces in the explicit task, mostly during imagination of avoiding these faces. The posterior cingulate cortex is an important node in the processing of social-affective stimuli (e.g., Amodio & Frith, 2006; Kross, Egner, Ochsner, Hirsch, & Downey, 2007; Northoff & Bermpohl, 2004), it is involved in emotional experience (Britton et al., 2006; Koenigsberg et al., 2009), emotion regulation, particularly distancing oneself from aversive images (Koenigsberg et al., 2010), and it has recently been shown to be essential when forming social preferences (Chen et al., 2010).

Regions which show activation irrespective of behavioral tendency and emotional expression

Irrespective of behavioral tendency and emotional expression, patients showed stronger reactivity of the primary visual cortex (calcarine gyrus, BA 17) than controls. Interestingly, greater activation of this region to negative facial expressions has been reported to be associated with a good clinical outcome in depression (Keedwell et al., 2010). Additionally, patients exhibited stronger activation of the postcentral gyrus (BA 4), which has been repeatedly observed in studies investigating neural dysfunctions during processing of emotional faces in depression (e.g., Beevers, Clasen, Stice, & Schnyer, 2010; Fu et al., 2004). Concerning emotion processing, Adolphs (2002) indicated that the somatosensory cortex is essential for emotional contagion. Hence, this elevated activation of primary visual and primary somatosensory cortices might reflect intensified neural processing at a very basic level. Probably when confronted with emotional expressions and instructed to approach or avoid these faces, depressed patients not only show perceptual biases yielding stronger visual responses but are also affected more strongly emotionally.

Besides its association with various motor functions and even speech perception and production (for review, see Ackermann, Mathiak, & Riecker, 2007), the cerebellum is also known to be involved in emotion processing (Schmahmann, 2000), emotional modulation of cognitive processing (Simpson et al., 2000), and emotional experience (Derntl et al., 2010; Hofer et al., 2006, 2007). We observed cerebellar activation during approach and avoidance in the implicit task in both groups; however, controls showed significantly stronger recruitment of this region irrespective of behavioral tendency. Together with further functional abnormalities (Liu et al., 2010; Naismith et al., 2010), one can speculate that this region plays a neglected role not only in emotion but also in the pathophysiology of depression.

Behavioral performance and self-report data

The behavioral performance partly corroborates previous results from our lab (Seidel et al., 2010a), where we also observed that patients showed stronger avoidance during the explicit task (i.e., conscious focusing on approach or avoidance of the presented face). However, during the implicit task, which rather prompts automatic behavioral tendencies, no group difference was apparent. As hypothesized, we observed significantly higher BIS and lower BAS scores in patients, thereby supporting previous findings (e.g., Kasch et al., 2002; McFarland et al., 2006; Seidel et al., 2010a).

Hence, we assume that the self-reported behavioral tendencies and the performance in the explicit task are prone to the perceptual and interpretative biases of depressed patients, who tend to consciously draw back from positive social context and hence putatively positive social experiences (Gable & Shean, 2000). Beevers (2005) postulated a dual process model of information processing in depression, proposing an associative mode, which acts automatically, and a reflective mode, which is effortful and consumes cognitive resources. This model suggests that depressive cognition is characterized by a negatively biased automatic processing not corrected by the reflective mode. Our data indicate that with respect to the aberrant approach and avoidance tendencies in depression, and thus social withdrawal, the reflective mode (explicit task) is disturbed. These disturbances in consciously controlled aspects of social interaction might therefore be accessible to cognitive behavioral therapy (CBT), and the dysfunctional behavioral tendencies to behavioral activation therapy (BAT). Hence, our data have a clinical implication which should be addressed in greater detail in future studies.

Regarding emotion recognition, we observed no significant group difference but a significant group-by-emotion interaction, indicating that patients showed higher accuracy for fearful faces. We did not expect this finding; however, if we look at the published studies reporting emotion-recognition deficits in depression, there are several differences in task design and sample characteristics that might explain the diversity in findings (for overview, see Bourke, Douglas & Porter, 2010). While we used an explicit emotion-recognition task showing 36 stimuli (6 per emotion and 6 neutral expression) with a forced-choice answering format, several other studies only showed a limited range of emotions including happiness, sadness, and neutral expressions (e.g., Gur et al., 1992; Mikhailova, Vladimirova, Iznak, Tsusulkovskaya, & Sushko, 1996), and some relied on the face in the crowd task (e.g., Suslow et al., 2001, 2004), which does not directly assess recognition accuracy, or presented schematic faces in an emotion-discrimination task (e.g., Bouhuys et al., 1996). Moreover, we investigated depressed patients without any comorbidity, in contrast to most other studies, which included depressed patients with a comorbid anxiety disorder (e.g., Bouhuys et al., 1997; Gilboa-Schechtman, Erhard-Weiss, & Jeczemien, 2002), or even merged depressed and bipolar patients (e.g., Gur et al., 1992; Rubinow & Post, 1992). In the latter, we reported emotion-recognition deficits (Derntl et al., 2009d). Moreover, recent results from Anderson and colleagues (2011) indicate no significant difference in recognition accuracy between controls and currently depressed patients but showed that remitted patients performed significantly better than the two other groups. Thus, these results support our data showing no significant difference in emotion-recognition performance between currently depressed patients and matched controls.

Limitations

Due to the small sample size, analysis of gender differences was not possible. However, depression in men and women differs in prevalence (e.g., Kessler et al., 2005), etiology (e.g., Piccinelli & Wilkinson, 2000), and severity and symptom presentation (e.g., Smith et al., 2008). Moreover, in a previous study from our lab applying the implicit and explicit task to measure approach and avoidance behavior in depression (Seidel et al., 2010a), we observed significantly stronger social withdrawal in female patients. Future studies should explore whether these behavioral differences are accompanied by distinct neural responses further characterizing female and male depression. Moreover, due to the small sample size and the exploratory nature of most of the corollary analyses, we did not apply an alpha correction for multiple correlations.

Considering the impact of medication on neural processing, it has recently been proposed that antidepressants modulate affective processing rather than directly affecting mood (Harmer, 2010; Harmer, Goodwin, & Cowen, 2009). Consequently, previous studies have repeatedly shown that antidepressant medication affects amygdala responses to emotional stimuli: while some observed reduced amygdala activation to negative stimuli (e.g., Fu et al., 2004; Harmer, Mackay, Reid, Cowen, & Goodwin, 2006; Norbury et al., 2007; Sheline et al., 2001), others observed enhanced amygdala activation to positive faces (e.g., Fu et al., 2007; Norbury, Mackay, Cowen, Goodwin, & Harmer, 2009; Schaefer, Putnam, Benca, & Davidson, 2006). We observed significantly less amygdala activation to happy and angry faces in the explicit task and a trend toward stronger amygdala activation irrespective of emotion in controls in the implicit joystick task. Thus, our data from medicated patients only partly support pervious findings. In light of the small number of fMRI studies addressing antidepressant treatment effects on the neural substrates of emotion processing in depressed patients, and considering our small sample size with mixed medication, it is hard to infer how medication influenced the current results. Therefore, future neuroimaging studies should highlight how and where antidepressants influence the neural correlates of emotional competencies, such as social approach and withdrawal.

The color of the frames (blue = pull, yellow = push) was not counterbalanced across subjects. Thus, we cannot rule out that color may represent a potential confound. Future studies should use counterbalanced designs to control for this issue.

Conclusion

This study investigated the behavioral and neural correlates of implicit and explicit social approach and withdrawal in patients suffering from major depression. We found stronger social withdrawal in depressed patients, and this was also reflected in stronger neural activation of a widespread network during avoidance of angry faces. Moreover, during approach, patients showed stronger activation of orbitofrontal and supramarginal gyri, regions that have been associated with anticipated negative outcome and memory retrieval. We also observed a significant decrease in amygdala activation, particularly during processing of happy faces in patients. Additionally, the significant correlations between psychopathology and amygdala activation, as well as behavioral data, support the notion that more pronounced depressive symptoms are accompanied by stronger neural dysfunctions and inadequate behavioral tendencies. This in turn may aggravate the disorder by negative social interactions contributing to isolation and reinforcing cognitive biases.

REFERENCES

Ackermann, H., Mathiak, K., & Riecker, A. (2007). The contribution of the cerebellum to speech production and speech perception: Clinical and functional imaging data. *Cerebellum, 6*(3), 202–213.

Adolphs, R. (2002). Recognizing emotion from facial expressions: Psychological and neurological mechanisms. *Behavioral and Cognitive Neuroscience Reviews, 1*(1), 21–62.

Amodio, D. M., & Frith, C. D. (2006). Meeting of minds: The medial frontal cortex and social cognition. *Nature Reviews Neuroscience, 7*(4), 268–277.

Amunts, K., Kedo, O., Kindler, M., Pieperhoff, P., Mohlberg, H., Shah, N. J., et al. (2005). Cytoarchitectonic mapping of the human amygdala, hippocampal region and entorhinal cortex: Intersubject variability and probability maps. *Anatomy and Embryology, 210*(5–6), 343–352.

Anderson, I. M., Shippen, C., Juhasz, G., Chase, D., Thomas, E., Downey, D., et al. (2011). State-dependent alteration in face emotion recognition in depression. *British Journal of Psychiatry, 198*, 302–308. doi:10.1192/bjp.bp.110.078139

APA.(1994). Diagnostic and statistical manual of mental disorders. Washington, DC: APA.

Ashburner, J., & Friston, K. J. (2003). Rigid body registration. In R. S. Frackowiak, K. J. Friston, C. Frith, R. Dolan, C. J. Price, S. Zeki, et al. (Eds.), *Human brain function* (pp. 635–655). London, UK: Academic Press.

Ashburner, J., & Friston, K. J. (2005). Unified segmentation. *NeuroImage, 26*(3), 839–851.

Balleine, B. W., Delgado, M. R., & Hikosaka, O. (2007). The role of the dorsal striatum in reward and decision-making. *Journal of Neuroscience, 27*(31), 8161–8165.

Beck, A. T. (2008). The evolution of the cognitive model of depression and its neurobiological correlates. *American Journal of Psychiatry, 165*, 969–977.

Beck, A. T., Erbaugh, J., Ward, C. H., Mock, J., & Mendelsohn, M. (1961). An inventory for measuring depression. *Archives of General Psychiatry, 4*, 561–571.

Beer, J. S., & Lombardo, M. V. (2007). Insights into emotion regulation from neuropsychology. In J. J. Gross (Ed.), *Handbook of emotion regulation* (pp. 69–86). New York, NY: Guilford.

Beevers, C. G. (2005). Cognitive vulnerability to depression: A dual process model. *Clinical Psychology Review, 25*(7), 975–1002.

Beevers, C. G., Clasen, P., Stice, E., & Schnyer, D. (2010). Depression symptoms and cognitive control of emotion cues: A functional magnetic resonance imaging study. *Neuroscience, 167*(1), 97–103.

Bouhuys, A. L., Geerts, E., & Mersch, P. P. (1997). Relationship between perception of facial emotions and anxiety in clinical depression: Does anxiety-related perception predict persistence of depression? *Journal of Affective Disorders, 43*(3), 213–223.

Bouhuys, A. L., Geerts, E., Mersch, P. P., & Jenner, J. A. (1996). Nonverbal interpersonal sensitivity and persistence of depression: Perception of emotions in schematic faces. *Psychiatry Research, 64*(3), 193–203.

Bourke, C., Douglas, K., & Porter, R. (2010). Processing of facial emotion expression in major depression: A review. *Australian and New Zealand Journal of Psychiatry, 44*(8), 681–696.

Britton, J. C., Phan, K. L., Taylor, S. F., Welsh, R. C., Berridge, K. C., & Liberzon, I. (2006). Neural correlates of social and nonsocial emotions: An fMRI study. *NeuroImage, 31*(1), 397–409.

Cacioppo, J. T., Priester, J. R., & Berntson, G. G. (1993). Rudimentary determinants of attitudes. II. Arm flexion and extension have differential effects on attitudes. *Journal of Personality and Social Psychology, 65*(1), 5–17.

Campbell-Sills, L., Liverant, G. I., & Brown, T. A. (2004). Psychometric evaluation of the behavioral inhibition/behavioral activation scales in a large sample of outpatients with anxiety and mood disorders. *Psychological Assessment, 16*(3), 244–254.

Carr, L., Iacoboni, M., Dubeau, M. C., Mazziotta, J. C., & Lenzi, G. L. (2003). Neural mechanisms of empathy in humans: A relay from neural systems for imitation to limbic areas. *Proceedings of the National Academy of Sciences of the United States of America, 100*(9), 5497–5502.

Carver, C. S., & White, T. I. (1994). Behavioral inhibition, behavioral activation, and affective responses to impending reward and punishment: The BIS/BAS scales. *Journal of Personality and Social Psychology, 67*, 319–333.

Cavanna, A. E., & Trimble, M. R. (2006). The precuneus: A review of its functional anatomy and behavioural correlates. *Brain: A Journal of Neurology, 129*(Pt 3), 564–583.

Ciaramelli, E., Grady, C. L., & Moscovitch, M. (2008). Top-down and bottom-up attention to memory: A hypothesis (AtoM) on the role of the posterior parietal cortex in memory retrieval. *Neuropsychologia, 46*(7), 1828–1851.

Chen, M., & Bargh, J. A. (1999). Consequences of automatic evaluation: Immediate behavioral predispositions to approach or avoid the stimulus. *Personality and Social Psychology Bulletin, 25*(2), 215–224.

Chen, A. C., Welsh, R. C., Liberzon, I., & Taylor, S. F. (2010). 'Do I like this person?' A network analysis of midline cortex during a social preference task. *NeuroImage, 51*(2), 930–939.

Coan, J. A., & Allen, J. J. B. (2003). Frontal EEG asymmetry and the behavioral activation and inhibition systems. *Psychophysiology, 40*(1), 106–114.

Collins, D. L., Neelin, P., Peters, T. M., & Evans, A. C. (1994). Automatic 3d intersubject registration of MR volumetric data in standardized Talairach space. *Journal of Computer-Assisted Tomography, 18*(2), 192–205.

Corr, P. J. (2008). *The reinforcement sensitivity theory of personality*. Cambridge University Press.

Cox, R. W. (1996). AFNI: Software for analysis and visualization of functional magnetic resonance neuroimages. *Computational and Biomedical Research, 29*, 162–173.

Dannlowski, U., Ohrmann, P., Bauer, J., Kugel, H., Arolt, V., et al. (2007). Amygdala reactivity to masked negative faces is associated with automatic judgmental bias in major depression: A 3 T fMRI study. *Journal of Psychiatry & Neuroscience: JPN, 32*(6), 423–429.

Dapretto, M., Davies, M. S., Pfeifer, J. H., Scott, A. A., Sigman, M., Bookheimer, S. Y., et al. (2006). Understanding emotions in others: Mirror neuron dysfunction in children with autism spectrum disorders. *Nature Neuroscience, 9*(1), 28–30.

Davidson, R. J. (1992). Anterior cerebral asymmetry and the nature of emotion. *Brain and Cognition, 20*(1), 125–151.

Davidson, R.J. (1998). Affective style and affective disorders: Perspectives from affective neuroscience. *Cognition and Emotion, 12*, 307-330.

den Ouden, H. E. M., Frith, U., Frith, C., & Blakemore, S. (2005). Thinking about intentions. *NeuroImage, 28*(4), 787–796.

Derntl, B., Finkelmeyer, A., Eickhoff, S., Kellermann, T., Falkenberg, D. I., Schneider, F., et al. (2010). Multidimensional assessment of empathic abilities: Neural correlates and gender differences. *Psychoneuroendocrinology, 35*(1), 67–82.

Derntl, B., Habel, U., Robinson, S., Windischberger, C., Kryspin-Exner, I., Gur, R. C., et al. (2009b). Amygdala activation during recognition of emotions in a foreign ethnic group is associated with duration of stay. *Social Neuroscience, 4*(4), 294–307.

Derntl, B., Habel, U., Windischberger, C., Robinson, S., Kryspin-Exner, I., Gur, R. C., et al. (2009a). General and specific responsiveness of the amygdala during explicit emotion recognition in females and males. *BMC Neuroscience, 10*, 91.

Derntl, B., Kryspin-Exner, I., Fernbach, E., Moser, E., & Habel, U. (2008). Emotion recognition accuracy in healthy young females is associated with cycle phase. *Hormones and Behavior, 53*(1), 90–95.

Derntl, B., Seidel, E., Kryspin-Exner, I., Hasmann, A., & Dobmeier, M. (2009d). Facial emotion recognition in patients with bipolar I and bipolar II disorder. *British Journal of Clinical Psychology, 48*(Pt 4), 363–375.

Derntl, B., Windischberger, C., Robinson, S., Kryspin-Exner, I., Gur, R. C., Moser, E., et al. (2009c). Amygdala activity to fear and anger in healthy young males is associated with testosterone. *Psychoneuroendocrinology*, *34*(5), 687–693.

Douglas, K. M., & Porter, R. J. (2010). Recognition of disgusted facial expressions in severe depression. *British Journal of Psychiatry*, *197*, 156–157.

Downar, J., Crawley, A. P., Mikulis, D. J., & Davis, K. D. (2002). A cortical network sensitive to stimulus salience in a neutral behavioral context across multiple sensory modalities. *Journal of Neurophysiology*, *87*(1), 615–620.

Duckworth, K. L., Bargh, J. A., Garcia, M., & Chaiken, S. (2002). The automatic evaluation of novel stimuli. *Psychological Science*, *13*(6), 513–519.

Eickhoff, S. B., Heim, S., Zilles, K., & Amunts, K. (2006). Testing anatomically specified hypotheses in functional imaging using cytoarchitectonic maps. *NeuroImage*, *32*(2), 570–582.

Eickhoff, S. B., Stephan, K. E., Mohlberg, H., Grefkes, C., Fink, G. R., Amunts, K., et al. (2005). A new SPM toolbox for combining probabilistic cytoarchitectonic maps and functional imaging data. *NeuroImage*, *25*(4), 1325–1335.

Elliot, A.J., & Covington, M.V. (2001). Approach and avoidance motivation. *Educational Psychology Review*, *13*(2), 73–92.

Evans, A. C., Marrett, S., Neelin, P., Collins, L., Worsley, K., Dai, W., et al. (1992). Anatomical mapping of functional activation in stereotactic coordinate space. *Neuroimage*, *1*(1), 43–53.

Fitzgerald, D. A., Angstadt, M., Jelsone, L. M., Nathan, P. J., & Phan, K. L. (2006). Beyond threat: Amygdala reactivity across multiple expressions of facial affect. *NeuroImage*, *30*(4), 1441–1448.

Fu, C. H. Y., Williams, S. C. R., Brammer, M. J., Suckling, J., Kim, J., Cleare, A. J., et al. (2007). Neural responses to happy facial expressions in major depression following antidepressant treatment. *American Journal of Psychiatry*, *164*(4), 599–607.

Fu, C. H. Y., Williams, S. C. R., Cleare, A. J., Brammer, M. J., Walsh, N. D., Kim, J., et al. (2004). Attenuation of the neural response to sad faces in major depression by antidepressant treatment: A prospective, event-related functional magnetic resonance imaging study. *Archives of General Psychiatry*, *61*(9), 877–889.

Fusar-Poli, P., Placentino, A., Carletti, F., Landi, P., Allen, P., Surguladze, S., et al. (2009). Functional atlas of emotional faces processing: A voxel-based meta-analysis of 105 functional magnetic resonance imaging studies. *Journal of Psychiatry & Neuroscience*, *34*(6), 418–432.

Gable, S. L., & Shean, G. D. (2000). Perceived social competence and depression. *Journal of Social and Personal Relationships*, *17*, 139–150.

Gilboa-Schechtman, E., Erhard-Weiss, D., & Jeczemien, P. (2002). Interpersonal deficits meet cognitive biases: Memory for facial expressions in depressed and anxious men and women. *Psychiatry Research*, *113*(3), 279–293.

Gray, J. A. (1972). The psychophysiological basis of introversion-extraversion: A modification of Eysenck's theory. In V. D. Nebylitsyn & J. A. Gray (Eds.), *Biological bases of individual behavior* (pp. 182–205). New York, NY: Academic Press.

Gray, J. A. (1982). *The neuropsychology of anxiety: An enquiry into the functions of the septo-hippocampal system*. Oxford, UK: Oxford University Press.

Gray, J. A. (1990). Brain systems that mediate both emotion and cognition. *Cognition & Emotion*, *4*(3), 269–288.

Gray, J. A. (1994). Framework for a taxonomy of psychiatric disorder. In S. H. M. van Goozen & N. E. Van de Poll (Eds.), *Emotions: Essays on emotion theory* (pp. 29–59). Mahwah, NJ: Lawrence Erlbaum Associates, Inc.

Gray, J. A., & McNaughton, N. (2000). *The neuropsychology of anxiety: An enquiry into the functions of the septo-hippocampal system* (2nd ed.). Oxford, UK: Oxford University Press.

Graybiel, A. M. (2005). The basal ganglia: Learning new tricks and loving it. *Current Opinion in Neurobiology*, *15*(6), 638–644.

Gur, R. C., Erwin, R. J., & Gur, R. E. (1992). Neurobehavioral probes for physiologic neuroimaging studies. *Archives of General Psychiatry*, *49*(5), 409–414.

Gur, R. C., Sara, R., Hagendoorn, M., Marom, O., Hughett, P., Macy, L., et al. (2002). A method for obtaining 3-dimensional facial expressions and its standardization for use in neurocognitive studies. *Journal of Neuroscience Methods*, *115*(2), 137–143.

Habel, U., Windischberger, C., Derntl, B., Robinson, S., Kryspin-Exner, I., Gur, R. C., et al. (2007). Amygdala activation and facial expressions: Explicit emotion discrimination versus implicit emotion processing. *Neuropsychologia*, *45*(10), 2369–2377.

Hamilton, M. (1960). A rating scale for depression. *Journal of Neurology, Neurosurgery, and Psychiatry*, *23*, 56–62.

Harmer, C. J. (2010). Antidepressant drug action: A neuropsychological perspective. *Depression and Anxiety*, *27*, 231–233.

Harmer, C. J., Goodwin, G. M., & Cowen, P. J. (2009). Why do antidepressants take so long to work? A cognitive neuropsychological model of antidepressant drug action. *British Journal of Psychiatry*, *195*, 102–108.

Harmer, C. J., Mackay, C. E., Reid, C. B., Cowen, P. J., & Goodwin, G. M. (2006). Antidepressant drug treatment modifies the neural processing of nonconscious threat cues. *Biological Psychiatry*, *59*(9), 816–820.

Harmon-Jones, E., & Allen, J. J. (1997). Behavioral activation sensitivity and resting frontal EEG asymmetry: Covariation of putative indicators related to risk for mood disorders. *Journal of Abnormal Psychology*, *106*(1), 159–163.

Hartig, J., & Moosbrugger, H. (2003). Die "ARES-Skalen" zur Erfassung der individuellen BIS- und BAS-Sensitivität: Entwicklung einer Lang- und einer Kurzfassung. *Zeitschrift für Differentielle und Diagnostische Psychologie*, *24*, 291–308.

Hasler, B. P., Allen, J. J. B., Sbarra, D. A., Bootzin, R. R., & Bernert, R. A. (2010). Morningness-eveningness and depression: Preliminary evidence for the role of the behavioral activation system and positive affect. *Psychiatry Research*, *176*(2–3), 166–173.

Hofer, A., Siedentopf, C. M., Ischebeck, A., Rettenbacher, M. A., Verius, M., Felber, S., et al. (2006). Gender differences in regional cerebral activity during the perception of emotion: A functional MRI study. *NeuroImage*, *32*(2), 854–862.

Hofer, A., Siedentopf, C. M., Ischebeck, A., Rettenbacher, M. A., Verius, M., Felber, S., et al. (2007). Sex differences in brain activation patterns during processing of positively and negatively valenced emotional words. *Psychological Medicine*, *37*(1), 109–119.

Horstmann, G. (2003). What do facial expressions convey: Feeling states, behavioral intentions, or action requests? *Emotion, 3*(2), 150–166.

Hsu, D. T., Langenecker, S. A., Kennedy, S. E., Zubieta, J., & Heitzeg, M. M. (2010). fMRI BOLD responses to negative stimuli in the prefrontal cortex are dependent on levels of recent negative life stress in major depressive disorder. *Psychiatry Research, 183*(3), 202–208.

Johnstone, T., van Reekum, C. M., Urry, H. L., Kalin, N. H., & Davidson, R. J. (2007). Failure to regulate: Counterproductive recruitment of top-down prefrontal-subcortical circuitry in major depression. *Journal of Neuroscience, 27*(33), 8877–8884.

Kasch, K. L., Rottenberg, J., Arnow, B. A., & Gotlib, I. H. (2002). Behavioral activation and inhibition systems and the severity and course of depression. *Journal of Abnormal Psychology, 111*(4), 589–597.

Keedwell, P. A., Drapier, D., Surguladze, S., Giampietro, V., Brammer, M., & Phillips, M. (2010). Subgenual cingulate and visual cortex responses to sad faces predict clinical outcome during antidepressant treatment for depression. *Journal of Affective Disorders, 120*(1–3), 120–125.

Kessler, R. C., Demler, O., Frank, R. G., Olfson, M., Pincus, H. A., Walters, E. E., et al. (2005). Prevalence and treatment of mental disorders, 1990 to 2003. *New England Journal of Medicine, 352*, 2515–2523.

Kjaer, T. W., Nowak, M., & Lou, H. C. (2002). Reflective self-awareness and conscious states: PET evidence for a common midline parietofrontal core. *NeuroImage, 17*(2), 1080–1086.

Koenigsberg, H. W., Fan, J., Ochsner, K. N., Liu, X., Guise, K. G., Pizzarello, S., et al. (2009). Neural correlates of the use of psychological distancing to regulate responses to negative social cues: A study of patients with borderline personality disorder. *Biological Psychiatry, 66*(9), 854–863.

Koenigsberg, H. W., Fan, J., Ochsner, K. N., Liu, X., Guise, K., Pizzarello, S., et al. (2010). Neural correlates of using distancing to regulate emotional responses to social situations. *Neuropsychologia, 48*(6), 1813–1822.

Kringelbach, M. L., & Rolls, E. T. (2004). The functional neuroanatomy of the human orbitofrontal cortex: Evidence from neuroimaging and neuropsychology. *Progress in Neurobiology, 72*(5), 341–372.

Kross, E., Egner, T., Ochsner, K., Hirsch, J., & Downey, G. (2007). Neural dynamics of rejection sensitivity. *Journal of Cognitive Neuroscience, 19*(6), 945–956.

Lamm, C., Batson, C. D., & Decety, J. (2007). The neural substrate of human empathy: Effects of perspective-taking and cognitive appraisal. *Journal of Cognitive Neuroscience, 19*(1), 42–58.

Lamm, C., & Singer, T. (2009). The social neuroscience of empathy. *Annals of the New York Academy of Sciences, 1156*, 81–96.

Lehrl, S. (1996). Der MWT– ein Intelligenztest für die ärztliche Praxis. *Praxis für Neurologie und Psychiatrie, 7*, 488–491.

Leppänen, J. M. (2006). Emotional information processing in mood disorders: A review of behavioral and neuroimaging findings. *Current Opinion in Psychiatry, 19*(1), 34–39.

Li, C. R., & Sinha, R. (2008). Inhibitory control and emotional stress regulation: Neuroimaging evidence for frontal-limbic dysfunction in psycho-stimulant addiction. *Neuroscience and Biobehavioral Reviews, 32*(3), 581–597.

Liu, Z., Xu, C., Xu, Y., Wang, Y., Zhao, B., Lv, Y., et al. (2010). Decreased regional homogeneity in insula and cerebellum: A resting-state fMRI study in patients with major depression and subjects at high risk for major depression. *Psychiatry Research, 182*(3), 211–215.

Lou, H. C., Luber, B., Crupain, M., Keenan, J. P., Nowak, M., Kjaer, T. W., et al. (2004). Parietal cortex and representation of the mental self. *Proceedings of the National Academy of Sciences of the United States of America, 101*(17), 6827–6832.

Marsh, A. A., Ambady, N., & Kleck, R. E. (2005). The effects of fear and anger facial expressions on approach- and avoidance-related behaviors. *Emotion, 5*(1), 119–124.

McFarland, B. R., Shankman, S. A., Tenke, C. E., Bruder, G. E., & Klein, D. N. (2006). Behavioral activation system deficits predict the six-month course of depression. *Journal of Affective Disorders, 91*(2–3), 229–234.

Mikhailova, E. S., Vladimirova, T. V., Iznak, A. F., Tsusulkovskaya, E. J., & Sushko, N. V. (1996). Abnormal recognition of facial expression of emotions in depressed patients with major depression disorder and schizotypal personality disorder. *Biological Psychiatry, 40*(8), 697–705.

Moser, E., Derntl, B., Robinson, S., Fink, B., Gur, R. C., & Grammer, K. (2007). Amygdala activation at 3T in response to human and avatar facial expressions of emotions. *Journal of Neuroscience Methods, 161*(1), 126–133.

Naghavi, H. R., & Nyberg, L. (2005). Common fronto-parietal activity in attention, memory, and consciousness: Shared demands on integration? *Consciousness and Cognition, 14*(2), 390–425.

Naismith, S. L., Lagopoulos, J., Ward, P. B., Davey, C. G., Little, C., & Hickie, I. B. (2010). Fronto-striatal correlates of impaired implicit sequence learning in major depression: An fMRI study. *Journal of Affective Disorders, 125*(1–3), 256–261.

Neumann, R., & Strack, F. (2000). Approach and avoidance: The influence of proprioceptive and exteroceptive cues on encoding of affective information. *Journal of Personality and Social Psychology, 79*(1), 39–48.

Norbury, R., Mackay, C. E., Cowen, P. J., Goodwin, G. M., & Harmer, C. J. (2007). Short-term antidepressant treatment and facial processing. Functional magnetic resonance imaging study. *British Journal of Psychiatry: The Journal of Mental Science, 190*, 531–532.

Norbury, R., Taylor, M. J., Selvaraj, S., Murphy, S. E., Harmer, C. J., & Cowen, P. J. (2009). Short-term antidepressant treatment modulates amygdala response to happy faces. *Psychopharmacology, 206*(2), 197–204.

Northoff, G., & Bermpohl, F. (2004). Cortical midline structures and the self. *Trends in Cognitive Sciences, 8*(3), 102–107.

Oldfield, R.C. (1971). The assessment and analysis of handedness: The Edinburgh inventory. *Neuropsychologia, 9*(1), 97–113.

Packard, M.G., & Knowlton, B.J.(2002). Learning and memory functions of the Based Ganglia. *Annual Review of Neuroscience, 25*, 563–593.

Phillips, M. L., Drevets, W. C., Rauch, S. L., & Lane, R. (2003). Neurobiology of emotion perception. II. Implications for major psychiatric disorders. *Biological Psychiatry, 54*(5), 515–528.

Piccinelli, M., & Wilkinson, G. (2000). Gender differences in depression. Critical review. *British Journal of*

Psychiatry: The Journal of Mental Science, 177, 486–492.

Pickering, A. D., & Gray, J. A. (1999). The neuroscience of personality. In L. Pervin & O. John (Eds.), *Handbook of Personality* (2nd ed., pp. 277–299). New York, NY: Guilford Press.

Puca, R. M., Rinkenauer, G., & Breidenstein, C. (2006). Individual differences in approach and avoidance movements: How the avoidance motive influences response force. *Journal of Personality, 74*(4), 979–1014.

Rauch, A. V., Reker, M., Ohrmann, P., Pedersen, A., Bauer, J., Dannlowski, U., et al. (2010). Increased amygdala activation during automatic processing of facial emotion in schizophrenia. *Psychiatry Research, 182*(3), 200–206.

Reitan, M. G. (1958). Validity of the Trial Making Test as an indicator of organic brain damage. *Perceptual and Motor Skills, 8*, 271–276.

Reuter, M., Stark, R., Hennig, J., Walter, B., Kirsch, P., Schienle, A., et al. (2004). Personality and emotion: Test of Gray's personality theory by means of an fMRI study. *Behavioral Neuroscience, 118*(3), 462–469.

Rubinow, D. R., & Post, R. M. (1992). Impaired recognition of affect in facial expression in depressed patients. *Biological Psychiatry, 31*(9), 947–953.

Schaefer, H. S., Putnam, K. M., Benca, R. M., & Davidson, R. J. (2006). Event-related functional magnetic resonance imaging measures of neural activity to positive social stimuli in pre- and post-treatment depression. *Biological Psychiatry, 60*(9), 974–986.

Scheuerecker, J., Meisenzahl, E.M., Koutsoulers, N., Roesner, M., Schöpf, V., Linn, J., et al.(2010). Orbitofrontal volume reductions during emotion recognition in patients with major depression. *Journal of Psychiatry and Neuroscience, 35*(5), 311–320.

Schmahmann, J. D. (2004). Disorders of the cerebellum: Ataxia, dysmetria of thought, and the cerebellar cognitive affective syndrome. *Journal of Neuropsychiatry and Clinical Neuroscience, 16*, 367–378.

Schulte-Rüther, M., Markowitsch, H. J., Fink, G. R., & Piefke, M. (2007). Mirror neuron and theory of mind mechanisms involved in face-to-face interactions: A functional magnetic resonance imaging approach to empathy. *Journal of Cognitive Neuroscience, 19*(8), 1354–1372.

Schulte-Rüther, M., Markowitsch, H. J., Shah, N. J., Fink, G. R., & Piefke, M. (2008). Gender differences in brain networks supporting empathy. *NeuroImage, 42*(1), 393–403.

Segrin, C. (2000). Social skills deficits associated with depression. *Clinical Psychology Review, 20*, 379–403.

Seidel, E., Habel, U., Finkelmeyer, A., Schneider, F., Gur, R. C., & Derntl, B. (2010a). Implicit and explicit behavioral tendencies in male and female depression. *Psychiatry Research, 177*(1–2), 124–130.

Seidel, E., Habel, U., Kirschner, M., Gur, R. C., & Derntl, B. (2010b). The impact of facial emotional expressions on behavioral tendencies in women and men. *Journal of Experimental Psychology. Human Perception and Performance, 36*(2), 500–507.

Seitz, R. J., Schäger, R., Scherfeld, D., Friederichs, S., Popp, K., Wittsack, H. J., et al. (2008). Valuating other people's emotional face expression: A combined functional magnetic resonance imaging and electroencephalography study. *Neuroscience, 152*(3), 713–722.

Sheline, Y. I., Barch, D. M., Donnelly, J. M., Ollinger, J. M., Snyder, A. Z., & Mintun, M. A. (2001). Increased amygdala response to masked emotional faces in depressed subjects resolves with antidepressant treatment: An fMRI study. *Biological Psychiatry, 50*(9), 651–658.

Simpson, J. R., Ongür, D., Akbudak, E., Conturo, T. E., Ollinger, J. M., Snyder, A. Z., et al. (2000). The emotionalmodulation of cognitive processing: An fMRI study. *Journal of Cognitive Neuroscience, 12*(Suppl 2), 157–170.

Singer, T., & Lamm, C. (2009). The social neuroscience of empathy. *Annals of the New York Academy of Sciences, 1156*, 81–96.

Smith, D. J., Kyle, S., Forty, L., Cooper, c., Walters, J., Russell, E., et al.(2008). Differences in depressive symptom profile between males and females. *Journal of Affective Disorders, 108*(3), 279–284.

Smith, K. L. W., Matheson, F. I., Moineddin, R., & Glazier, R. H. (2007). Gender, income and immigration differences in depression in Canadian urban centres. *Canadian Journal of Public Health. Revue Canadienne De Santé Publique, 98*(2), 149–153.

Surguladze, S. A., El-Hage, W., Dalgleish, T., Radua, J., Gohier, B., & Phillips, M. L. (2010). Depression is associated with increased sensitivity to signals of disgust: A functional magnetic resonance imaging study. *Journal of Psychiatric Research, 44*, 894–902. doi:10.1016/j.jpsychires.2010.02.010

Surguladze, S. A., Young, A. W., Senior, C., Brébion, G., Travis, M. J., & Phillips, M. L. (2004). Recognition accuracy and response bias to happy and sad facial expressions in patients with major depression. *Neuropsychology, 18*(2), 212–218.

Suslow, T., Dannlowski, U., Lalee-Mentzel, J., Donges, U., Arolt, V., & Kersting, A. (2004). Spatial processing of facial emotion in patients with unipolar depression: A longitudinal study. *Journal of Affective Disorders, 83*(1), 59–63.

Suslow, T., Junghanns, K., & Arolt, V. (2001). Detection of facial expressions of emotions in depression. *Perceptual and Motor Skills, 92*(3 Pt 1), 857–868.

Suslow, T., Konrad, C., Kugel, H., Rumstadt, D., Zwitserlood, P., Schöning, S., et al. (2010). Automatic mood-congruent amygdala responses to masked facial expressions in major depression. *Biological Psychiatry, 67*(2), 155–160.

Tse, W. S., & Bond, A. J. (2004). The impact of depression on social skills. *Journal of Nervous and Mental Disease, 192*(4), 260–268.

Vogeley, K., & Fink, G. R. (2003). Neural correlates of the first-person-perspective. *Trends in Cognitive Sciences, 7*(1), 38–42.

Von Aster, M., Neubauer, A., & Horn, R. (2006). *Hamburg-Wechsler-Intelligenz-Test für Erwachsene III*. Frankfurt, Germany: Harcourt.

Wagner, A. D., Shannon, B. J., Kahn, I., & Buckner, R. L. (2005). Parietal lobe contributions to episodic memory retrieval. *Trends in Cognitive Sciences, 9*(9), 445–453.

Whalen, P. J., Shin, L. M., Somerville, L. H., McLean, A. A., & Kim, H. (2002). Functional neuroimaging studies of the amygdala in depression. *Seminars in Clinical Neuropsychiatry, 7*(4), 234–242.

Wittchen, H.-U., Zaudig, M., & Fydrich, T. (1997). *Strukturiertes Klinisches Interview für DSM-IV*. Göttingen, Germany: Hogrefe.

Are you really angry? The effect of intensity on facial emotion recognition in frontotemporal dementia

Fiona Kumfor[1,2], Laurie Miller[3,4], Suncica Lah[4], Sharpley Hsieh[1,2], Sharon Savage[1], John R. Hodges[1,2], and Olivier Piguet[1,2]

[1]Neuroscience Research Australia, Sydney, Australia
[2]School of Medical Sciences, University of New South Wales, Sydney, Australia
[3]Neuropsychology Unit, Royal Prince Alfred Hospital, Sydney, Australia
[4]School of Psychology, University of Sydney, Sydney, Australia

Frontotemporal dementia (FTD) is a neurodegenerative brain disorder that affects the frontal and temporal lobes predominantly. Impaired emotion recognition has been reported in two FTD subtypes: behavioral-variant FTD (bvFTD) and semantic dementia (SD), but has not been investigated in the third subtype: progressive nonfluent aphasia (PNFA).
Methods: Recognition of six basic facial emotions (anger, disgust, fear, sadness, surprise, and happiness) was investigated in 41 FTD patients (bvFTD = 16; SD = 12; PNFA = 13) and 37 age- and education-matched controls, using two tests. In one task, intensity of emotional expression was increased to identify cognitive components contributing to emotion recognition performance.
Results: All patient groups demonstrated impaired overall facial emotion recognition compared to controls. Performance, however, improved with increased emotion intensity in bvFTD and PNFA groups, the effect of intensity on emotion recognition being particularly pronounced for negative emotions. In contrast, increased intensity of facial emotion did not change performance in SD.
Conclusions: Patients with SD demonstrate a primary emotion processing impairment, whereas PNFA and bvFTD patients' emotional disturbance is in part mediated by attentional deficits. These findings indicate that a subset of FTD patients may benefit from enhanced emotional intensity that will facilitate facial emotion recognition.

Keywords: Caricatures; Behavioral-variant FTD; Semantic dementia; Progressive nonfluent aphasia.

Recognition of emotion is a critical component of human social interaction. This aspect of social cognition, underlies the ability to perceive the intentions and dispositions of others (Brothers, 1990), based on their facial expression, prosody, body language, or a combination of these cues. Emotion recognition is impaired in frontotemporal dementia (FTD) (Lavenu, Pasquier, Lebert, Petit, & Van der Linden, 1999; Neary, et al., 1998), a progressive neurodegenerative brain disorder that affects the frontal and temporal lobes predominantly. Disturbances of social cognition, especially emotion recognition, may contribute to the changes in personality and behavior observed in FTD. Understanding the disturbance in emotion recognition in FTD subtypes will provide insight into the deficits experienced in these clinical syndromes and enhance clinical diagnosis and management.

Correspondence should be addressed to: Olivier Piguet, Neuroscience Research Australia, Barker Street, Randwick NSW 2031, Australia. E-mail: o.piguet@neura.edu.au

This project was supported by a National Health and Medical Research Council (NHMRC) of Australia Project Grant (no. 510106). F.K. is supported by an Australian Postgraduate Award (APA). J.R.H. is supported by an Australian Research Council Federation Fellowship (no. FF0776229). O.P. is supported by an NHMRC Clinical Career Development Award fellowship (no. 510184).

http://dx.doi.org/10.1080/17470919.2011.620779

CLINICAL PRESENTATION OF FTD

Frontotemporal dementia is the second most common neurodegenerative disorder affecting individuals under the age of 65 (Ratnavalli, Brayne, Dawson, & Hodges, 2002). Three clinical subtypes are generally described: behavioral-variant FTD (bvFTD), semantic dementia (SD), and progressive nonfluent aphasia (PNFA) (Hodges & Patterson, 2007; Neary et al., 1998; Piguet, Hornberger, Mioshi, & Hodges, 2011; Rascovsky et al., 2007). Diagnosis is based on the predominant clinical features at initial presentation, although the subtypes tend to merge with disease progression (Gorno-Tempini et al., 2011; Kertesz, McMonagle, Blair, Davidson, & Munoz, 2005). Early presentation in bvFTD is characterized by changes in personality and behavior, and executive dysfunction on cognitive testing. In contrast, the primary presenting features in SD and PNFA are language disturbances: a fluent speech output characterized by a lack of content words secondary to loss of semantic knowledge in SD (Hodges & Patterson, 2007), and nonfluent, effortful, and labored speech with relatively well-preserved comprehension in PNFA (Neary et al., 1998).

Cognitive and behavioral changes tend to reflect the pattern of underlying brain pathology. In bvFTD, greatest atrophy is observed in the frontal lobes bilaterally, most particularly affecting the orbitofrontal and medial frontal cortices (Seeley et al., 2008). In SD, marked atrophy is present in the anterior temporal lobe. This atrophy is typically asymmetrical, more commonly greater on the left than on the right (Hodges, Patterson, Oxbury, & Funnell, 1992). In PNFA, atrophy is less widespread and typically constrained to the inferior posterior frontal region (i.e., Broca's area) and the insula (Nestor et al., 2003).

A range of deficits in social cognition, including emotion recognition is present in FTD (Kipps & Hodges, 2006; Lavenu et al., 1999; Lough et al., 2006). Patients with FTD tend to show greater difficulty in recognizing facial emotions than both healthy controls and patients with Alzheimer's disease. Findings are, however, somewhat inconsistent: Some studies report a recognition deficit that is specific to negative emotions (e.g., anger, disgust, fear, sadness) (Kessels et al., 2007), although others have found impaired recognition of both positive (e.g., happiness, surprise) and negative facial expressions (Bediou et al., 2009; Snowden et al., 2008). In these studies, however, FTD was investigated as a single disease entity, and FTD subtypes were not analyzed separately, and this may account for the disparity in results reported across studies.

NEURAL SUBSTRATES OF EMOTION RECOGNITION IN FTD

Ekman and colleagues proposed a small set of "basic" emotions (anger, disgust, fear, sadness, surprise, and happiness), each characterized by a distinct facial expression, physiology, and evolutionary history (Ekman, 1992; Ekman, Friesen, & Ellsworth, 1972). Subsequent lesion and functional neuroimaging studies have demonstrated specific neural substrates that underlie the recognition of these basic emotions (e.g., Calder, Keane, Lawrence, & Manes, 2004; Calder, Keane, Manes, Antoun, & Young, 2000a; Calder et al., 1996; Ekman, 1992; Kesler West et al., 2001; LaBar, Gatenby, Gore, LeDoux, & Phelps, 1998). These results led to the hypothesis that specific emotions are processed by discrete neural substrates, referred to as the multimodal system model of emotion recognition. To date, specific neural substrates have been identified for a number of basic emotions, with the amygdala associated with fear recognition (Adolphs, Tranel, Damasio, & Damasio, 1995; Allman & Brothers, 1994), the insula associated with disgust recognition (Calder et al., 2000a; Phillips et al., 1997, 1998; Sprengelmeyer et al., 1996), and the ventral striatum associated with anger recognition (Blair, Morris, Frith, Perrett, & Dolan, 1999; Calder et al., 2004).

The pattern of emotion recognition deficits in subtypes of FTD, however, matches only loosely what would be expected from predictions of the multimodal system model of emotion recognition. For example, the consistent impairment of anger recognition in bvFTD (Keane, Calder, Hodges, & Young, 2002; Kipps, Nestor, Acosta-Cabronero, Arnold, & Hodges, 2009b; Lough et al., 2006; Rosen et al., 2004) fits the predominant atrophy observed in the orbitofrontal and anterior cingulate cortices, which play an important role in processing this type of emotional stimulus. Nevertheless, these patients' impairment is not emotion specific, as deficits in recognizing other emotions are also reported (Keane et al., 2002; Rosen et al., 2004). Similarly, despite the significant amygdala atrophy observed in SD, patients exhibit a broad deficit in recognizing negative emotions, rather than a recognition deficit limited to fear alone (Rosen et al., 2004; Rosen et al., 2002b). Emotion recognition in PNFA has been investigated in one study only (Rankin et al., 2009), and a specific or disproportionate impairment in recognition of disgust has not been reported.

The discrepancy between the multimodal system model predictions and current findings in FTD may be due to factors other than deficits in emotion processing. Patients with FTD may perform poorly because of perceptual or attentional difficulties that interfere with emotion processing, resulting in pervasive rather than specific deficits. Despite an increased interest in this domain, understanding of the mechanisms underlying facial emotion recognition remains relatively limited.

The goal of this investigation was to identify the mechanisms contributing to emotion processing disturbance in FTD. In this study, we manipulated the perceptual and attentional demands by exaggerating the emotional expression of the stimuli in one of the facial emotion-recognition tasks. This process is achieved by digitally manipulating the position of numerous geographic points of difference, between an emotional and a neutral facial expression (e.g., corner of the mouth, eyebrow frown, eye opening) (Calder et al., 2000b). In this way, the intensity or salience of the emotional expression can be either enhanced (known as caricatures) or degraded (known as anti-caricatures). Previous studies have found that caricatured emotional expressions are more quickly identified (Calder, Young, Rowland, & Perrett, 1997), are rated as more intensely expressing the emotion (Benson, Campbell, Harris, Frank, & Tovée, 1999; Calder et al., 2000b), and are identified as well as, or better than, non-exaggerated "real-life" expressions. Increased intensity of facial expressions has also been associated with activation of the same specific neural regions associated with recognition of real-life emotions, and this activation amplifies with increasing intensity (Morris et al., 1998; Phillips et al., 1997). Thus, the main hypothesis of this study was that, in some instances, the use of caricatures would improve emotion recognition by reducing perceptual and attentional task demands.

In addition, how emotion recognition deficits vary across FTD subtypes is not well specified. The second aim of this study was to investigate facial emotion recognition in PNFA, which has not been extensively examined to date, and compare performance with bvFTD and SD profiles. It was hypothesized that the deficits observed in these groups would be consistent with their predominant regions of atrophy. Hence anger recognition was expected to be impaired in bvFTD, fear recognition impaired in SD, and disgust recognition impaired in PNFA.

METHODS

Participants

Forty-one patients meeting clinical diagnostic criteria for FTD (bvFTD = 16; SD = 12; PNFA = 13) were recruited from the Frontotemporal Dementia Research Group in Sydney. All patients were seen by the same experienced behavioral neurologist (J.R.H), and diagnosis was based on clinical assessment, comprehensive neuropsychological assessment, and presence of brain atrophy on structural magnetic resonance imaging (MRI). Patients presenting with behavioral features of bvFTD in the absence of brain atrophy or progression, also known as phenocopy bvFTD, were excluded (Davies et al., 2006). Three SD patients with predominant right-sided temporal lobe atrophy were also excluded, as this small number of cases prevented separate analyses for this group. Within the PNFA group, patients exhibiting impaired repetition and comprehension for sentences and reduced word span were excluded. This presentation, known as logopenic progressive aphasia, is strongly associated with Alzheimer's disease (Gorno-Tempini et al., 2008). Other exclusion criteria included presence of other types of dementia or other neurological disease; diagnosis of major depression, schizophrenia, obsessive compulsive disorder or substance abuse; and Mini-Mental State Examination (MMSE) score below 19/30.

Thirty-seven healthy, age- and education-matched controls were recruited from the patients' families and friends, or from the local area. Participants were excluded if any of the following criteria were present: significant history of psychiatric or neurological condition, history of substance abuse or medications with CNS side effects, and MMSE score below 27/30.

The Southeastern Sydney and Illawarra Area Health Service and the University of New South Wales ethics committees approved the study. Participants, or their person responsible, provided informed consent. All participants volunteered their time but were reimbursed for travel costs.

Materials and procedure

Ekman 60 Task

This task assesses recognition of 60 facial expressions across six basic emotions (anger, disgust, fear, happiness, sadness, and surprise), using stimuli from the Pictures of Facial Affect series (Ekman & Friesen, 1976). These images are not manipulated, and emotion is expressed at a normal or unchanged level of intensity, referred to here as 100% intensity. Images are pseudorandomly presented on a computer screen, and participants view each stimulus for 5 s and select the label that best describes the emotional expression.

The labels are present throughout testing, and selection is untimed. Participants respond either by using the mouse to click the appropriate label, pointing to the label, or saying their response (depending on their preference) for the researcher to record. No feedback on response accuracy is provided.

Ekman Caricatures

This task is similar to the Ekman 60 with the same procedure, but uses only two models (one male, one female). In this task, faces have been digitally manipulated to increase the intensity of emotional expression by altering critical facial features (see Appendix 1 for example stimuli) (Young, Perrett, Calder, Sprengelmeyer, & Ekman, 2002). In total, 48 images are presented across the same six basic emotions. These emotions vary across four levels of intensity, with emotional expression increased from 100% emotional expression by +15%, +30%, +50%, or +75%.

Hence, comparison of overall performance between the Ekman 60 and the Ekman Caricatures will identify the contribution of attentional and perceptual processes to emotion recognition. Investigations of each emotion in the Ekman Caricatures separately will identify the effect of manipulation of intensity on recognition of specific emotions.

Neuropsychological and behavioral measures

In addition to the MMSE (Folstein, Folstein, & McHugh, 1975) and the Addenbrooke's Cognitive Examination-Revised (ACE-R) (Mioshi, Dawson, Mitchell, Arnold, & Hodges, 2006), all participants were given tests of attention and working memory (Digit Span subtest of the Wechsler Adult Intelligence Scale) (Wechsler, 1997), semantic knowledge (Boston Naming Test, 15-item version) (Mack, Freed, White Williams, & Henderson, 1992), processing speed (Trail Making Test) (Tombaugh, 2004), verbal fluency (Controlled Oral Word Association Test (COWAT) (Spreen & Strauss, 1998), and inhibition of prepotent response (Hayling Test) (Burgess & Shallice, 1997) (Table 2). These tests were administered as part of a larger evaluation of cognitive functions.

Statistical analyses

Data were analyzed by PASWS 19.0 (IBM, Inc., Chicago, IL, USA). First, variables were plotted and checked for normal distribution by Kolmogorov-Smirnov tests. Given the non-normal distribution of the individual emotion scores on the Ekman 60 and the Ekman Caricatures, these variables were analyzed by nonparametric tests.

Between-group comparisons for sociodemographic and neuropsychological tests were performed by univariate analysis of variance (ANOVA) followed by post-hoc tests where appropriate. The effect of intensity on overall emotion recognition was investigated with a 2×4 repeated-measures ANOVA with Intensity (Ekman 60, Ekman Caricatures) as the within-subjects variable and Diagnosis (control, bvFTD, SD, PNFA) as the between-subjects variable. Follow up t-tests comparing each group's performance on the Ekman Caricatures with the Ekman 60 performance were then conducted. Between-group comparisons for the six basic emotions on the Ekman 60, and the Ekman Caricatures were investigated by Kruskal–Wallis analyses of variance, followed by post-hoc Mann–Whitney tests comparing each patient group with controls. Finally, to investigate the effect of emotion intensity, an index of change was calculated as based on the difference in performance between +15% and +75% for each basic emotion. Again, between-group differences were investigated by Kruskal–Wallis analyses of variance for each emotion, followed by post-hoc Mann–Whitney tests comparing each patient group with controls. Statistical significance was set at .05, and all analyses were corrected for multiple comparisons where appropriate.

RESULTS

Demographic data

Groups were well matched for age, $F(3, 72) = 2.26$, $p = .09$, and level of education, $F(3, 72) = .75, p = .53$. Within the FTD subgroups, disease duration differed significantly, $F(2, 38) = 4.74$, $p = .02$, with longer duration observed in the SD than PNFA group ($p = .01$). This difference probably reflects the reported longer time to diagnosis from disease onset in SD patients (Table 1).

Neuropsychological performance

All patient groups were significantly impaired on both cognitive screening measures (MMSE and ACE-R), compared to controls (Table 2). Across patient groups, PNFA performed significantly worse than bvFTD on the MMSE ($p = .046$), and SD performed significantly worse than bvFTD ($p < .001$) and PNFA ($p = .01$)

TABLE 1
Demographic characteristics of the study sample

	Controls	bvFTD	SD	PNFA	F	Post-hoc Patient vs. controls	Post-hoc Between subtypes
n	37	16	12	13	–		
Male/female	24/13	12/4	9/3	9/4	–		
Age	64.6 (4.5)	61.5 (9.7)	62.4 (8.8)	65.5 (11.4)	0.9	N/A	N/A
Education (years)	13.8 (2.5)	12.1 (3.3)	12.5 (3.4)	13.0 (3.3)	1.4	N/A	N/A
Disease duration (months)	–	42 (24.2)	58 (25.5)	30 (16.1)	4.7*	N/A	SD > PNFA*

Notes: Scores are means (SD); $*p < .05$. N/A = not applicable.

TABLE 2
Neuropsychological variables across healthy controls and three FTD subgroups

	Controls	bvFTD	SD	PNFA	F	Impaired compared to controls	Post-hoc Between subtypes
MMSE	29.4 (0.9)	26.8 (2.6)	24.0 (3.1)	23.5 (5.1)	22.4	bvFTD* SD** PNFA**	bvFTD > PNFA*
ACE-R	95.6 (3.0)	82.8 (6.8)	59.7 (11.7)	72.1 (18.3)	52.6	bvFTD** SD** PNFA**	bvFTD > SD** bvFTD > PNFA* PNFA > SD*
Digit Span Forwards	7.0 (1.3)	5.6 (0.7)	6.3 (1.3)	4.4 (1.3)	14.3	bvFTD* PNFA**	SD > PNFA**
Digit Span Backwards	5.5 (1.3)	4.1 (0.8)	4.5 (1.1)	3.2 (0.9)	15.6	bvFTD** PNFA**	SD > PNFA*
Boston – 15	14.7 (0.6)	13.6 (1.4)	2.4 (1.7)	11.4 (3.3)	181.4	SD** PNFA**	bvFTD > SD** bvFTD > PNFA* PNFA > SD**
Trails A	32.4 (12.7)	59.6 (38.9)	38.7 (18.0)	63.5 (45.6)	6.2	bvFTD** PNFA**	ns
Trails B	81.9 (31.9)	145.1 (88.2)	107.3 (47.0)	142.7 (66.8)	6.3	bvFTD** PNFA*	ns
COWAT	44.2 (12.1)	24.5 (11.9)	25.2 (10.9)	16.8 (12.5)	21.7	bvFTD** SD** PNFA**	ns
Hayling[#]	6.4 (.7)	3.0 (2.3)	1.0 (.0)	3.0 (2.5)	28.9	bvFTD** SD** PNFA**	ns

Notes: Scores are mean (SD); $*p < .05$, $**p < .01$; *ns*: not significant. MMSE (Mini-Mental State Examination): total score /30; ACE-R (Addenbrooke's Cognitive Examination-Revised): total score /100; Digit Span Forwards and Digit Span Backwards: maximum span raw score; Boston – 15 total: total score /15; Trails A and Trails B: time to complete (seconds); COWAT (Controlled Oral Word Association Test): total number of words produced (F, A, S); Hayling: overall scaled score /10 (average score = 6). [#] Only two SD and six PNFA patients were assessed on this task.

on the ACE-R, reflecting the relatively high language demands of this test. Significant differences between patient groups and controls were also observed on most cognitive tasks. Post-hoc tests showed that the cognitive profiles and deficits were consistent with the clinical presentation generally associated with each of the subtypes (Table 2). The bvFTD group was significantly impaired on tasks of attention (Digit Span Forwards), working memory (Digit Span Backwards), and executive functioning (COWAT and Hayling), whereas the SD group was significantly impaired on the semantic task (Boston Naming Test – 15) and two language-based executive functioning tasks (COWAT and Hayling) compared to controls, bvFTD and PNFA groups (all $ps < .001$). The PNFA group demonstrated impaired performance on all neuropsychological tests compared to controls and showed variable performance compared to the other FTD subgroups, performing poorer than bvFTD on the MMSE, ACE-R, and Boston Naming Test – 15, and worse than the SD group for Digit Span Forwards and Backwards. These deficits likely reflect the overall reduced verbal output in these patients compared to the other patient groups (Table 2).

Overall emotion recognition

A significant interaction between Diagnosis and Intensity was present, $F(3, 73) = 3.52$, $p = .019$, indicating that the effect of intensity differed across the various diagnostic groups. A main effect of Diagnosis $F(3, 73) = 19.211$, $p < .001$, was significant, with all patient groups performing worse than controls (all $ps < .001$) and a trend for the SD group to perform worse than the bvFTD group ($p = .058$). A significant main effect of Intensity was observed, with performance better on the Caricatures than on the Ekman 60 task $F(1, 73) = 50.80$, $p < .001$. Post-hoc t-tests indicated that increased intensity on the Caricatures task significantly improved emotion recognition for control ($p < .001$), bvFTD ($p < .001$), and PNFA ($p = .006$), but not SD ($p > .05$) groups (Figure 1).

Recognition of basic emotions: Ekman 60

An overall effect of group was present for each of the four negative emotions on the Ekman 60 task (anger:

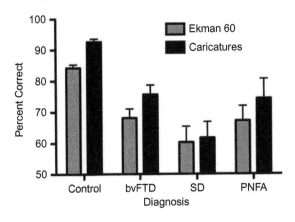

Figure 1. Percentage of correctly recognized emotions on the Ekman 60 and Ekman Caricatures tasks in healthy controls and FTD subtypes. Error bars represent *SEM*.

$H(3) = 22.556$, $p < .001$; disgust: $H(3) = 24.497$, $p < .001$; fear: $H(3) = 14.937$, $p = .002$; and sadness: $H(3) = 14.713$, $p = .002$), but not for the positive emotions (Figure 2). Post-hoc tests indicated that bvFTD performed below controls for all negative emotions (anger: $U = 123.0$, $z = -3.172$, $p = .002$; disgust: $U = 125.5$, $z = -3.152$, $p = .002$; fear: $U = 122.0$, $z = -3.161$, $p = .002$; and sadness: $U = 158.0$, $z = -2.466$, $p = .014$), as did the SD group (anger: $U = 76.0$, $z = -3.448$, $p = .001$; disgust: $U = 29.0$, $z = -4.580$, $p < .001$; fear: $U = 113.0$, $z = -2.552$, $p = .011$; and sadness: $U = 108.0$, $z = -2.720$, $p = .007$). The PNFA group performed below controls for anger: $U = 86.5$, $z = -3.458$, $p = .001$; fear: $U = 129.5$, $z = -2.473$, $p = .013$; and sadness: $U = 106.5$, $z = -3.043$, $p = .002$, but not for disgust.

Recognition of basic emotions: Ekman Caricatures

On the Caricatures task, a similar profile of performance emerged with an overall effect of group observed for negative emotions only (anger: $H(3) = 22.738$, $p < .001$; disgust: $H(3) = 31.242$, $p < .001$; fear: $H(3) = 23.582$, $p < .001$; and sadness: $H(3) = 23.595$, $p < .001$) (Figure 2). Post-hoc tests indicated that the bvFTD group performed worse than controls for recognition on all negative emotions, collapsed across all intensity levels (anger: $U = 123.0$, $z = -3.172$, $p = .002$; disgust: $U = 183.0$, $z = -2.491$, $p = .013$; fear: $U = 134.5$, $z = -4.140$, $p < .001$; and sadness: $U = 137.5$, $z = -3.428$, $p = .001$). SD patients showed a similar pattern of performance (anger: $U = 44.0$, $z = -4.553$, $p < .001$; disgust: $U = 33.0$, $z = -5.178$, $p < .001$; fear: $U = 65.0$, $z = -3.767$, $p < .001$; and sadness: $U = 51.5$, $z = -4.375$, $p < .001$). The PNFA group's performance was worse than controls for anger: $U = 141.5$, $z = -2.517$,

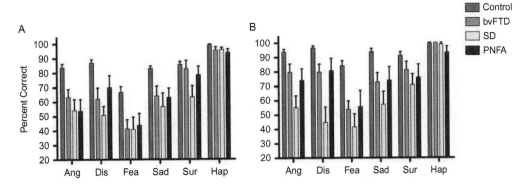

Figure 2. Percentage of correctly recognized emotions on the Ekman 60 (A) and Ekman Caricatures (B) for each of the six basic emotions (Ang: anger, Dis: disgust, Fea: fear, Sad: sadness, Sur: surprise, and Hap: happiness) in healthy controls and FTD subtypes. Errors bars represent *SEM*.

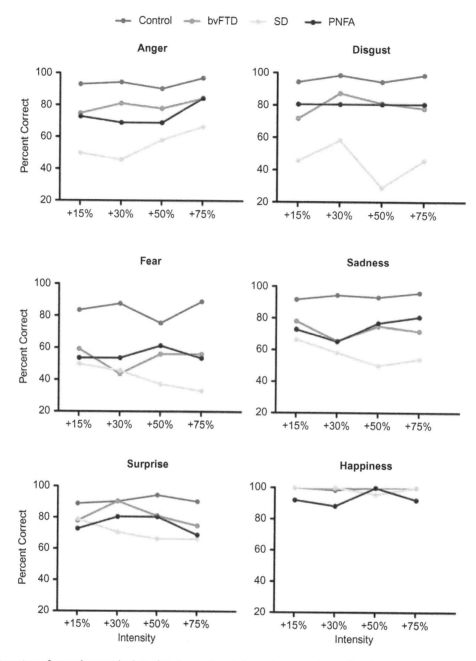

Figure 3. Percentage of correctly recognized emotions (anger, disgust, fear, sadness, surprise, and happiness) across the four levels of intensity in healthy controls and FTD subtypes.

$p = .012$; disgust: $U = 147.5$, $z = -2.202$; and sadness: $U = 142.5$, $z = -2.497$, $p = .013$, but not fear.

The effect of intensity on emotion recognition

Examination of change scores for each negative emotion, revealed an overall effect of group for disgust: $H(3) = 11.013$, $p = .012$; fear: $H(3) = 15.345$, $p = .002$; and sadness: $H(3) = 10.348$, $p = .016$, indicating that, for these three emotions, change in performance from low to high intensity differed across groups (Figure 3). In contrast, change in performance for anger stimuli did not differ across groups. Post-hoc tests showed that bvFTD performance was not significantly different from controls for these three basic emotions, indicating that bvFTD patients benefited from increasing intensity to a similar degree as controls. In contrast, the SD group's change score was significantly less than controls for disgust ($U = 108.0$, $z = -1.594$, $p = .001$), fear ($U = 73$, $z = -3.731$, $p < .001$), and sadness ($U = 113.5$, $z = -2.926$,

$p = .003$), indicating minimal benefit of increasing intensity compared to controls. PNFA patients generally improved to a similar extent as controls except for the emotion fear ($U = -2.461$, $z = .014$, $p = .014$). Change scores for positive emotions are not reported given the absence of between-group differences for these emotions in the previous analyses.

DISCUSSION

This study sought to explore emotion recognition performance in FTD and to identify the mechanisms responsible for impaired performance in each FTD subtype by modulating the emotional intensity of the stimuli. This study reveals a number of novel findings regarding emotion processing in the three subtypes of FTD, findings that have clinical implications for the management of these patients. Using two tasks of facial emotion recognition, our study revealed significant impairment in emotion recognition in all three subtypes of FTD, including in PNFA, indicating that contrary to current assumptions, PNFA patients do have demonstrable deficits in emotion recognition. Results from this study further illustrated an effect of emotion intensity on performance accuracy. Importantly, the performance improvement observed with increased intensity of emotion was present in bvFTD and PNFA groups only and was absent in SD, indicating differences in the mechanisms underlying emotion-detection deficits across subtypes of FTD.

Profiles of emotion-recognition deficits

We predicted that deficits in emotion recognition would differ across FTD subtypes, the deficits reflecting the predominant locus of brain atrophy of each subtype (i.e., impaired anger recognition in bvFTD, impaired fear recognition in SD, and impaired disgust recognition in PNFA). Contrary to our predictions, deficits in emotion recognition were more widespread than expected.

Overall, bvFTD patients were impaired for recognition of all negative emotions, including anger, whether intensity of emotion was modulated or not, a finding in keeping with previous studies (Kipps, Mioshi, & Hodges, 2009a; Kipps et al., 2009b; Lough et al., 2006). Impaired recognition of anger in patients with bvFTD is consistent with the predictions of the model and reflects the orbitofrontal cortex atrophy reported in these patients (Krill, Macdonald, Patel, Png, & Halliday, 2005; Piguet et al., 2011; Seeley et al., 2008). Although impairment in the recognition of

sadness, fear, and disgust, was not predicted, these emotion-recognition deficits can be accounted for by the multimodal system model, in that the discrete neural substrates necessary for these basic emotions undergo pathological changes in bvFTD. Preserved ability to recognize sadness has been associated with integrity of the subcallosal cingulate cortex (Phan, Wager, Taylor, & Liberzon, 2002). Atrophy in this region has been implicated in the disturbance of empathy (Rankin et al., 2006) and is the site of significant neuropathological changes in bvFTD (Schroeter, Raczka, Neumann, & Yves von Cramon, 2008; Seeley et al., 2008). Similarly, atrophy of the anterior insula, which is involved in disgust recognition, has been shown to be affected to some extent in bvFTD (Rosen et al., 2002a; Schroeter et al., 2008; Seeley, 2010; Seeley et al., 2008). Atrophy of the amygdala is also present in bvFTD (Barnes et al., 2006; Brambati et al., 2007) and most likely contributes to the reduced fear recognition in this group.

In SD, a severe impairment in fear recognition was observed with other negative emotions (anger, sadness, and disgust) affected to a lesser degree. This disproportionate impairment of fear is consistent with the multimodal system theory's prediction of reduced fear recognition in the presence of anterior temporal lobe, particularly amygdala, atrophy (Adolphs et al., 1995; Ekman, 1992). Recognition of other negative emotions, however, was also impaired. This finding confirms prior reports of a global impairment for recognition of negative emotions in SD (Rosen et al., 2002b, 2004). Although it is possible that these patients performed poorly due to loss of semantic knowledge or difficulty in reading the label, this explanation does not fully account for their performance. These patients performed above chance for each of the emotions, indicating sufficient word knowledge to perform the task to some extent. Thus, the observed pattern of deficits more likely reflects a profound involvement of the amygdala in processing emotion. Theorists suggest the amygdala may be essential for processing ambiguous stimuli (Whalen, 1999), or for processing emotions related to behavioral withdrawal (Anderson, Spencer, Fulbright, & Phelps, 2000). With disease progression, atrophy in SD extends beyond the anterior temporal region to involve other brain regions necessary for recognition of other basic emotions (such as the insula, and striatal regions). Although the SD group included in this study was relatively mild, presence of atrophy in other regions cannot be discounted.

Contrary to the prediction of the model, a disproportionate impairment of recognition of disgust was not observed in PNFA at 100% intensity. In fact, recognition of disgust in PNFA was similar to controls. Although the insula has been implicated in disgust recognition, only some regions of the insula are involved in emotion processing, with other areas specialized for other functions such as speech production (Ackermann & Riecker, 2004; Adolphs, 2002a). It is plausible that in PNFA, regions necessary for speech are disproportionately damaged compared to those areas necessary for recognition of disgust. Alternatively, some researchers have suggested that the insula may be necessary not only for processing of disgust but also for interoceptive awareness and the processing of "physical emotions" (Craig, 2009; Critchley, Wiens, Rotshtein, Öhman, & Dolan, 2004; Damasio et al., 2000). Consistent with this hypothesis, the PNFA group did demonstrate impaired recognition of anger, sadness, and fear at 100% intensity.

Emotion recognition impairment in PNFA has not been reported before (Rankin et al., 2009). The impairment observed in this study is unlikely to be due to greater disease severity in the selected PNFA patients, as these participants had a similar cognitive performance to the other patient groups. Rather, we suggest that using a larger group and a sensitive measure of emotion recognition has allowed us to uncover deficits not previously reported. The observed impaired recognition of sadness, anger, and fear is likely due to pathological changes in regions beyond the insula, such as the anterior cingulate and striatal regions, which may have resulted in the impaired sadness and anger recognition observed in this study (Fukui & Kertesz, 2000). Disconnection among brain regions involved in emotion recognition may also contribute to the observed deficits, with white matter changes of the superior longitudinal fasciculus and uncinate fasciculus previously reported in this FTD subtype (Whitwell et al., 2010).

Taken together, our results suggest that the multimodal system model of emotion accounts to some extent for emotion-recognition deficits in FTD, although its predictions tend to be more specific than the deficits reported in this study. The multimodal system model has evolved from studies involving patients with discrete lesions and studies using functional imaging in healthy controls. As such, the model appears less able to account for the deficits arising from diffuse brain changes and disconnections across brain regions, such as those seen in neurodegenerative disorders.

Multiple processes of emotion recognition: The effect of intensity

In this study, mechanisms underlying poor performance on emotion-recognition tasks were investigated by manipulating the intensity of facial emotional expressions. Both overall and individual emotion recognition significantly improved in bvFTD and PNFA with increased intensity, whereas the SD group showed limited response to that manipulation. Importantly, the bvFTD and PNFA groups were significantly impaired on neuropsychological measures of attention and working memory. Increasing the salience of the emotions, and therefore reducing the attentional and perceptual demands of the task, saw an improvement in performance in these two groups. These findings suggest that emotion-recognition disturbance in bvFTD and PNFA may be attributable in part to attentional deficits. In contrast, the lack of improvement in SD indicates a primary emotion-processing deficit in this group that cannot be overcome with changes in the attentional and perceptual demands of the task.

Emotional stimuli tend to be processed preferentially over nonemotional stimuli (Dolan, 2002; Öhman, Flykt, & Esteves, 2001). This preferential processing is mediated by two networks that are thought to work in parallel: Bottom-up, pre-attentive processing occurs in regions from the amygdala to cortical visual areas (Vuilleumier, Armony, Driver, & Dolan, 2001), whereas top-down, preferential allocation of spatial attention to emotional stimuli is mediated by prefrontal attentional networks (Hopfinger, Buonocore, & Mangun, 2005; LaBar et al., 1998; LeDoux, 1998; Palermo & Rhodes, 2007; Vuilleumier, 2005).

The top-down network relies on frontoparietal cortical regions, regions which overlap with the atrophy observed in bvFTD and to some extent PNFA (dorsolateral prefrontal cortex, anterior cingulate gyrus, ventromedial prefrontal cortex, superior temporal sulcus, and intraparietal sulcus) (Palermo & Rhodes, 2007). At low levels of intensity, emotional salience may be insufficient to attract preferential attentional resources in bvFTD and PNFA. With increasing intensity, however, as emotional salience becomes greater, attentional resources become preferentially allocated to the stimulus by the top-down network, resulting in accurate emotion detection. In contrast, in SD, the amygdala is compromised. This area is critical for emotion analysis and receives feedback both from the bottom-up and top-down pathways (Adolphs, 2002b). Hence, amygdala damage results in widespread emotion recognition impairment, which cannot be overcome by increasing emotional salience. This position is supported by a recent study that demonstrated that perceptual deficits contribute to impaired performance on emotion recognition tasks in bvFTD, but not SD (Miller et al., in press).

Notably, the effect of intensity was largest for negative emotions. From an evolutionary perspective, detection of some emotions is more critical than others to assist survival. As such, anger detection is important for identifying potential aggression in others and preparation for defense. Similarly, fear detection is important for avoidance of dangerous situations, and detection of disgust is important for avoiding potentially hazardous foods and environments (Darwin, 1872). From this perspective, processing of these emotions, which are critical for survival, is likely to be more robust and amenable to compensation than other emotions, such as surprise, that are relevant to social situations. Accordingly, a stronger response to modulation of intensity would also be anticipated in those emotions that are linked to survival than in those associated with social situations, as was generally observed in this study.

It is also possible that increasing salience has a specific effect on arousal, with increase in arousal sufficient for a critical threshold to be met, and recognition of the emotional stimulus to occur. In this study, highly arousing negative emotions responded to changes in intensity to a significant extent. Investigating whether there are circumstances under which manipulation of intensity is effective for low arousing emotions will help to distinguish these theoretical accounts.

Clinical implications

The emotion-recognition deficits observed in this study illustrate extensive disruption to emotion processing across all subtypes of FTD. The impact of these deficits on appropriate social interactions and the ability to form and maintain relationships is substantial. Thus, the potential for rehabilitation of emotion recognition is an important target for intervention. The current results suggest that increasing the emotional salience of environmental stimuli may be sufficient for attentional resources to be allocated to these stimuli and improve emotion recognition in PNFA and bvFTD, at least for negative emotions.

It can only be speculated whether increasing emotional salience might also offer an effective intervention in the later stages of the disease, and for other emotions. Investigating the effect of salience longitudinally may provide some insight into the effects of perceptual and attentional demands on emotion processing, and help determine whether increasing emotional salience is an effective intervention later in the disease course.

Summary

These results demonstrate significant impairment in recognition of emotions across all subtypes of FTD, including PNFA. Furthermore, they demonstrate that

the underlying mechanism causing impaired performance on emotion-recognition tasks differs across subtypes. The results suggest that while SD patients perform poorly because of an underlying primary emotion-processing deficit, bvFTD and PNFA patients appear to be impaired because of inattention or perceptual difficulties, or due to partially degraded emotion-recognition structures that respond only to highly salient emotional material. Investigation of the neural correlates of emotion detection in these patients by structural or functional imaging will help understand what structures are necessary for recognition of specific emotions and help determine what effect manipulation of intensity is having in this population. Clinically, these results suggest that compensation of impaired emotion-recognition performance may be possible in the bvFTD and PNFA subtypes of FTD. Given the striking behavioral changes in these patients that alter their interpersonal conduct and ability to form and maintain relationships, simple interventions such as these are likely to have significant impact on these patients' functioning.

REFERENCES

Ackermann, H., & Riecker, A. (2004). The contribution of the insula to motor aspects of speech production: A review and a hypothesis. *Brain and Language, 89,* 320–328.

Adolphs, R. (2002a). Neural systems for recognizing emotion. *Current Opinion in Neurobiology, 12,* 169–177.

Adolphs, R. (2002b). Recognizing emotion from facial expressions: Psychological and neurological mechanisms. *Behavioral and Cognitive Neuroscience Reviews, 1,* 21–62.

Adolphs, R., Tranel, D., Damasio, H., & Damasio, A. (1995). Fear and the human amygdala. *Journal of Neuroscience, 15,* 5879–5891.

Allman, J., & Brothers, L. (1994). Faces, fear and the amygdala. *Nature, 372,* 613–614.

Anderson, A., Spencer, D. D., Fulbright, R. K., & Phelps, E. A. (2000). Contribution of the anteromedial temporal lobes to the evaluation of facial emotion. *Neuropsychology, 14,* 526–536.

Barnes, J., Whitwell, J. L., Frost, C., Josephs, K. A., Rossor, M. N., & Fox, N. C. (2006). Measurements of the amygdala and hippocampus in pathologically confirmed Alzheimer disease and frontotemporal lobar degeneration. *Archives of Neurology, 63,* 1434–1439.

Bediou, B., Ryff, I., Mercier, B., Milliery, M., Henaff, M.-A., D'Amato, T., et al. (2009). Impaired social cognition in mild Alzheimer disease. *Journal of Geriatric Psychiatry and Neurology, 22,* 130–140.

Benson, P. J., Campbell, R., Harris, T., Frank, M. G., & Tovée, M. J. (1999). Enhancing images of facial expressions. *Perception and Psychology, 61,* 259–274.

Blair, J. R., Morris, J. S., Frith, C. D., Perrett, D. I., & Dolan, R. J. (1999). Dissociable neural responses to facial expressions of sadness and anger. *Brain, 122,* 883–893.

Brambati, S. M., Renda, N. C., Rankin, K. P., Rosen, H., Seeley, W. W., Ashburner, J., et al. (2007). A tensor based morphometry study of longitudinal gray matter contraction in FTD. *NeuroImage, 35*, 998–1003.

Brothers, L. (1990). The social brain: A project for integrating primate behaviour and neurophysiology in a new domain. *Concepts in Neuroscience, 1*, 27–51.

Burgess, P., & Shallice, T. (1997). The Hayling and Brixton tests. Bury St Edmonds, UK: Thames Valley Test Company.

Calder, A. J., Keane, J., Lawrence, A. D., & Manes, F. (2004). Impaired recognition of anger following damage to the ventral striatum. *Brain, 127*, 1958–1969.

Calder, A. J., Keane, J., Manes, F., Antoun, N., & Young, A. W. (2000a). Impaired recognition and experience of disgust following brain injury. *Nature Neuroscience, 3*, 1077–1079.

Calder, A. J., Rowland, D., Young, A. W., Nimmo-Smith, I., Keane, J., & Perrett, D. I. (2000b). Caricaturing facial expressions. *Cognition, 76*, 105–146. doi: 10.1016/S0010–0277(00)00074–3

Calder, A. J., Young, A. W., Rowland, D., & Perrett, D. (1997). Computer-enhanced emotion in facial expressions. *Proceedings of the Royal Society London. Series B, Biological Sciences, 264*, 919–925.

Calder, A. J., Young, A. W., Rowland, D., Perrett, D. I., Hodges, J. R., & Etcoff, N. L. (1996). Facial emotion recognition after bilateral amygdala damage: Differentially severe impairment of fear. *Cognitive Neuropsychology, 13*, 699–745.

Craig, A. D. (2009). How do you feel – now? The anterior insula and human awareness. *Nature Reviews Neuroscience, 10*, 59–70.

Critchley, H. D., Wiens, S., Rotshtein, P., Öhman, A., & Dolan, R. J. (2004). Neural systems supporting interoceptive awareness. *Nature Neuroscience, 7*, 189–195.

Damasio, A. R., Grabowski, T. J., Bechara, A., Damasio, H., Ponto, L. L. B., Parvizi, J., & Hichwa, R. D. (2000). Subcortical and cortical brain activity during the feeling of self-generated emotions. *Nature Neuroscience, 3*, 1049–1056.

Darwin, C. (1872). *The expression of the emotions in man and animals*. London: John Murray.

Davies, R. R., Kipps, C. M., Mitchell, J., Kril, J. J., Halliday, G. M., & Hodges, J. R. (2006). Progression in frontotemporal dementia: Identifying a benign behavioural variant by magnetic resonance imaging. *Archives of Neurology, 63*, 1627–1631.

Dolan, R. J. (2002). Emotion, cognition, and behavior. *Science, 298*, 1191–1194.

Ekman, P. (1992). An argument for basic emotions. *Cognition and Emotion, 6*, 169–200.

Ekman, P., & Friesen, W. V. (1976). *Pictures of facial affect*. San Francisco: CA: Consulting Psychologists Press.

Ekman, P., Friesen, W. V., & Ellsworth, P. (1972). *Emotion in the human face: Guidelines for research and integration of findings*. New York, NY: Pergamon (Eds.).

Folstein, M. F., Folstein, S. E., & McHugh, P. R. (1975). "Mini-mental state": A practical method for grading the mental state of patients for clinicians. *Journal of Psychiatry Research, 12*, 189–198.

Fukui, T., & Kertesz, A. (2000). Volumetric study of lobar atrophy in Pick complex and Alzheimer's disease. *Journal of the Neurological Sciences, 174*, 111–121.

Gorno-Tempini, M. L., Brambati, S. M., Ginex, V., Ogar, J. M., Dronkers, N. F., Marcone, A., et al. (2008). The logopenic/phonological variant of primary progressive aphasia. *Neurology, 71*, 1227–1234.

Gorno-Tempini, M. L., Hillis, A., Weintraub, S., Kertesz, A., Mendez, M., Cappa, S. F., et al. (2011). Recommendations for the classification of primary progressive aphasia. *Neurology, 76*, 1006–1014.

Hodges, J. R., & Patterson, K. (2007). Semantic dementia: A unique clinicopathological syndrome. *Lancet Neurology, 6*, 1004–1014.

Hodges, J. R., Patterson, K., Oxbury, S., & Funnell, E. (1992). Semantic dementia: Progressive fluent aphasia with temporal lobe atrophy. *Brain, 115*, 1783–1806.

Hopfinger, J. B., Buonocore, M. H., & Mangun, G. R. (2005). The neural mechanisms of top-down attentional control. *Nature Neuroscience, 3*, 284–291.

Keane, J., Calder, A. J., Hodges, J. R., & Young, A. W. (2002). Face and emotion processing in frontal variant frontotemporal dementia *Neuropsychologia, 40*, 655–665.

Kertesz, A., McMonagle, P., Blair, M., Davidson, W., & Munoz, D. G. (2005). The evolution and pathology of frontotemporal dementia. *Brain, 218*, 1996–2005.

Kesler West, M. L., Andersen, A. H., Smith, C. D., Avison, M. J., Davis, C. E., Kryscio, R. J., et al. (2001). Neural substrates of facial emotion processing using fMRI. *Cognitive Brain Research, 11*, 213–226.

Kessels, R. P. C., Gerritsen, L., Montagne, B., Ackl, N., Diehl, J., & Danek, A. (2007). Recognition of facial expression of different emotional intensities in patients with frontotemporal lobar degeneration. *Behavioural Neurology, 18*, 31–36.

Kipps, C. M., & Hodges, J. R. (2006). Theory of mind in frontotemporal dementia. *Social Neuroscience, 1*, 235–244.

Kipps, C. M., Mioshi, E., & Hodges, J. R. (2009a). Emotion, social functioning and activities of daily living in frontotemporal dementia. *Neurocase, 15*, 182–189.

Kipps, C. M., Nestor, P. J., Acosta-Cabronero, J., Arnold, R., & Hodges, J. R. (2009b). Understanding social dysfunction in the behavioural variant of frontotemporal dementia: The role of emotion and sarcasm processing. *Brain, 132*, 592–603.

Krill, J. J., Macdonald, V., Patel, S., Png, F., & Halliday, G. (2005). Distribution of brain atrophy in behavioural variant frontotemporal dementia. *Journal of the Neurological Sciences, 232*, 83–90.

LaBar, K. S., Gatenby, J. C., Gore, J. C., LeDoux, J. E., & Phelps, E. A. (1998). Human amygdala activation during conditioned fear acquisition and extinction: A mixed-trial fMRI study. *Neuron, 20*, 937–945.

Lavenu, I., Pasquier, F., Lebert, F., Petit, H., & Van der Linden, M. (1999). Perception of emotion in frontotemporal dementia and Alzheimer disease. *Alzheimer Disease & Associated Disorders, 13*, 96–101.

LeDoux, J. E. (1998). Fear and the brain: Where have we been, and where are we going? *Biological Psychiatry, 44*, 1229–1238.

Lough, S., Kipps, C. M., Treise, C., Watson, P., Blair, J. R., & Hodges, J. R. (2006). Social reasoning, emotion and empathy in frontotemporal dementia. *Neuropsychologia, 44*, 950–958.

Mack, W. J., Freed, D. M., White Williams, B., & Henderson, V. W. (1992). Boston naming test: Shortened versions for use in Alzheimer's disease. *Journal of Gerontology, 47*, 154–158.

Miller, B. L., Hsieh, S., Lah, S., Savage, S., Hodges, J. R., & Piguet, O. (in press). One size does not fit all: Face emotion processing impairments in semantic dementia, behavioural-variant frontotemporal dementia and Alzheimer's disease are mediated by distinct cognitive deficits. *Behavioural Neurology*.

Mioshi, E., Dawson, K., Mitchell, J., Arnold, R., & Hodges, J. R. (2006). The Addenbrooke's Cognitive Examination Revised (ACE-R): A brief cognitive test battery for dementia screening. *International Journal of Geriatric Psychiatry, 21*, 1078–1085.

Morris, J. S., Friston, K. J., Buchel, C., Frith, C. D., Young, A. W., Calder, A. J., et al. (1998). A neuromodulatory role for the human amygdala in processing emotional facial expressions. *Brain, 121*, 47–57.

Neary, D., Snowden, J. S., Gustafson, L., Passant, U., Stuss, D., Black, S., et al. (1998). Frontotemporal lobar degeneration: A consensus on clinical diagnostic criteria. *Neurology, 51*, 1546–1554.

Nestor, P. J., Graham, N. L., Fryer, T. D., Williams, G. B., Patterson, K., & Hodges, J. R. (2003). Progressive non-fluent aphasia is associated with hypometabolism centred on the left anterior insula. *Brain, 126*, 2406–2418.

Öhman, A., Flykt, A., & Esteves, F. (2001). Emotion drives attention: Detecting the snake in the grass. *Journal of Experimental Psychology: General, 130*, 466–478.

Palermo, R., & Rhodes, G. (2007). Are you always on my mind? A review of how face perception and attention interact. *Neuropsychologia, 45*, 75–92.

Phan, K. L., Wager, T., Taylor, S. F., & Liberzon, I. (2002). Functional neuroanatomy of emotion: A meta-analysis of emotion activation studies in PET and fMRI. *NeuroImage, 16*, 331–348.

Phillips, M. L., Young, A. W., Scott, S. K., Calder, A. J., Andrew, C., Giampietro, V., et al. (1998). Neural responses to facial and vocal expressions of fear and disgust. *Proceedings of the Royal Society. Series B, Biological Sciences, 365*, 1809–1817.

Phillips, M. L., Young, A. W., Senior, C., Brammer, M., Andrew, C., Calder, A. J., et al. (1997). A specific neural substrate for perceiving facial expressions of disgust. *Nature, 389*, 495–498.

Piguet, O., Hornberger, M., Mioshi, E., & Hodges, J. R. (2011). Behavioural-variant frontotemporal dementia: Diagnosis, clinical staging, and management. *Lancet Neurology, 10*, 162–172.

Rankin, K. P., Gorno-Tempini, M. L., Allison, S., Stanley, C. M., Glenn, S., Weiner, M. W., et al. (2006). Structural anatomy of empathy in neurodegenerative disease. *Brain, 129*, 2945–2956.

Rankin, K. P., Salazar, A., Gorno-Tempini, M. L., Sollberger, M., Wilson, S. M., Pavlic, D., et al. (2009). Detecting sarcasm from paralinguistic cues: Anatomic and cognitive correlates in neurodegenerative disease. *NeuroImage, 47*, 2005–2015.

Rascovsky, K., Hodges, J. R., Kipps, C. M., Johnson, J. K., Seeley, W. W., Mendez, M., et al. (2007). Diagnostic criteria for the behavioural variant of frontotemporal dementia: Current limitations and future directions.

Alzheimer Disease & Associated Disorders, 21, S14–S18.

Ratnavalli, E., Brayne, C., Dawson, K., & Hodges, J. R. (2002). The prevalence of frontotemporal dementia. *Neurology, 58*, 1615–1621.

Rosen, H. J., Gorno-Tempini, M. L., Goldman, W. P., Perry, R. J., Schuff, N., Weiner, M., et al. (2002a). Patterns of brain atrophy in frontotemporal dementia and semantic dementia. *Neurology, 58*, 198–208.

Rosen, H. J., Pace-Savitsky, K., Perry, R. J., Kramer, J. H., Miller, B. L., & Levenson, R. L. (2004). Recognition of emotion in the frontal and temporal variants of frontotemporal dementia. *Dementia and Geriatric Cognitive Disorder, 17*, 277–281.

Rosen, H. J., Perry, R. J., Murphy, J., Kramer, J. H., Mychack, P., Schuff, N., et al. (2002b). Emotion comprehension in the temporal variant of frontotemporal dementia. *Brain, 125*, 2286–2295.

Schachter, S. (1964). The interaction of cognitive and physiological determinants of emotional state. In L. Berkowitz (Ed.), *Advances in experimental social psychology* (vol. 1). New York, NY: Academic Press.

Schroeter, M. L., Raczka, K., Neumann, J., & Yves von Cramon, D. (2008). Neural networks in frontotemporal dementia – a meta-analysis. *Neurobiology of Aging, 29*, 418–426.

Seeley, W. W. (2010). Anterior insula degeneration in frontotemporal dementia. *Brain Structure & Function, 214*, 465–475.

Seeley, W. W., Crawford, R., Rascovsky, K., Kramer, J. H., Weiner, M., Miller, B. L., et al. (2008). Frontal paralimbic network atrophy in very mild behavioural variant frontotemporal dementia. *Archives of Neurology, 65*, 249–255.

Snowden, J. S., Austin, N. A., Sembi, S., Thompson, J. C., Craufurd, D., & Neary, D. (2008). Emotion recognition in Huntington's disease and frontotemporal dementia. *Neuropsychologia, 46*, 2638–2649.

Spreen, O., & Strauss, E. A. (1998). *A compendium of neuropsychological tests: Administration, norms and commentary*, 2nd edn. New York, NY: Oxford University Press.

Sprengelmeyer, R., Young, A. W., Calder, A. J., Karnat, A., Lange, H., Hömberg, V., et al. (1996). Loss of disgust: Perception of faces and emotions in Huntington's disease. *Brain, 119*, 1647–1665.

Tombaugh, T. N. (2004). Trail Making Test A and B: Normative data stratified by age and education. *Archives of Clinical Neuropsychology, 19*, 203–214.

Vuilleumier, P. (2005). How brains beware: Neural mechanisms of emotional attention. *Trends in Cognitive Sciences, 9*, 585–594.

Vuilleumier, P., Armony, J. L., Driver, J., & Dolan, R. J. (2001). Effects of attention and emotion on face processing in the human brain: An event-related fMRI study. *Neuron, 30*, 829–841.

Wechsler, D. (1997). *WAIS-III administration and scoring manual*. San Antonio, TX: Psychological Corporation.

Whalen, P. J. (1999). Fear, vigilance, and ambiguity: Initial neuroimaging studies of the human amygdala. *Psychological Science, 7*, 177–187.

Whitwell, J. L., Avula, R., Senjem, M. L., Kantarci, K., Weigand, S. D., Samikoglu, A., et al. (2010). Gray and white matter water diffusion in the syndromic variants of frontotemporal dementia. *Neurology, 74*, 1279–1287.

Young, A. W., Perrett, D., Calder, A. J., Sprengelmeyer, R., & Ekman, P. (2002). *Facial expressions of emotion – stimuli and tests (FEEST)*. Thames Valley Test Company, Thurston, Suffolk.

APPENDIX 1

Example stimuli from the Ekman 60 task (100%) and the Ekman Caricatures task for Model MO (A) for the emotion disgust and Model JJ (B) for the emotion sadness, across the four levels of intensity (+15%, +30%, +50%, +75%).

(A)

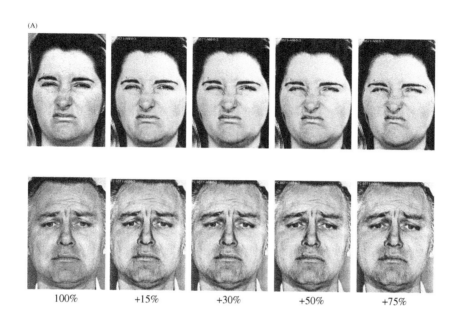

100%   +15%   +30%   +50%   +75%

Perception of emotion in psychiatric disorders: On the possible role of task, dynamics, and multimodality

Patricia Garrido-Vásquez[1*], Sarah Jessen[1,2*], and Sonja A. Kotz[1]

[1]Minerva Research Group "Neurocognition of Rhythm in Communication," Max Planck Institute for Human Cognitive and Brain Sciences, Leipzig, Germany
[2]Cluster of Excellence "Languages of Emotion," Free University of Berlin, Germany

Experimental evidence suggests an impairment in emotion perception in numerous psychiatric disorders. The results to date are primarily based on research using static displays of emotional facial expressions. However, our natural environment is dynamic and multimodal, comprising input from various communication channels such as facial expressions, emotional prosody, and emotional semantics, to name but a few. Thus, one critical open question is whether alterations in emotion perception in psychiatric populations are confirmed when testing patients in dynamic and multimodal naturalistic settings. Furthermore, the impact task demands may exert on results also needs to be reconsidered. Focusing on schizophrenia and depression, we review evidence on how emotions are perceived from faces and voices in these disorders and examine how experimental task demands, stimulus dynamics, and modality may affect study results.

Keywords: Emotion perception; Multimodal; Task effects; Depression; Schizophrenia.

Accurately perceiving others' emotions is crucial for successful interpersonal interactions and may be a key factor in social deficits observed in many neuropsychiatric disorders. During recent decades, a great deal of evidence on emotion perception in psychiatric populations has accumulated. Even though this research has considerably advanced our understanding of how emotions are perceived in people with pathological mental states, we argue that inconsistencies in the results need to be viewed in light of the experimental procedures applied. As we will show, these procedures can be improved by considering higher ecological validity of stimuli, as, for instance, by using multimodal stimuli (i.e., the combination of face and voice). While an emotional expression in one modality is often enough to determine a person's emotional state, these information sources normally occur simultaneously, providing seemingly redundant information. However, simultaneous perception of congruent emotional information is by no means redundant, but rather offers a number of benefits to the perceiver as will be outlined in this review. These benefits may be especially high when information-processing resources are limited, as is often apparent in psychiatric disorders. Along these lines, dynamic rather than static stimuli provide a more ecologically valid approach to test emotion perception. Furthermore, implicit rather than explicit task instructions may affect how emotional expressions are perceived as well as reduce confounds due to cognitive task demands.

Correspondence should be addressed to: Sonja A. Kotz, Max Planck Institute for Human Cognitive and Brain Sciences, P.O. Box 500355, 04303 Leipzig, Germany. E-mail: kotz@cbs.mpg.de

*These authors contributed equally to the paper.

The authors would like to thank three anonymous reviewers for helpful comments on an earlier version of this manuscript. We also thank Rosie Wallis for carefully checking our English. We gratefully acknowledge the funding of the Canadian Institute of Health Research (CIHR-MPO:62867: Marc D. Pell and S.A.K.) and the German Science Foundation (DFG-FOR-499: S.A.K.).

The present review aims to promote a more naturalistic approach to emotion perception in patient investigations. To this end, we first present a general introduction to underline the benefits offered by multimodal, dynamic stimulus presentations and implicit task instructions. Second, we introduce models of how emotions are perceived from speech and faces. We then exemplify by focusing on two mental disorders, schizophrenia and major depression, and summarize what has been done so far to understand emotion perception in these two disorders. Finally, we draw disorder-specific and general conclusions from the current state of the art in research. The motivation for focusing on these specific disorders is that they are assumed to tap into different stages of emotion processing: While schizophrenia has been shown to already affect early sensory processes (Johnston, Stojanov, Devir, & Schall, 2005), major depression is considered to alter later processing stages (Gotlib & Joormann, 2010). Likewise, modality, dynamics, and task manipulations may have different effects in these two psychiatric populations, as we try to elucidate in the present review. Our literature review concerning schizophrenia and depression is limited to the perception of social emotional cues (i.e., vocal and facial expressions).

MULTIMODAL EMOTION PROCESSING

Several studies have shown facilitated emotional information processing when information is simultaneously presented in several modalities (Giard & Peronnet, 1999; Kreifelts, Ethofer, Grodd, Erb, & Wildgruber, 2007; Paulmann, Jessen, & Kotz, 2009). Emotions are recognized faster (Giard & Peronnet, 1999) and more accurately (Giard & Peronnet, 1999; Kreifelts et al., 2007) when simultaneously conveyed by face and voice rather than when presented unimodally.

Furthermore, multimodal perception allows fine-tuning of emotional expression that cannot be achieved unimodally. The voice can have a strong influence on how we perceive the face and vice versa (de Gelder & Vroomen, 2000). If, for instance, an ambiguous facial expression is presented with a happy voice, participants tend to perceive the facial expression as happy, while if the same facial expression is accompanied by sad prosody, the facial expression is also perceived as sad (de Gelder & Vroomen, 2000).

In addition, multimodal perception can ensure that our assessment of someone's emotional state is clear if one of the modalities fails us. The less reliable information from one modality becomes, the more we benefit from information provided in another modality (Werner & Noppeney, 2010). Consider, for example,

being in the dark when you can hardly see your conversational partner but very clearly hear his or her voice.

For these reasons, multisensory perception is an essential mechanism in emotion communication. Yet multisensory perception of emotions is largely understudied in the field of neuropsychiatric disorders. However, it seems crucial to study multisensory emotion perception in patients who suffer from emotional as well as social deficits. It is an open question whether multimodal perception offers the same benefit to these patients as it does to healthy people, or, in contrast, may hamper emotion perception.

MODELS OF PROCESSING EMOTION FROM SPEECH AND FACES

One important information source in emotional communication is a person's voice. Three basic steps have been suggested for the processing of emotional speech (Schirmer & Kotz, 2006): a first sensory processing phase, followed by the integration of acoustic cues to form a salient percept, and finally cognitive processes operating on such percepts. In the first step, auditory input is processed in the primary auditory cortices, projecting to the bilateral superior temporal sulcus (STS). These primary features are then integrated in the superior temporal gyrus as well as sulcus, moving along to anterior portions of the STS. Based on these processes, different steps occur depending on the context, the task, and the exact specification of the stimulus. In emotional speech, emotional prosodic content is accompanied by semantic content. Thus, both types of emotion information need to be compared and aligned. This particular processing step engages the left inferior frontal gyrus. On the other hand, if participants have just to process and label prosodic information, the right inferior frontal gyrus, as well as the orbitofrontal cortex, seem to play an essential role (Schirmer & Kotz, 2006). A similar model was proposed by Wildgruber, Ackermann, Kreifelts, and Ethofer (2006), positing three steps: the extraction of supra-segmental cues, the representation of supra-segmental sequences, and the explicit judgment of emotional information. These steps correspond essentially to the three steps introduced above, and also involve mostly the same brain regions.

Regarding the processing of emotions from facial expressions, Adolphs (2002) proposes a two-step model, dividing the processing into a perception and a recognition part. However, the second step can be further divided into a recognition and a conceptual knowledge aspect, where the recognized emotion is associated with previously known information about the person and the general context. In the first, sensory step, simple as well as highly salient features are processed. This encoding encompasses subcortical regions such as the superior colliculus and the pulvinar nucleus of the thalamus as well as primary sensory cortices, such as the striate cortex. Following this first step, more detailed emotion-processing mechanisms that can be subsumed under the term *recognition* are employed. Relevant facial features are analyzed in visual association cortices. If motion information is contained, the middle and middle superior temporal areas are involved. This information is passed on to the fusiform and the superior temporal cortices, where emotional content and social relevance is processed. The amygdala and the orbitofrontal cortex are involved in then guiding various aspects of further emotion processing. First, feedback is directed back to regions implicated in earlier processing stages, allowing fine-tuned processing of the ongoing input. Second, connections to various cortical regions, such as the somatosensory and the prefrontal cortex, as well as the hippocampus, allow the perceived emotions to be put into emotional context based on previous knowledge. Finally, a simulation of another person's emotional state is facilitated by connections to motor structures such as the basal ganglia and the frontal operculum, as well as brainstem nuclei, enabling us to "feel with" somebody (Adolphs, 2002). Besides these elaborate emotion-processing pathways, the pulvinar-amygdalar pathway provides a second, faster route for processing emotion information (Adolphs, 2002; Vuilleumier & Pourtois, 2007). Here, coarse stimulus features, in particular, are used to quickly determine potentially dangerous situations in order to allow a fast and adequate response.

Overall, processing of auditory as well as visual information can thus be divided into three essential steps: namely, early feature processing, integration of these features, and finally evaluation of the percept. Different regions in the brain have been proposed to be involved in the integration of auditory and visual emotional information, such as the right posterior STS (Ethofer, Pourtois, & Wildgruber, 2006), the right posterior insula (Ethofer et al., 2006), the amygdala (Ethofer et al., 2006; de Gelder, Vroomen, & Pourtois, 2004; O'Doherty, Rolls, & Kringelbach, 2004), and the orbitofrontal cortex (O'Doherty et al., 2004). However, no specific model exists which accounts for the integration of multisensory emotional information, and it remains unclear to what extent an integration process is emotion specific or, rather, draws on mechanisms involved in multisensory perception, such as audiovisual speech perception, in general.

DYNAMIC VERSUS STATIC EMOTIONAL STIMULI

While speech and facial expressions are both inherently dynamic, facial expressions are often tested in static displays. However, our environment is dynamic, and the use of dynamic stimuli in emotion research is hence a much more naturalistic approach for studying emotion perception than from photos or drawings of facial expressions.

In fact, dynamic face stimuli are more easily recognized, especially in the case of non-prototypical emotion displays (Ambadar, Schooler, & Cohn, 2005; Bould & Morris, 2008), and emotional deviants are easier to detect in a dynamic visual search task (Horstmann & Ansorge, 2009). The advantage of dynamic stimuli is further underlined in a steady-state, visual evoked-potential study (Mayes, Pipingas, Silberstein, & Johnston, 2009). This method, in which stimuli are presented together with a flicker while the electroencephalogram (EEG) is recorded, provides information on cortical activation patterns and their latencies, as well as on inhibitory and excitatory processes. Applying this method, the authors found that processing of dynamic facial stimuli proved to be faster and yielded more activity at temporal electrodes and less activity at frontal electrodes than static ones. This may indicate more efficient processing of dynamic stimuli (Mayes et al., 2009). Furthermore, neuroimaging studies report more extended activation patterns for dynamic than for static facial stimuli. This concerns activation of the middle temporal gyrus, inferior and superior temporal gyri, visual areas, most notably the middle occipital and fusiform gyri, and frontal regions, especially the inferior frontal gyrus (Kilts, Egan, Gideon, Ely, & Hoffman, 2003; LaBar, Crupain, Voyvodic, & McCarthy, 2003; Sato, Kochiyama, Yoshikawa, Naito, & Matsumura, 2004; Trautmann, Fehr, & Herrmann, 2009). Increased middle occipital activation, however, is also observed when viewing dynamic mosaics (Sato et al., 2004) whereas the effects in temporal brain regions are connected to the social relevance of stimuli.

EXPLICIT VERSUS IMPLICIT TASK INSTRUCTIONS

In emotion perception experiments, task instructions can be generally divided into two categories: explicit and implicit. Explicit tasks are aimed at the emotional content of stimuli and comprise, among others, emotion categorization (assigning a verbal emotion label to a stimulus), matching of emotional stimuli in terms of the emotion they express, or valence ratings. There is also a wealth of implicit tasks that are not directed toward the emotionality of a stimulus. An example for an implicit measure is the dot-probe paradigm, in which a neutral and an emotional stimulus are presented simultaneously at different spatial locations and one stimulus is subsequently replaced by a dot probe to which subjects have to react as quickly as possible. Response latencies in this task can reveal attentional biases, as reactions should be faster when the probe appears in a previously attended location. Other examples of implicit tasks are decisions on physical stimulus characteristics, such as color, or gender decisions.

Some studies have directly compared how identical emotional stimulus material is processed under different task instructions. For example, event-related potential (ERP) studies have shown that explicit and implicit task instructions lead to quantitative differences in face processing. Some researchers reported different scalp distributions and latencies of an N400-like component when comparing identity- and emotion-matching tasks, as well as longer reaction times and higher error rates in an explicit task (Bobes, Martin, Olivares, & Valdés-Sosa, 2000; Münte et al., 1998), supporting the view that information processing is facilitated under implicit task instructions.

In neuroimaging research, increased amygdala activations in explicit compared to implicit tasks (Gorno-Tempini et al., 2001; R. C. Gur et al., 2002) and also the opposite pattern (Critchley et al., 2000) have been reported. In fact, the amygdala seems to be more sensitive to the absence of a task (i.e., passive viewing) than to differences between implicit and explicit tasks (Costafreda, Brammer, David, & Fu, 2008; but see Fusar-Poli et al., 2009). Furthermore, an absence of task-modulated amygdala effects but heightened responses in the somatosensory cortex have been reported during explicit face processing (Winston, O'Doherty, & Dolan, 2003).

ERP studies on emotional prosody perception also suggest task-dependent differences. Wambacq, Shea-Miller, and Abubakr (2004) reported an early differentiation (P200) between neutral and negative emotional prosody only when emotion was not task-relevant, while the explicit condition yielded a similar but later effect around 360 ms after stimulus onset. In a cross-modal priming study, Schirmer, Kotz, and Friederici (2002) presented an emotional-prosodic stimulus followed by a visually presented word or non-word, with participants required to make a lexical decision. On incongruent trials, the authors found an N400 effect that differed between women and men as a function of the interstimulus interval. In contrast,

when the task was to determine the valence of a word, no sex effects were found (Schirmer, Kotz, & Friederici, 2005). Kotz and Paulmann (2007) reported comparable effects when assessing ERP responses to emotional-prosodic and combined prosodic-semantic expectancy violations under implicit and explicit processing instructions. However, ERP amplitudes were overall more positive-going and larger in the explicit task.

Brain activation patterns revealed by neuroimaging methods during emotional prosody processing also depend considerably on study designs, among them task instructions (Kotz, Meyer, & Paulmann, 2006). In fMRI studies, explicit versus implicit emotional prosody processing has been related to increased orbitofrontal and inferior frontal activations (Ethofer et al., 2009; Imaizumi et al., 1997; Wildgruber et al., 2004, 2005). Enhanced activations in the right superior temporal region have also been observed under explicit versus implicit task instructions when processing emotional prosody (Ethofer et al., 2009; Wildgruber et al., 2005). To sum up, task demands appear to have a considerable impact on study results, both at the neural level and behaviorally. In many cases, reaction times and/or error rates are lower for implicit compared to explicit tasks, for face and prosody processing (Critchley et al., 2000; Ethofer et al., 2009; Gorno-Tempini et al., 2001; Kotz & Paulmann, 2007; Münte et al., 1998; Wildgruber et al., 2005). However, depending on the nature of the task assigned, an implicit task might be more difficult than one that is explicit (R. C. Gur et al., 2002). Thus, a simple distinction between implicit and explicit task dimensions may be too coarse, but choosing a simple implicit task could significantly reduce task demands. This may be a critical issue when testing patients who suffer from executive dysfunction. High task demands, as often occur in explicit tasks, can confound executive dysfunction with emotional processing deficits. This could lead to false assumptions about emotional processing deficits in patients (e.g., Adolphs, Schul, & Tranel, 1998). On the other hand, a very simple task, or even more so, the absence of a task, may lead to increased distraction in psychiatric populations and could also alter study results. These facts need to be considered when designing patient studies or interpreting their outcomes.

SCHIZOPHRENIA

Schizophrenia is a complex neuropsychiatric disorder comprising numerous symptoms such as hallucinations, delusions, incoherent thought, and blunting of affect, which can occur in basically any combination. While disturbances in the dopaminergic system have been commonly associated with schizophrenia, other pathological changes in the brain also seem to play an important role in the neuronal basis of schizophrenia.

Among the numerous deficits associated with schizophrenia, one often observes abnormalities in early sensory processing and deficits in emotion perception. It has been frequently reported that patients with schizophrenia seem to be unable to filter out relevant sensory information from irrelevant information in a stream of complex sensory input. This phenomenon has been described as a deficit in sensory gating (Freedman et al., 1987), and is also reflected in the electrophysiological brain response. When healthy people hear two consecutive sounds, an early auditory evoked potential (P50) in response to the second sound is reduced. This is commonly interpreted as reflecting the above-mentioned gating mechanism. In individuals with schizophrenia, no such reduction can be observed (Bramon, Rabe-Hesketh, Sham, Murray, & Frangou, 2004). Hence, schizophrenic patients seem to process sensory information in a different way than healthy controls.

At the same time, patients with schizophrenia appear to have difficulty in processing emotional information (Chan, Li, Cheung, & Gong, 2010). However, emotional information is usually presented in one or the other sensory modality, raising the question of to what degree emotion perception in patients with schizophrenia is influenced by a general sensory processing deficit, and what role different modalities play in emotion perception. Do the reported deficits in emotion perception originate from an early, sensory deficit, from a deficit in early emotion processing, or rather at a later stage in emotion processing? Furthermore, differences at several of these stages are conceivable, and may have a differential effect dependent on the precise experimental setup. An overview of selected emotion perception studies in schizophrenia is provided in Table 1.

Schizophrenia: Explicit versus implicit tasks

Various tasks, such as emotion identification (e.g., Bach et al., 2009; Hempel, Hempel, Schönknecht, Stippich, & Schröder, 2003; Kucharska-Pietura, David, Masiak, & Phillips, 2005; Quintana, Wong, Ortiz-Portillo, Marder, & Mazziotta, 2003) and emotion matching (e.g., Hempel et al., 2003; Martin, Baudouin, Tiberghien, & Franck, 2005; Quintana et al., 2003; Salgado-Pineda, Fakra, Delaveau,

Hariri, & Blin, 2010), have been employed to investigate explicit emotion perception in patients with schizophrenia. While these tasks themselves differ in that one requires explicit labeling while the other only requires a comparison between a number of stimuli, a recent meta-analysis by Kohler, Walker, Martin, Healey, and Moberg (2010) shows that no difference in impairment can be found by these tasks. This suggests that the observed deficits, also in emotion-identification tasks, are likely to arise from emotional or at least stimulus-processing deficits rather than an impairment in correctly labeling emotions. Regarding implicit emotion processing, gender decision tasks (Johnston et al., 2005; Lepage et al., 2011; Williams et al., 2007), identity matching (Martin et al., 2005; Quintana et al., 2003), and age discrimination (R. E. Gur et al., 2002) have been used primarily in imaging studies, while behavioral studies have focused on the priming paradigm (Höschel & Irle, 2001; Rauch et al., 2010) or the Stroop task (Roux, Christophe, & Passerieux, 2010).

Considering behavioral studies, large differences in emotion processing between patients and controls are observed in explicit and implicit tasks. For instance, in an emotion-matching and also an identity-matching task with patients, Martin et al. (2005) reported a performance decrease that was especially severe if both emotion and identity varied. Interestingly, this suggests that patients had difficulty in processing only one stimulus dimension while ignoring the other. Furthermore, the performance deficit was larger in the emotion than in the identity task. A similar picture, namely a deficit in both tasks, albeit larger in an emotion-identification task, was also found by several other authors (e.g., Sachs, Steger-Wuchse, Kryspin-Exner, Gur, & Katschnig, 2004). While these results suggest a combination of general and emotion-specific impairments, other studies provide results pointing clearly in one or the other direction. Kerr and Neale (1993) reported no emotion-specific processing deficit, while Edwards, Pattison, Jackson, and Wales (2001) described a selective impairment in emotional processing. Evidence of both a general and a specific impairment, has been reported for auditory and visual stimuli (Kerr & Neale, 1993; Edwards et al., 2001).

While the pattern of results remains unclear for studies employing explicit tasks, a clearer picture emerges when considering implicit processing tasks. In an affective priming paradigm, Höschel and Irle (2001) demonstrated that both healthy controls and patients with schizophrenia show increased priming by negative stimuli. However, this effect was even more strongly increased in the patient group, suggesting enhanced automatic processing of (negative)

TABLE 1

Comparison of selected studies investigating emotion perception in schizophrenia.

Authors	Participants		Medication		Method	Stimuli	Emotions							Task	Results
	Patients	H. Controls	Medicated	Unmedicated			Anger	Fear	Sadness	Happiness	Disgust	Neutral	Other		
de Gelder et al. (2005)	13	13	13	0	Behav	AV			+	+				e	Weaker influence of voice but stronger influence of face
de Jong et al. (2009)	55[1]	50	52	0	Behav	AV		+	+	+				e	Weaker influence of face
de Jong et al. (2010)	55[1]	50	52	0	Behav	AV	+	+	+	+				e,i	Stronger integration when auditory distractor present, no impairment by visual distractor
Linden et al. (2010)	34	34	30	4	Behav	V	+			+		+		e,i	No difference for i task but impairment for e task
Suslow et al. (2005)	88[2]	30	88	0	Behav	V		+	+	+		+		e	Differences only for patients with affective symptoms
Tomlinson et al. (2006)	16	24	?	?	Behav	V[3]	+	+	+	+	+		+	e	Motion benefit for both groups, but smaller for patients
van den Stock et al. (in press)	31	21	28	3	Behav	V[4]	+	+	+					e	Deficits for perception of body expressions
Das et al. (2007)	16	16	16	0	Behav	AV[4]		+		+				e	Increased audiovisual integration
	14	14	10	4	fMRI	V	+	+				+		p	Reduced activation/differences in connectivity of AMG
Johnston et al. (2005)	11	15	11	0	fMRI/EEG	V	+	+	+	+	+	+	+	e, i	ERP: reduction of VPP and P3; fMRI: reduced activity of FFG
Salgado-Pineda et al. (2010)	14	14	14	0	fMRI	V	+	+	+	+				e	Reduced activation of AMG for sustained stimulation
An et al. (2003)	20[5]	20	15	11	EEG	V						+	+	i	Reduced P3
Turetsky et al. (2007)	16	16	10	6	EEG	V			+	+		+		e	Larger N170
Wynn et al. (2008)	26	27	26	0	EEG	V	+	+	+	+		+	+	e, i	Reduced N250

Notes: H. Controls: healthy controls; Behav: behavioral; V: visual; A: auditory (prosody); e: explicit; i: implicit; p: passive viewing; FFG: fusiform gyrus; AMG: amygdala. 1: 53 paranoid, 2 residual; 2: 30 schizophrenic patients with flat affect, 30 schizophrenic patients with anhedonia, 28 schizophrenic patients without affective symptoms; 3: point-light faces; 4: body language; 5: 16 paranoid, 4 undifferentiated.

emotional information. In contrast, other studies did not find any difference in implicit emotion processing (Demily et al., 2010; Linden et al., 2010; Roux et al., 2010), or found an impairment only for patients suffering from negative affective symptoms (Suslow, Droste, Roestel, & Arolt, 2005).

In sum, behavioral studies yield a rather heterogeneous picture regarding a possible emotion-processing deficit in schizophrenia. The fact that implicit emotion processing seems less affected than explicit processing may suggest that pre-attentive processing is relatively spared. The reported deficits in explicit tasks may arise from rather late evaluative processing. Furthermore, the ability to correctly discriminate emotional facial expressions under explicit conditions correlates with general cognitive performance in individuals with schizophrenia (Kohler, Bilker, Hagendoorn, Gur, & Gur, 2000; Sachs et al., 2004), suggesting a link between the performance in emotion discrimination and other tasks.

In order to shed light on the underlying mechanisms and potential deficits of emotion perception, numerous neuroimaging studies have been conducted in recent years. Overall, the results suggest deficits at multiple levels. Several studies have shown that patients with schizophrenia show a decrease in activation in the basal-limbic system, including the amygdala (R. E. Gur et al., 2002; Hempel et al., 2003; Johnston et al., 2005; Leitman et al., 2007; Schneider et al., 1998; Williams et al., 2007), the hippocampus (R. E. Gur et al., 2002; Hempel et al., 2003), the anterior cingulate cortex (Hempel et al., 2003; Williams et al., 2007), and the medial prefrontal cortex (MPFC) (Das et al., 2007; Williams et al., 2007). Based on these findings, Williams and colleagues suggest a breakdown in the amygdala–MPFC connection to be related to the observed deficits in the processing of emotional, especially of fearful, stimuli. As abnormalities in functional integration of these regions were observed irrespective of whether fearful stimuli were processed consciously or subconsciously, this suggests deficits in early, fast processing as well as disturbed frontal control mechanisms affecting late evaluative processing.

An interesting differentiation between explicit and implicit tasks can be seen when considering a meta-analysis by Li, Chan, McAlonan, and Gong (2010), who report an overall decrease in activation particularly in the amygdala and in the fusiform gyrus in emotion perception. However, while changes in amygdala activation were observed irrespective of task, the fusiform gyrus was less activated only in explicit tasks. Similar results are reported in a recent study by Quintana et al. (2011), who report underactivation

in the fusiform gyrus only when attention is directed to emotional features of a stimulus. These findings further support the notion that at least two separate systems are impaired in emotion processing in patients with schizophrenia: a fast, pre-attentive system, involving the amygdala and its surrounding network, and, at least in visual emotion perception, an attention-modulated system, which also seems deficient but is only involved when participants have to consciously process facial features.

Some studies also report differences in sensory areas, such as the middle occipital gyrus (Johnston et al., 2005), in visual emotion perception, both in explicit and implicit tasks, and primary auditory areas for explicit emotion perception from voices (Leitman et al., 2007). These findings suggest a third aspect that may also affect emotion processing at later stages, namely deficits in an early encoding phase.

A differentiation between an early, purely sensory, and later, more evaluative processing stages is also in line with electrophysiological evidence specifically investigating the time course of visual emotion perception. In addition to an fMRI study, Johnston et al. (2005) conducted an ERP study with the same patient group and paradigm. They reported a reduced vertex positive potential (VPP) amplitude in patients, associated with deficits in the encoding of emotion information. Furthermore, they observed differences in the P3a, which they interpreted as being associated with the encoding deficits at earlier processing stages. Hence, these results support a more general view in which the encoding of facial features is disturbed, irrespective of specific emotional content. Similar results were observed by Turetsky et al. (2007), who reported N170 amplitude differences in an emotion-recognition task as indicative of deficient structural encoding, which also affects differences at the P3. Support for the hypothesis that early processing deficits are at least partly responsible for emotion-processing deficits comes from a study by Kee, Kern, and Green (1998), who found a strong correlation with deficits in visual (and auditory) emotion recognition and measures of neurocognitive function assessing early deficits in perceptual processing. Other studies, however, did not report any N170 differences, but only differences at later emotion processing stages, irrespective of task specification (An et al., 2003; Wynn, Lee, Horan, & Green, 2008).

Overall, behavioral and neuroimaging results suggest deficits in at least two different sub-processes of emotion processing. Explicit and implicit tasks lead to differences in performance and point to separate brain structures underlying the behaviorally observed deficits. Hence, they provide a valuable tool in assessing the interplay between the observed networks.

Schizophrenia: Dynamic versus static stimuli

Another potentially important factor influencing visual emotion perception concerns the role of dynamic information; what happens if stimuli are not static pictures but dynamic videos, a situation much more comparable to emotion perception in an everyday life? It has been reported that healthy controls show an improvement in emotion recognition for dynamic over static point-light displays (Atkinson, Dittrich, Gemmell, & Young, 2004). Furthermore, imaging studies have shown that static stimuli activate different brain regions than dynamic stimuli (Kilts et al., 2003). In particular, static stimuli activate areas associated with motor imagery, such as motor and parietal areas (Decety & Grèzes, 1999), suggesting that static stimuli require motion simulation to be processed completely. Dynamic stimuli therefore present an interesting case when we investigate emotion perception in patients with schizophrenia. On the one hand, we could expect a decrease in performance, as the processing of dynamic information is an additional processing load and information needs to be integrated correctly in order to be beneficial rather than detrimental. Indeed, Archer, Hay, and Young (1994) showed that patients performed worse in an emotion-identification task when stimuli were dynamic rather than static. On the other hand, one can argue that dynamic stimuli are more natural, utilize different brain networks, and lead to a processing benefit in healthy participants, and hence may also be beneficial to patients. Evidence in support of this view comes from a study by Tomlinson, Jones, Johnston, Meaden, and Wink (2006), who observed that patients with schizophrenia showed an overall worse performance in emotion identification, but nevertheless improved when dynamic stimuli were presented. In contrast to Archer et al. (1994), Tomlinson et al. (2006) used point-light stimuli and were thus able to investigate the perception of motion information selectively. Therefore, motion itself seems to provide a benefit to patients with schizophrenia, and the deficits observed by Archer and colleagues are likely to arise from a different stimulus aspect. Determining the interplay between different aspects, and establishing which aspects prove beneficial and which detrimental is particularly necessary when investigating early, sensory-driven emotion perception.

Schizophrenia and multimodal emotion perception

One aspect that needs investigation is the perception of emotional information from multiple sensory modalities at the same time. On the one hand, multimodal presentation of congruent emotional information in healthy controls usually leads to facilitated perception (de Gelder & Vroomen, 2000); on the other hand, it increases the processing demands in terms of sorting out which piece of information from one processing stream is associated with which piece from the other processing stream. As patients with schizophrenia have trouble integrating different sensory streams (Magnée, Oranje, van Engeland, Kahn, & Kemner, 2009), multimodal perception may therefore do more harm than good. Previous studies have shown that patients with schizophrenia seem to be able to accomplish audiovisual integration for very simple sensory input but show strong deficits in more complex, social situations, such as audiovisual speech perception (de Gelder, Vroomen, Annen, Masthof, & Hodiamont, 2003).

A few studies have investigated the multimodal perception of emotional information in patients with schizophrenia (de Jong, Hodiamont, & de Gelder, 2010; de Jong, Hodiamont, van den Stock, & de Gelder, 2009; de Gelder et al., 2005; van den Stock, de Jong, Hodiamont, & de Gelder, in press). Using photos of emotional facial expressions accompanied by matching or mismatching prosodic information, all these studies describe an anomalous integration between auditory and visual information, but no emotion-specific effects. Regarding the exact nature of this anomalous integration, the evidence provided by the studies is inconclusive. When patients are instructed to determine the emotion conveyed by a facial expression and ignore simultaneously presented auditory information, reduced integration compared to healthy controls is observed (de Gelder et al., 2005). When participants are instructed to attend to the auditory modality while ignoring the visual information, the reverse pattern is seen; patients with schizophrenia then show stronger integration than healthy controls (de Gelder et al., 2005). In the study by de Jong et al. (2009), a slightly modified paradigm was used; again participants were instructed to attend to the auditory modality while ignoring the visual information. This time, decreased integration was found. If, however, an auditory distractor was included in the paradigm, integration increased in patients while it decreased in healthy controls (de Jong et al., 2010). In a fourth study by the same group, increased integration of emotional body language and emotional vocalizations was observed in patients instructed to pay attention to visual information (van den Stock et al., in press).

Overall, the studies described here seem to show that there is an abnormality in the multimodal integration of emotional information, but its exact nature is still unclear. As no comparison to neutral stimuli was made, it is still an open question whether the observed effects are specific to emotions, or are deficits in integration per se.

Emotion perception in schizophrenia: Appraisal and outlook

Task, stimulus dynamics, and multimodality are three factors that are likely to influence emotion perception in patients with schizophrenia. The heterogeneous pattern of results in explicit tasks and the relatively better performance in implicit tasks suggest several underlying mechanisms that may be differentially impaired. If this observation of intact implicit processing can be confirmed, it would necessitate a new discussion about what exactly emotion discrimination tasks are measuring, and whether it can still be claimed that patients with schizophrenia are impaired in emotion perception per se. This assumption is further supported by brain-imaging results; at least two separate aspects, namely an early sensory and a late cognitive processing aspect, seem to be affected to different degrees. Further light can be cast on the interaction and impairment of these mechanisms by investigating dynamic and multimodal stimuli, two features that in healthy participants lead to facilitated processing. Movement information seems to result in a processing benefit, providing evidence of an intact integration of motion information, probably at early processing steps. Furthermore, multimodal perception seems to be affected. However, the exact extent and emotion specificity remain unclear.

In summary, it is of relevance for future studies to disentangle these different components in order to gain a clearer understanding of emotion-processing deficits in patients with schizophrenia.

DEPRESSION

Major depressive disorder (MDD) is one of the most prevalent neuropsychiatric disorders in Western society. For example, around one-sixth of the American population suffers from MDD at least once during life (Kessler et al., 2003). There is evidence that MDD leads to biased emotional processing, which is a maintaining factor of the disorder and may also be present in people at risk or in remission. Understanding how emotion is processed in MDD or MDD-prone individuals could thus be important in preventing and treating the disorder.

It should first be considered that depression severely alters cognitive functions. Problems in thinking and concentrating are among the diagnostic criteria for MDD, according to the *Diagnostic and Statistic Manual of Mental Disorders* (DSM-IV, American Psychiatric Association, 1994). MDD patients exhibit slowed information processing, as evidenced by reaction times (e.g., Leyman, de Raedt, Schacht, & Koster, 2007) and enhanced latencies in the ERP, such as the P300 (Vandoolaeghe, Hunsel, Nuyten, & Maes, 1998). MDD patients are also more interference prone than healthy controls, as most clearly manifested in increased error rates (Elliott et al., 1997). These deficits need to be taken into account when running experiments with this patient population, as they could considerably influence study results.

Generally, there are three views on emotion processing in depression:

(1) Negative potentiation posits that MDD patients are biased toward negatively valenced stimuli (e.g., Beck, 1967).
(2) Positive attenuation considers that processing of positively valenced information is impaired in MDD (e.g., Clark & Watson, 1991).
(3) General blunting of emotional processing irrespective of valence is also suggested (emotion context insensitivity; Rottenberg & Gotlib, 2004).

Importantly, points (1) and (2) do not exclude each other and can be present in combination, while (3) is in line with (2), but additionally posits blunted processing of negatively valenced stimuli. An overview of selected studies assessing emotion perception in MDD is provided in Table 2.

Depression: Explicit versus implicit tasks

A negativity bias in MDD is widely supported by studies applying explicit categorization of facial emotion, as patients tend to classify neutral or ambiguous facial expressions as sad or negative (Douglas & Porter, 2010; Gollan, Pane, McCloskey, & Coccaro, 2008; R. C. Gur et al., 1992; Hale, 1998; Hale, Jansen, Bouhuys, & van den Hoofdakker, 1998; Leppänen, Milders, Bell, Terriere, & Hietanen, 2004; Luck & Dowrick, 2004; Naranjo et al., 2010), and happy faces as neutral (R. C. Gur et al., 1992; Surguladze et al.,

TABLE 2

Comparison of selected studies testing emotion perception in major depression. Note that due to a very high number of facial emotion categorization studies in the literature, they are treated exclusively in the text. In the case of treatment studies, only results at baseline are reported

Authors	Participants		Medication		Method	Stimuli	Emotions							Task	Results
	Patients	Controls	Medicated	Unmedicated			anger	fear	sadness	happiness	disgust	neutral	other		
Fritzsche et al. (2010)	20	20	2	18	Behav	V			+	+		+		i	Attentional bias toward sad and away from happy faces
Goeleven et al. (2006)	20	20	?	?	Behav	V			+	+				e	Reduced negative priming of sad faces, indicating reduced inhibition
Gotlib, Kasch et al. (2004)	88	55	?	?	Behav	V	+		+	+		+		i	Attentional bias toward sad faces
Gotlib, Kransnoperova et al. (2004)	19	16	8	11	Behav	V	+		+	+		+		i	Attentional bias toward sad faces
Joormann et al. (2007)	26	19	16	10	Behav	V			+	+		+		i	Attentional bias toward sad and away from happy faces
Kan et al. (2004)	16	20	15	1	Behav	A	+	+	+	+	+		+	e	Tendency to categorize surprise as negative
Leyman et al. (2007)	20	20	16	4	Behav	V	+					+		i	Reduced attentional disengagement from angry faces
Luck et al. (2004)	49	30	0	49	Behav	A	+		+	+		+		e	More negative classifications
Naranjo et al. (2010)	23	23	23	0	Behav	A	+	+	+	+	+	+	+	e	Tendency to categorize neutral and surprise as negative
Uekermann et al. (2008)	29	29	12	17	Behav	A	+	+	+	+		+		e	Impaired for all subtests except for stimuli with congruent semantics and prosody; all categories except sadness affected
Almeida et al. (2010)	15	15	13	2	fMRI	V		+	+	+		+		e	No group differences
Dannlowski et al. (2008)	28	28	28	0	fMRI	V¹	+		+	+		+		p	No group differences; increased AMG response in risk allele carriers

Study					Method	Stim					Type	Findings
Frodl et al. (2009)	12	12	8	4	fMRI	V	+		+	+	e, i	Higher neural response, reduced with explicit task
Fu et al. (2004)	19	19	0	19	fMRI	V[2]	+		+	+	i	Stronger neural response, slower RT
Fu et al. (2007)	19	19	0	19	fMRI	V[2]		+	+	+	i	Range of neural response to different intensities reduced in limbic-subcortical and visual areas
Fu et al. (2008)	16	16	0	16	fMRI	V[2]	+		+	+	i	Higher overall AMG-hippocampus activity, stronger reaction to intensity increase
Gotlib et al. (2005)	18	18	9	9	fMRI	V	+		+	+	i	Stronger neural response to sad and happy
Lawrence et al. (2004)	9	11	9	0	fMRI	V[2]	+	+	+	+	i	Weaker neural response to all faces, subcortically and frontally
Lee et al. (2008)	21	15	10	11	fMRI	V	+		+	+	p	Weaker neural response, subcortically and frontally
Sheline et al. (2001)	11	11	0	11	fMRI	V[1]		+		+	p	Greater AMG activation to emotional faces
Surguladze et al. (2005)	16	14	16	0	fMRI	V[2]	+		+	+	i	Happy: reduced activation in FFG and putamen; sad: increased activation in FFG, putamen, parahippocampal gyrus/AMG
Suslow et al. (2010)	30	26	30	0	fMRI	V[1]	+		+	+	e	Stronger AMG reaction to sad primes; reduced to happy
van Wingen et al. (2010)	18	30	0	18	fMRI	V	+	+	+		e[3]	Increased activation in AMG, IFG, and ACC to labeling task
Victor et al. (2010)	22	25	0	22	fMRI	V[1]	+		+	+	i	Stronger AMG reaction to sad primes; reduced to happy; faster RT to sad primes
Chang et al. (2010)	15	15	0	15	EEG	V	+		+	+	i	No difference between neutral and emotional stimuli from around 220 ms after onset
Dai et al. (2011)	19	20	?	?	EEG/Behav	V	+		+	+	e	Enhanced P1 to sad face preceded by sad target; no negative priming for sad faces
Deldin et al. (2001)	19	15	14	5	EEG	V	+		+	+	i	Abolished P300 difference between previously seen sad and happy faces
Deveney & Deldin (2004)	17	17	6	11	EEG	V	+		+	+	i	No slow wave amplitude reduction for sad faces
Foti et al. (2010)	19	25	0	19	EEG	V	+	+	+	+	p	No difference of emotional versus neutral faces in late positive potential

Notes: H. Controls: healthy controls; Behav: behavioral; V: visual; A: auditory (prosody); e: explicit; i: implicit; p: passive viewing; ACC: anterior cingulate cortex; IFG: inferior frontal gyrus; FFG: fusiform gyrus; AMG: amygdala; RT: reaction time. 1: subliminally presented, masked faces; 2: morphed to different emotional intensities; 3: two explicit tasks: matching of facial expressions, matching of facial expression to verbal label.

2004). In the latter study by Surguladze et al. (2004), this was limited to happy expressions of 50% intensity, and only at longer stimulus durations (2000 ms instead of 100 ms). Thus, biases in MDD may benefit from long presentation times and non-prototypical emotion expressions providing more scope for an individual's own interpretations. This speaks in favor of biases at later, evaluative processing stages. This evaluative bias fits with the finding that morphed sad faces need less intensity to be categorized as sad by MDD patients, while the intensity of happy facial expressions needs to be higher in order to be categorized as happy (Joormann & Gotlib, 2006). Even healthy subjects have been shown to perceive more sadness or rejection in ambiguous line drawings of faces after depressive mood induction (Bouhuys, Bloem, & Groothuis, 1995). Thus, evaluative biases can also occur in the absence of MDD; mood itself may play an essential role. A positivity bias in controls not present in patients has also been reported, reflected by an increased tendency of controls to classify neutral faces as happy (Douglas & Porter, 2010; Gollan et al., 2008; Surguladze et al., 2004).

Results from negative priming, a method in which a former distractor becomes task-relevant in a subsequent trial, revealed deficient inhibition in MDD patients when the distractor was a sad face (Dai, Feng, & Koster, 2011; Goeleven, de Raedt, Baert, & Koster, 2006). This suggests that patients allocated attention to the sad face even though it had to be ignored, while the happy face had to be categorized.

The negative bias in MDD previously outlined for faces is also reported for the categorization of vocal stimuli (Kan, Mimura, Kamijima, & Kawamura, 2004; Luck & Dowrick, 2004; Naranjo et al., 2010). For example, MDD patients judge negatively valenced stimuli as more negative than healthy controls (Naranjo et al., 2010). Importantly, stimulus characteristics may play a pivotal role in MDD. Uekermann, Abdel-Hamid, Lehmkämper, Vollmoeller, and Daum (2008) tested emotional prosody categorization and found broad impairments in MDD patients for almost all prosodic stimuli (with neutral semantics and with mismatching semantics). However, no group effects were found when semantics matched the emotional tone conveyed in a sentence. Additionally, the participants' executive functions were related to task performance. Thus, convergent semantic information may largely reduce between-group differences. Reduced impact on executive functions and richer information availability could account for this effect. Likewise, in a study by Emerson, Harrison, and Everhart (1999) examining schoolboys, it was shown

that although all participants were better at emotional categorization when semantics and prosody of a sentence matched, the performance decline for mismatching sentences was much more pronounced in depressed than in healthy participants.

Preliminary ERP data from our laboratory on a group of elderly participants without clinically relevant depression symptoms indicate that increased depression scores affect the integration of emotional-prosodic and semantic information, as reflected in smaller differences in the ERP response to pure prosodic (unintelligible) and normal speech. Based on these results, one may argue that depressive symptoms hamper the integration of information channels, and that the benefit MDD patients showed for semantically and prosodically congruent stimuli in Uekermann et al. (2008) rather reflects richer information availability and/or the absence of distracting information than integration per se.

Taken together, the results on emotional prosody perception in depression confirm the presence of a negative evaluation bias and thus extend the findings from facial expressions to speech, although stimulus characteristics may play an additional role in these findings.

Apart from explicit tasks, implicit methods have also been used at the behavioral level, most notably the dot-probe task (described in the introduction). At face presentation times of 1000 ms, an attentional bias toward sad faces has been reported for MDD patients (Fritzsche et al., 2010; Gotlib, Kasch et al., 2004; Gotlib, Krasnoperova, Yue, & Joormann, 2004; Joormann & Gotlib, 2007), and this bias was preserved in remitted patients (Fritzsche et al., 2010; Joormann & Gotlib, 2007). Even in never-depressed daughters of depressed mothers, this effect was found after negative mood induction (Joormann, Talbot, & Gotlib, 2007; Kujawa et al., 2011). A related method, spatial cueing, revealed prolonged attending to angry versus neutral faces in MDD compared to controls, suggesting that the bias extends beyond expressions of sadness when participants are instructed to attend to the faces (Leyman et al., 2007). By contrast, in dot-probe studies including anger, the bias appears to be sadness-specific (Gotlib, Krasnoperova et al., 2004; Gotlib, Kasch et al., 2004); thus, instructions that direct attention may be significant. Some dot-probe studies also report a positivity bias in healthy controls not present in MDD or MDD-prone individuals (Fritzsche et al., 2010; Joormann & Gotlib, 2007; Joormann et al., 2007). These studies demonstrate that depression leads to preferred attention to sad, or more generally, negatively valenced faces and away from positive expressions at late processing stages, even when emotion is not task-relevant.

In short, both explicit and implicit behavioral studies revealed biased emotion processing in depression, and not only evaluative but also attentional biases are supported by the literature.

Electrophysiology has also yielded interesting results: Dai et al. (2011) reported abnormal ERP responses to sad faces in MDD already occurring around 100 ms after stimulus onset when the face was preceded by a sad target that had to be categorized. Thus, after judging a face as sad, this kind of stimulus is processed abnormally in MDD, as can be observed during early sensory processing. Importantly, sad faces preceded by a happy target did not provoke such early effects, speaking against an early sensory deficit or bias. In an electromyographic study (Sloan, Bradley, Dimoulas, & Lang, 2002), students who were dysphoric revealed reduced activity of the zygomaticus, a facial muscle which responds to positive stimuli, during the explicit categorization of happy faces. Instead, they showed more corrugator activity for happy faces, which is normally correlated with processing negative stimuli. In fact, positive and negative facial expressions were processed as if they were all negative in dysphorics, even though categorization performance was comparable to controls, and thus there was no evaluative bias. This study, which temporally covers the seconds range, is complemented by an ERP study with millisecond resolution, in which students who were depressed exhibited reduced processing especially of mildly happy faces, starting around 350 ms after face onset (Cavanagh & Geisler, 2006).

Taken together, results from electrophysiological studies using explicit tasks indicate that subclinical depression leads to reduced positive reactivity, and that sad stimuli are processed abnormally in MDD when attention has previously been drawn to them. The time course suggests that biases already emerge before evaluative decisions, but no early, sensory processing deficits are confirmed.

ERP studies applying implicit tasks also support altered processing of facial emotion in MDD. Deldin, Keller, Gergen, and Miller (2000) observed a generally attenuated N200 component at right-posterior electrode sites in MDD, an effect which was strongest in response to happy facial expressions. In another study, MDD patients did not differ in their P300 response to previously seen happy versus sad faces while controls did (Deldin, Keller, Gergen, & Miller, 2001). The mismatch negativity (MMN), a measure of pre-attentive change detection, was not evident in MDD patients from around 220 ms post-stimulus onset onward, as their ERP response to sad, happy, and neutral schematic faces did not differ. However, the MMN was present before this time window, as neutral faces initially elicited brain responses that differed from the two emotional categories (Chang, Xu, Shi, Zhang, & Zhao, 2010). This means that early perceptual encoding including the face-related N170 component may be largely intact, and that blunting takes effect at a stage in which emotional significance is supposed to be extracted. Foti, Olvet, Klein, and Hajcak (2010) reported a reduced differentiation of fearful and angry faces from neutral ones in the ERP in MDD at a late processing stage (late positive potential). This further supports the view of emotional blunting in MDD, although it may also reflect the fact that MDD patients are more likely to perceive neutral faces as negative. Blunting is also supported in a study by Kayser, Bruder, Tenke, Stewart, and Quitkin (2000), who presented unpleasant (wounded) and neutral (intact) faces and reported a lack of a differentiation of these two stimulus types in MDD in a comparable time window, suggesting generally reduced emotional reactivity. Deveney and Deldin (2004) assessed memory retention of emotional faces during several seconds (slow-wave). While this component was increasingly reduced for sad relative to neutral and happy faces in healthy participants, no differences emerged in the depressed group, suggesting that MDD patients fail to disengage from sad faces over time. This is nicely complemented by an fMRI study on memory retention of positively and negatively valenced pictures: Successful retention of the latter was associated with enhanced amygdala activation during encoding in MDD patients (Hamilton & Gotlib, 2008); thus, amygdala hyperactivity might play a role in the enhanced retention of negative stimuli.

To sum up, electrophysiological studies indicate that a few hundred milliseconds from the onset of a face, emotion perception is blunted in MDD, as reflected in reduced neural differentiation between positive, negative, and neutral faces. Negative potentiation, in contrast, seems to occur in the seconds range and may reflect a lack of attentional deployment from negative stimuli. This can be observed with explicit and implicit tasks.

In fMRI studies, there is some indication that explicit compared to implicit tasks may reduce group effects at the neural level. Studies using implicit tasks or passive viewing provide ample evidence of negative potentiation—that is, heightened neural responses in MDD compared to controls when individuals are presented with negative facial expressions (Frodl et al., 2009; Fu et al., 2004, 2008; Gotlib et al., 2005; Sheline et al., 2001; Surguladze et al., 2005; Victor, Furey, Fromm, Ohman, & Drevets, 2010, but see Lawrence et al., 2004; Lee et al., 2008). This extends to remitted patients (Neumeister et al., 2006) and individuals at genetic risk of MDD (Dannlowski et al., 2008; Hariri et al., 2005). Positive attenuation has also been corroborated by neuroimaging studies using implicit tasks (Fu et al., 2007; Lawrence et al., 2004; Surguladze et al., 2005; Victor et al., 2010).

In contrast, a study using emotional categorization reported no significant activation differences between MDD and control participants, for either positive or negative emotional faces (Almeida, Versace, Hassel, Kupfer, & Phillips, 2010). This fits well with Frodl et al. (2009), who observed that negative potentiation was greater when using a gender matching instead of an emotion matching task. It is also in line with a study by Monk et al. (2008) comparing passive viewing with different tasks in adolescents at risk of depression. While the data obtained from passive viewing support negative potentiation and positive attenuation, the data from the constrained attention conditions do not. The findings from Monk et al. also suggest that a challenging implicit task may diminish group differences. Thus, what has to be evaluated is probably not the explicit-implicit distinction but rather the potential cognitive load introduced by the task. Stronger frontal activations in the risk compared to the non-risk group when performing the tasks support this notion (Monk et al., 2008). Along these lines, an easy explicit task may also give rise to negative potentiation in MDD, compared to a more challenging one (van Wingen et al., 2011).

The amygdala, in particular, seems to play a central role in MDD, as most of the studies reporting negative potentiation find enhanced activations in this structure. Moreover, Dannlowski et al. (2007) observed that amygdala activation to sad and angry faces correlated positively with behaviorally manifested negative biases in MDD. Connectivity studies with depressed individuals have shown abnormal interactions of prefrontal regions with brain structures attributed to emotional processing. More specifically, connections between the amygdala and prefrontal cortex appear to be impaired in MDD (Dannlowski et al., 2009; Siegle, Thompson, Carter, Steinhauer, & Thase, 2007), a connection which has been implicated in successful emotion regulation (Banks, Eddy, Angstadt, Nathan, & Phan, 2007). Low cognitive load, as in easy implicit tasks, may exacerbate the impact of this disturbed circuit, enabling excessive amygdala activations in response to negative stimuli. Findings from the cognitive domain indicate that tasks which strongly bind attention and suppress rumination, a core feature of the disorder, have been associated with reduced between-group differences (for a review, see Gotlib & Joormann, 2010).

Depression: Dynamic versus static stimuli

Categorization studies using dynamic facial expressions have yielded promising results. Kan et al. (2004) tested videos of six emotions (happiness, surprise, anger, disgust, fear, and sadness) and found no group effects in recognition performance, a result which is in line with a recent study (Schaefer, Baumann, Rich, Luckenbaugh, & Zarate, 2010). One possible explanation for these results is that the information provided by dynamic stimuli is richer than from static displays and makes it easier for patients to recognize them, as discussed by the authors of both studies. Another explanation is that emotional information provided in moving stimuli that is continuously changing prevents MDD patients from focusing on a facial expression and starting to ruminate about it or project negativity onto it. This suggestion is in line with intact early processing of emotional cues and biases at later, more cognitive processing stages. In any case, these study results are promising and will have to be corroborated by further investigations.

However, as outlined in the previous section, impairments and negative biases have been reported in studies on emotional prosody categorization in MDD (Kan et al., 2004; Luck & Dowrick, 2004; Naranjo et al., 2010; Uekermann et al., 2008), and prosody is also dynamic in nature. In the realm of an apparent dissociation between dynamic speech and face stimuli, it must be considered that recognition performance is generally higher for emotions conveyed by faces than by voices (e.g., Kan et al., 2004). Likewise, cognitive impairment associated with MDD could potentiate challenging task demands in explicit tasks (i.e., prosody categorization), while congruent semantic information may reduce cognitive demands and decrease group differences (Emerson et al., 1999; Uekermann et al., 2008).

To sum up, dynamic facial expressions, as well as prosodic stimuli with matching semantics, have led to promising results in MDD. However, even in the absence of behavioral differences, there may still be alterations at the neural level (e.g., Sloan et al., 2002), as explicit categorization is not very informative about earlier processing steps. More research is needed to shed light on these open issues.

Depression and multimodal emotion perception

To our knowledge, so far there are no studies on multimodal emotion perception in MDD patients. There is, however, one ERP experiment by Campanella et al. (2010) that assesses a group of students displaying elevated but subclinical anxiety and depression scores. In an emotional target-detection oddball paradigm, they reported a reduced P300 amplitude when compared with students with low scores in these measures. Interestingly, this was the case only in the multimodal (prosodic speech cue and static facial expression) condition irrespective of emotion, and not in the unimodal conditions. Moreover, the P300 amplitude was negatively correlated with depression scores. This suggests that under multimodal input conditions, emotion-processing deficits may be more likely to take effect than under unimodal input. However, this is but one experiment in a subclinical population, and one can only speculate about multimodal emotion processing in MDD. Furthermore, in the experiment, a static facial expression was combined with a prosodic stimulus. Even though these were matched in terms of the emotion they expressed, the static-dynamic combination represents a mismatch, and mismatching information may affect depressive individuals more strongly than controls (Emerson et al., 1999; Uekermann et al., 2008). As long as multimodal information is congruent, MDD patients should benefit from it, at least when it comes to explicit recognition tasks, while neural correlates of multimodal stimuli still need to be elucidated in MDD.

Emotion perception in depression: Appraisal and outlook

While a considerable amount of work targeting facial emotion processing in MDD has accumulated in both implicit and explicit tasks, studies on emotional prosody or multimodal emotion displays are largely or completely missing to date. Furthermore, face stimuli have been presented statically rather than dynamically in the vast majority of studies.

As outlined in the introduction, emotion processing is not a unitary process but rather involves different processing steps in which different tasks are accomplished and different neural correlates may be engaged. Studies with MDD patients have helped to shape our understanding of how depression may influence emotion processing from faces. ERP studies indicate that processing alterations in MDD may

start around a few hundred milliseconds after stimulus onset, where there is evidence of general blunting, as neural response patterns fail to distinguish stimuli according to their emotional content (Chang et al., 2010; Foti et al., 2010). At around one second after face onset, patients have directed their attention to sad stimuli (Fritzsche et al., 2010; Gotlib, Krasnoperova, et al., 2004) and away from happy stimuli (Joormann & Gotlib, 2007; Joormann et al., 2007) even when faces are not task-relevant. This attentional focus then persists, as suggested by two ERP studies looking at late processing stages (Dai et al., 2011; Deveney & Deldin, 2004). When prompted to attend to the emotionality of a face, happy expressions may be processed as if they were negative (Sloan et al., 2002), a result which could explain biases evident in explicit categorization experiments. Thus, it seems that depression does not lead to preferential attention to negative or attention deployment from positive stimuli, but once negative stimuli have captured a patient's attention, the patient cannot disengage from them (Gotlib & Joormann, 2010). This may be especially the case when an experimental task is not challenging (Frodl et al., 2009) and does not provide sufficient distraction to suppress rumination.

Even though there is no convincing evidence so far that MDD biases early perceptual encoding of facial emotion, support for early alterations has been provided by three recent fMRI studies using subliminally presented, masked face primes, which revealed group-related activation differences in the amygdala (Sheline et al., 2001; Suslow et al., 2010; Victor et al., 2010). Another study reported similar results only when dividing the sample into genetic risk and non-risk groups rather than diagnostic groups (Dannlowski et al., 2008). The effects may correspond to the fast emotional face-processing pathway directly feeding into the amygdala (Adolphs, 2002; Vuilleumier & Pourtois, 2007). However, neutral faces were used to mask the emotional faces in these studies. Due to the low temporal resolution of fMRI, it cannot be ruled out that the activations reflect the processing of the neutral mask, influenced by emotional prime valence. Experiments using non-facial masks and the application of techniques providing higher temporal resolution, such as EEG, may be useful tools to further address this issue. ERP studies, despite their excellent temporal resolution, have so far failed to find alterations in early sensory processing.

In sum, at this point, we do not know whether depression affects early sensory processing. Evidence on later processing stages is much clearer, although the fact that the vast majority of studies are based on static face stimuli must be considered.

GENERAL DISCUSSION

Accurately perceiving others' emotions is an essential component of successful social interaction, a skill, which is impaired in many if not all psychiatric disorders. However, emotion perception is not a single-step process but rather a complex mechanism involving several interacting subcomponents.

As elaborated in the introduction, at least three such substeps can be distinguished: (1) sensory processing, (2) integration of sensory cues, and (3) evaluation of the perceived cues (Adolphs, 2002; Schirmer & Kotz, 2006). In principle, all of these steps are vulnerable to malfunction, raising the question of at which processing stage the observed deficits in a given neuropsychiatric disorder may occur.

Here, schizophrenia and depression are of particular interest. While a common finding in patients with schizophrenia is the presence of early sensory processing deficits, depression is characterized by deficits at later, cognitive processing stages. This dissociation in general deficits suggests that the processing of emotional information expressed in faces and voices may also result in specific processing deficits. Thus, while at first glance, emotion perception appears disturbed in both disorders, the underlying mechanisms leading to these alterations may be different. It is therefore conceivable that deficits arise in several processing steps, or from a different mechanism altogether, such as cognitive deficits affecting response behavior. Furthermore, it is open to discussion to what extent observed deficits are modality-specific or supramodal, and whether multimodal emotion perception can alleviate symptoms or, in contrast, lead to a decrease in performance. All these questions cannot be fully answered by the current state of the art in psychiatric research on emotion expressions, but the evidence reviewed here offers the possibility of some justified speculation, motivating further research.

Current studies suggest that emotion-processing deficits occur at various processing levels. For schizophrenia, in particular, studies show very early processing differences (Johnston et al., 2005; Leitman et al., 2007), suggesting that deficits already occur at the level of sensory encoding. Considering the well-known problems in sensory gating that are often described as one characteristic finding in patients with schizophrenia, this raises the question of to what extent these deficits are emotion-specific or rather a result of pathological early sensory encoding per se. In depression, evidence on deficits in early perceptual encoding is, to date, not convincing (e.g., Gotlib & Joormann, 2010).

Regarding later processing stages, emotional processing deficits have been reported for both patient populations (An et al., 2003; Dai et al., 2011; Deveney & Deldin, 2004; Wynn et al., 2008). While emotional valence appears to play a role in depression, as positively valenced material is attenuated and negatively valenced material is processed in an enhanced manner, no such distinction can be described for schizophrenia patients. These patterns suggest that the observed deficits in emotion processing arise from different underlying mechanisms. The observations regarding MDD speak in favor of an emotional bias, as processing differences are found at later stages and appear to be valence-driven. In schizophrenic patients, however, the picture is not that clear. While there is evidence for deficits at later evaluative steps, disturbances in early sensory processes are also commonly observed. On the one hand, this complex pattern may indicate that two separate processing mechanisms are affected. On the other hand, the late processing deficits could also be interpreted as a consequence of deficits at earlier stages. Thus, alterations in emotion processing per se seem more likely in depression than in schizophrenia.

To shed more light on emotional processing in these two disorders, it seems important to examine more closely several factors influencing emotion perception. Here, we focused on the three aspects—task, stimulus dynamics, and multimodality—as they seem to be especially relevant to emotion perception (e.g., Atkinson et al., 2004; Giard & Peronnet, 1999). Distinguishing between explicit and implicit tasks is important when investigating patient populations, as explicit task instructions may introduce differences between patient and healthy populations that are unrelated to emotional processing, but rather reflect cognitive deficits. In fact, schizophrenia patients generally perform worse in emotional as well as control tasks, suggesting a strong effect of cognitive factors (Kerr & Neale, 1993; Pomarol-Clotet et al., 2010). This is supported by the finding that difficulties are more pronounced in explicit than in implicit emotional processing tasks (Linden et al., 2010; Suslow et al., 2005). Patients with depression seem to be less affected by task settings and may even benefit from challenging tasks at the neural level (Frodl et al., 2009).

A second aspect to be taken into consideration is multisensory emotion perception. In healthy participants, providing congruent emotional information via several modalities usually leads to facilitated processing (Giard & Peronnet, 1999; Paulmann et al., 2009). Can similar benefits be observed in patient populations? Or do deficits rather increase, as multisensory stimuli are more complex than unisensory, thus requiring more elaborate processing? While investigating multisensory emotion perception may thus provide crucial insights into the relation between sensory and affective deficits, this dimension has hardly been investigated in patient populations. While no studies are known that have investigated multimodal emotion perception in MDD patients, a few studies have addressed this issue at the behavioral level in patients with schizophrenia (de Gelder et al., 2005; de Jong et al., 2009, 2010; van den Stock et al., in press). However, no clear picture results from these studies with respect to whether these patients can benefit from the multimodal perception of emotional information. For both disorders, it remains an open question whether multimodal emotional input leads to improved processing or introduces new problems.

A last aspect, also improving emotion perception in healthy controls, is the use of dynamic visual stimuli. Dynamic visual stimuli have higher ecological validity and provide more information than static pictures. In relation to the face stimuli used, it has been shown that, just like healthy controls, patient groups benefit from dynamic information (Atkinson et al., 2004; Kan et al., 2004; Schaefer et al., 2010; Tomlinson et al., 2006). Nevertheless, they are not widely used but should be considered for further research aiming at more naturalistic stimulus displays.

To conclude, emotion perception in patient populations is influenced by numerous factors, such as task specificity and stimulus dynamics. However, it seems that patients with schizophrenia and depression are affected differentially; while in schizophrenic patients, emotion-processing deficits occur at early as well as later stages in processing in patients with schizophrenia, MDD patients show impairments primarily at later evaluative stages. Acknowledging this fact could not only affect our understanding of a given disorder but also play a crucial role in the correct assessment and intervention.

REFERENCES

Adolphs, R. (2002). Recognizing emotion from facial expressions: Psychological and neurological mechanisms. *Behavioral and Cognitive Neuroscience Reviews*, *1*(1), 21–62.

Adolphs, R., Schul, R., & Tranel, D. (1998). Intact recognition of facial emotion in Parkinson's disease. *Neuropsychology*, *12*(2), 253–258.

Almeida, J. R. C., Versace, A., Hassel, S., Kupfer, D., & Phillips, M. (2010). Elevated amygdala activity to sad facial expressions: A state marker of bipolar but not unipolar depression. *Biological Psychiatry*, *67*(5), 414–421.

Ambadar, Z., Schooler, J. W., & Cohn, J. F. (2005). Deciphering the enigmatic face. *Psychological Science*, *16*(5), 403–410.

An, S. K., Lee, S. J., Lee, C. H., Cho, H. S., Lee, P. G., Lee, C., et al. (2003). Reduced P3 amplitudes by negative facial emotional photographs in schizophrenia. *Schizophrenia Research*, *64*(2–3), 125–135.

Archer, J., Hay, D. C., & Young, A. W. (1994). Movement, face processing and schizophrenia: Evidence of a differential deficit in expression analysis. *British Journal of Clinical Psychology*, *33*(Pt 4), 517–528.

Atkinson, A. P., Dittrich, W. H., Gemmell, A. J., & Young, A. W. (2004). Emotion perception from dynamic and static body expressions in point-light and full-light displays. *Perception*, *33*(6), 717–746.

Bach, D. R., Herdener, M., Grandjean, D., Sander, D., Seifritz, E., & Strik, W. K. (2009). Altered lateralisation of emotional prosody processing in schizophrenia. *Schizophrenia Research*, *110*(1–3), 180–187.

Banks, S. J., Eddy, K. T., Angstadt, M., Nathan, P. J., & Phan, K. L. (2007). Amygdala–frontal connectivity during emotion regulation. *Social Cognitive and Affective Neuroscience*, *2*(4), 303–312.

Beck, A. (1967). *Depression: Clinical, experimental, and theoretical aspects*. New York, NY: Hoeber Medical Division, Harper & Row.

Bobes, M. A., Martin, M., Olivares, E., & Valdés-Sosa, M. (2000). Different scalp topography of brain potentials related to expression and identity matching of faces. *Cognitive Brain Research*, *9*(3), 249–260.

Bouhuys, A. L., Bloem, G. M., & Groothuis, T. G. G. (1995). Induction of depressed and elated mood by music influences the perception of facial emotional expressions in healthy subjects. *Journal of Affective Disorders*, *33*(4), 215–226.

Bould, E., & Morris, N. (2008). Role of motion signals in recognizing subtle facial expressions of emotion. *British Journal of Psychology*, *99*(Pt 2), 167–189.

Bramon, E., Rabe-Hesketh, S., Sham, P., Murray, R. M., & Frangou, S. (2004). Meta-analysis of the P300 and P50 waveforms in schizophrenia. *Schizophrenia Research*, *70*(2–3), 315–329.

Campanella, S., Bruyer, R., Froidbise, S., Rossignol, M., Joassin, F., Kornreich, C., et al. (2010). Is two better than one? A cross-modal oddball paradigm reveals greater sensitivity of the P300 to emotional face–voice associations. *Clinical Neurophysiology*, *121*(11), 1855–1862.

Cavanagh, J., & Geisler, M. W. (2006). Mood effects on the ERP processing of emotional intensity in faces: A P3 investigation with depressed students. *International Journal of Psychophysiology*, *60*(1), 27–33.

Chan, R. C. K., Li, H., Cheung, E. F. C., & Gong, Q.-Y. (2010). Impaired facial emotion perception in schizophrenia: A meta-analysis. *Psychiatry Research*, *178*(2), 381–390.

Chang, Y., Xu, J., Shi, N., Zhang, B., & Zhao, L. (2010). Dysfunction of processing task-irrelevant emotional faces in major depressive disorder patients revealed by expression-related visual MMN. *Neuroscience Letters*, *472*(1), 33–37.

Clark, L. A., & Watson, D. (1991). Tripartite model of anxiety and depression: Psychometric evidence and taxonomic implications. *Journal of Abnormal Psychology*, *100*(3), 316–336.

Costafreda, S. G., Brammer, M. J., David, A. S., & Fu, C. H. Y. (2008). Predictors of amygdala activation during the processing of emotional stimuli: A meta-analysis of 385 PET and fMRI studies. *Brain Research Reviews, 58*(1), 57–70.

Critchley, H., Daly, E., Phillips, M., Brammer, M., Bullmore, E., Williams, S., et al. (2000). Explicit and implicit neural mechanisms for processing of social information from facial expressions: A functional magnetic resonance imaging study. *Human Brain Mapping, 9*(2), 93–105.

Dai, Q., Feng, Z., & Koster, E. H. W. (2011). Deficient distracter inhibition and enhanced facilitation for emotional stimuli in depression: An ERP study. *International Journal of Psychophysiology, 79*(2), 249–258.

Dannlowski, U., Ohrmann, P., Bauer, J., Deckert, J., Hohoff, C., Kugel, H., et al. (2008). 5-HTTLPR biases amygdala activity in response to masked facial expressions in major depression. *Neuropsychopharmacology, 33*(2), 418–424.

Dannlowski, U., Ohrmann, P., Bauer, J., Kugel, H., Arolt, V., Heindel, W., et al. (2007). Amygdala reactivity to masked negative faces is associated with automatic judgmental bias in major depression: A 3 T fMRI study. *Journal of Psychiatry & Neuroscience, 32*(6), 423–429.

Dannlowski, U., Ohrmann, P., Konrad, C., Domschke, K., Bauer, J., Kugel, H., et al. (2009). Reduced amygdala–prefrontal coupling in major depression: Association with MAOA genotype and illness severity. *International Journal of Neuropsychopharmacology, 12*(1), 11–22.

Das, P., Kemp, A. H., Flynn, G., Harris, A. W. F., Liddell, B. J., Whitford, T. J., et al. (2007). Functional disconnections in the direct and indirect amygdala pathways for fear processing in schizophrenia. *Schizophrenia Research, 90*(1–3), 284–294.

Decety, J., & Grèzes, J. (1999). Neural mechanisms subserving the perception of human actions. *Trends in Cognitive Sciences, 3*(5), 172–178.

de Gelder, B., & Vroomen, J. (2000). The perception of emotion by ear and by eye. *Cognition and Emotion, 14*(3), 289–311.

de Gelder, B., Vroomen, J., Annen, L., Masthof, E., & Hodiamont, P. (2003). Audio-visual integration in schizophrenia. *Schizophrenia Research, 59*(2–3), 211–218.

de Gelder, B., Vroomen, J., de Jong, J. J., Masthoff, E. D., Trompenaars, F. J., & Hodiamont, P. (2005). Multisensory integration of emotional faces and voices in schizophrenics. *Schizophrenia Research, 72*(2–3), 195–203.

de Gelder, B., Vroomen, J., & Pourtois, G. (2004). Multisensory perception of emotion, its time course, and its neural basis. In G. Calvert, C. Spence, & B. Stein (Eds.), *The handbook of multisensory processes* (pp. 581–596). Cambridge, MA: MIT Press.

de Jong, J. J., Hodiamont, P. P. G., & de Gelder, B. (2010). Modality-specific attention and multisensory integration of emotions in schizophrenia: Reduced regulatory effects. *Schizophrenia Research, 122*(1–3), 136–143.

de Jong, J. J., Hodiamont, P. P. G., van den Stock, J., & de Gelder, B. (2009). Audiovisual emotion recognition in schizophrenia: Reduced integration of facial and vocal affect. *Schizophrenia Research, 107*(2–3), 286–293.

Deldin, P. J., Keller, J., Gergen, J. A., & Miller, G. A. (2000). Right-posterior face processing anomaly in depression. *Journal of Abnormal Psychology, 109*(1), 116–121.

Deldin, P. J., Keller, J., Gergen, J. A., & Miller, G. A. (2001). Cognitive bias and emotion in neuropsychological models of depression. *Cognition & Emotion, 15*(6), 787–802.

Demily, C., Attala, N., Fouldrin, G., Czernecki, V., Ménard, J.-F., Lamy, S., et al. (2010). The emotional Stroop task: A comparison between schizophrenic subjects and controls. *European Psychiatry, 25*(2), 75–79.

Deveney, C. M., & Deldin, P. J. (2004). Memory of faces: A slow wave ERP study of major depression. *Emotion, 4*(3), 295–304.

Douglas, K. M., & Porter, R. J. (2010). Recognition of disgusted facial expressions in severe depression. *British Journal of Psychiatry, 197*(2), 156–157.

Edwards, J., Pattison, P. E., Jackson, H. J., & Wales, R. J. (2001). Facial affect and affective prosody recognition in first-episode schizophrenia. *Schizophrenia Research, 48*(2–3), 235–253.

Elliott, R., Baker, S. C., Rogers, R. D., O'Leary, D. A., Paykel, E. S., Frith, C. D., et al. (1997). Prefrontal dysfunction in depressed patients performing a complex planning task: A study using positron emission tomography. *Psychological Medicine, 27*(4), 931–942.

Emerson, C. S., Harrison, D. W., & Everhart, D. E. (1999). Investigation of receptive affective prosodic ability in school-aged boys with and without depression. *Neuropsychiatry, Neuropsychology, and Behavioral Neurology, 12*(2), 102–109.

Ethofer, T., Kreifelts, B., Wiethoff, S., Wolf, J., Grodd, W., Vuilleumier, P., et al. (2009). Differential influences of emotion, task, and novelty on brain regions underlying the processing of speech melody. *Journal of Cognitive Neuroscience, 21*(7), 1255–1268.

Ethofer, T., Pourtois, G., & Wildgruber, D. (2006). Investigating audiovisual integration of emotional signals in the human brain. *Progress in Brain Research, 156,* 345–361.

Foti, D., Olvet, D. M., Klein, D. N., & Hajcak, G. (2010). Reduced electrocortical response to threatening faces in major depressive disorder. *Depression and Anxiety, 27*(9), 813–820.

Freedman, R., Adler, L. E., Gerhardt, G. A., Walkdo, M., Baker, N., Rose, G. M., et al. (1987). Neurobiological studies of sensory gating in schizophrenia. *Schizophrenia Bulletin, 13*(4), 669–679.

Fritzsche, A., Dahme, B., Gotlib, I. H., Joormann, J., Magnussen, H., Watz, H., et al. (2010). Specificity of cognitive biases in patients with current depression and remitted depression and in patients with asthma. *Psychological Medicine, 40*(5), 815–826.

Frodl, T., Scheuerecker, J., Albrecht, J., Kleemann, A., Müller-Schunk, S., Koutsouleris, N., et al. (2009). Neuronal correlates of emotional processing in patients with major depression. *World Journal of Biological Psychiatry, 10*(3), 202–208.

Fu, C. H. Y., Williams, S. C. R., Brammer, M. J., Suckling, J., Kim, J., Cleare, A. J., et al. (2007). Neural responses to happy facial expressions in major depression following antidepressant treatment. *American Journal of Psychiatry, 164*(4), 599–607.

Fu, C. H. Y., Williams, S. C. R., Cleare, A. J., Brammer, M. J., Walsh, N. D., Kim, J., et al. (2004). Attenuation of the neural response to sad faces in major depression by antidepressant treatment: A prospective, event-related functional magnetic resonance imaging study. *Archives of General Psychiatry, 61*(9), 877–889.

Fu, C. H. Y., Williams, S. C. R., Cleare, A. J., Scott, J., Mitterschiffthaler, M. T., Walsh, N. D., et al. (2008). Neural responses to sad facial expressions in major depression following cognitive behavioral therapy. *Biological Psychiatry, 64*(6), 505–512.

Fusar-Poli, P., Placentino, A., Carletti, F., Landi, P., Allen, P., Surguladze, S., et al. (2009). Functional atlas of emotional faces processing: A voxel-based meta-analysis of 105 functional magnetic resonance imaging studies. *Journal of Psychiatry & Neuroscience, 34*(6), 418–432.

Giard, M., & Peronnet, F. (1999). Auditory-visual integration during multimodal object recognition in humans: A behavioral and electrophysiological study. *Journal of Cognitive Neuroscience, 11*(5), 473–490.

Goeleven, E., de Raedt, R., Baert, S., & Koster, E. H. W. (2006). Deficient inhibition of emotional information in depression. *Journal of Affective Disorders, 93*(1–3), 149–157.

Gollan, J. K., Pane, H. T., McCloskey, M. S., & Coccaro, E. F. (2008). Identifying differences in biased affective information processing in major depression. *Psychiatry Research, 159*(1–2), 18–24.

Gorno-Tempini, M. L., Pradelli, S., Serafini, M., Pagnoni, G., Baraldi, P., Porro, C., et al. (2001). Explicit and incidental facial expression processing: An fMRI study. *NeuroImage, 14*(2), 465–473.

Gotlib, I. H., & Joormann, J. (2010). Cognition and depression: Current status and future directions. *Annual Review of Clinical Psychology, 6*, 285–312.

Gotlib, I. H., Kasch, K. L., Traill, S., Arnow, B. A., Johnson, S. L., & Joormann, J. (2004). Coherence and specificity of information-processing biases in depression and social phobia. *Journal of Abnormal Psychology, 113*(3), 386–398.

Gotlib, I. H., Krasnoperova, E., Yue, D. N., & Joormann, J. (2004). Attentional biases for negative interpersonal stimuli in clinical depression. *Journal of Abnormal Psychology, 113*(1), 127–135.

Gotlib, I. H., Sivers, H., Gabrieli, J. D. E., Whitfield-Gabrieli, S., Goldin, P., Minor, K. L., et al. (2005). Subgenual anterior cingulate activation to valenced emotional stimuli in major depression. *Neuroreport, 16*(16), 1731–1734.

Gur, R. C., Erwin, R. J., Gur, R. E., Zwil, A. S., Heimberg, C., & Kraemer, H. C. (1992). Facial emotion discrimination. II. Behavioral findings in depression. *Psychiatry Research, 42*(3), 241–251.

Gur, R. E., McGrath, C., Chan, R. M., Schroeder, L., Turner, T., Turetsky, B. I., et al. (2002). An fMRI study of facial emotion processing in patients with schizophrenia. *American Journal of Psychiatry, 159*(12), 1992–1999.

Gur, R. C., Schroeder, L., Turner, T., McGrath, C., Chan, R. M., Turetsky, B. I., et al. (2002). Brain activation during facial emotion processing. *NeuroImage, 16*(3), 651–662.

Hale, W. W. (1998). Judgment of facial expressions and depression persistence. *Psychiatry Research, 80*(3), 265–274.

Hale, W. W., Jansen, J. H. C., Bouhuys, A. L., & van den Hoofdakker, R. H. (1998). The judgment of facial expressions by depressed patients, their partners and controls. *Journal of Affective Disorders, 47*(1–3), 63–70.

Hamilton, J. P., & Gotlib, I. H. (2008). Neural substrates of increased memory sensitivity for negative stimuli in major depression. *Biological Psychiatry, 63*(12), 1155–1162.

Hariri, A., Drabant, E., Munoz, K., Kolachana, B., Mattay, V., Egan, M., et al. (2005). A susceptibility gene for affective disorders and the response of the human amygdala. *Archives of General Psychiatry, 62*(2), 146–152.

Hempel, A., Hempel, E., Schönknecht, P., Stippich, C., & Schröder, J. (2003). Impairment in basal limbic function in schizophrenia during affect recognition. *Psychiatry Research, 122*(2), 115–124.

Horstmann, G., & Ansorge, U. (2009). Visual search for facial expressions of emotions: A comparison of dynamic and static faces. *Emotion, 9*(1), 29–38.

Höschel, K., & Irle, E. (2001). Emotional priming of facial affect identification in schizophrenia. *Schizophrenia Bulletin, 27*(2), 317–327.

Imaizumi, S., Mori, K., Kiritani, S., Kawashima, R., Sugiura, M., Fukuda, H., et al. (1997). Vocal identification of speaker and emotion activates different brain regions. *Neuroreport, 8*(12), 2809–2812.

Johnston, P. J., Stojanov, W., Devir, H., & Schall, U. (2005). Functional MRI of facial emotion recognition deficits in schizophrenia and their electrophysiological correlates. *European Journal of Neuroscience, 22*(5), 1221–1232.

Joormann, J., & Gotlib, I. H. (2006). Is this happiness I see? Biases in the identification of emotional facial expressions in depression and social phobia. *Journal of Abnormal Psychology, 115*(4), 705–714.

Joormann, J., & Gotlib, I. H. (2007). Selective attention to emotional faces following recovery from depression. *Journal of Abnormal Psychology, 116*(1), 80–85.

Joormann, J., Talbot, L., & Gotlib, I. H. (2007). Biased processing of emotional information in girls at risk for depression. *Journal of Abnormal Psychology, 116*(1), 135–143.

Kan, Y., Mimura, M., Kamijima, K., & Kawamura, M. (2004). Recognition of emotion from moving facial and prosodic stimuli in depressed patients. *Journal of Neurology, Neurosurgery & Psychiatry, 75*(12), 1667–1671.

Kayser, J., Bruder, G. E., Tenke, C. E., Stewart, J. E., & Quitkin, F. M. (2000). Event-related potentials (ERPs) to hemifield presentations of emotional stimuli: Differences between depressed patients and healthy adults in P3 amplitude and asymmetry. *International Journal of Psychophysiology, 36*(3), 211–236.

Kee, K. S., Kern, R. S., & Green, M. F. (1998). Perception of emotion and neurocognitive functioning in schizophrenia: What's the link? *Psychiatry Research, 81*(1), 57–65.

Kerr, S. L., & Neale, J. M. (1993). Emotion perception in schizophrenia: Specific deficit or further evidence of generalized poor performance? *Journal of Abnormal Psychology, 102*(2), 312–318.

Kessler, R. C., Berglund, P., Demler, O., Jin, R., Koretz, D., Merikangas, K. R., et al. (2003). The epidemiology of major depressive disorder: Results from the National Comorbidity Survey Replication (NCS-R). *Journal of the American Medical Association, 289*(23), 3095–3105.

Kilts, C. D., Egan, G., Gideon, D. A., Ely, T. D., & Hoffman, J. M. (2003). Dissociable neural pathways are involved in the recognition of emotion in static and dynamic facial expressions. *NeuroImage, 18*(1), 156–168.

Kohler, C. G., Bilker, W., Hagendoorn, M., Gur, R. E., & Gur, R. C. (2000). Emotion recognition deficit in schizophrenia: Association with symptomatology and cognition. *Biological Psychiatry, 48*(2), 127–136.

Kohler, C. G., Walker, J. B., Martin, E. A., Healey, K. M., & Moberg, P. J. (2010). Facial emotion perception in schizophrenia: A meta-analytic review. *Schizophrenia Bulletin, 36*(5), 1009–1019.

Kotz, S. A., Meyer, M., & Paulmann, S. (2006). Lateralization of emotional prosody in the brain: An overview and synopsis on the impact of study design. *Progress in Brain Research, 156*, 285–294.

Kotz, S. A., & Paulmann, S. (2007). When emotional prosody and semantics dance cheek to cheek: ERP evidence. *Brain Research, 1151*, 107–118.

Kreifelts, B., Ethofer, T., Grodd, W., Erb, M., & Wildgruber, D. (2007). Audiovisual integration of emotional signals in voice and face: An event-related fMRI study. *NeuroImage, 37*(4), 1445–1456.

Kucharska-Pietura, K., David, A. S., Masiak, M., & Phillips, M. L. (2005). Perception of facial and vocal affect by people with schizophrenia in early and late stages of illness. *British Journal of Psychiatry, 187*, 523–528.

Kujawa, A. J., Torpey, D., Kim, J., Hajcak, G., Rose, S., Gotlib, I. H., et al. (2011). Attentional biases for emotional faces in young children of mothers with chronic or recurrent depression. *Journal of Abnormal Child Psychology, 39*(1), 125–135.

LaBar, K. S., Crupain, M. J., Voyvodic, J. T., & McCarthy, G. (2003). Dynamic perception of facial affect and identity in the human brain. *Cerebral Cortex, 13*(10), 1023–1033.

Lawrence, N. S., Williams, A. M., Surguladze, S., Giampietro, V., Brammer, M. J., Andrew, C., et al. (2004). Subcortical and ventral prefrontal cortical neural responses to facial expressions distinguish patients with bipolar disorder and major depression. *Biological Psychiatry, 55*(6), 578–587.

Lee, B. T., Seok, J. H., Lee, B. C., Cho, S. W., Yoon, B. J., Lee, K. U., et al. (2008). Neural correlates of affective processing in response to sad and angry facial stimuli in patients with major depressive disorder. *Progress in Neuro-Psychopharmacology and Biological Psychiatry, 32*(3), 778–785.

Leitman, D. I., Hoptman, M. J., Foxe, J. J., Saccente, E., Wylie, G. R., Nierenberg, J., et al. (2007). The neural substrates of impaired prosodic detection in schizophrenia and its sensorial antecedents. *American Journal of Psychiatry, 164*(3), 474–482.

Lepage, M., Sergerie, K., Benoit, A., Czechowska, Y., Dickie, E., & Armony, J. L. (2011). Emotional face processing and flat affect in schizophrenia: Functional and structural neural correlates. *Psychological Medicine, 41*, 1833–1844.

Leppänen, J. M., Milders, M., Bell, J. S., Terriere, E., & Hietanen, J. K. (2004). Depression biases the recognition of emotionally neutral faces. *Psychiatry Research, 128*(2), 123–133.

Leyman, L., de Raedt, R., Schacht, R., & Koster, E. H. W. (2007). Attentional biases for angry faces in unipolar depression. *Psychological Medicine, 37*(3), 393–402.

Li, H., Chan, R. C. K., McAlonan, G. M., & Gong, Q. Y. (2010). Facial emotion processing in schizophrenia: A meta-analysis of functional neuroimaging data. *Schizophrenia Bulletin, 36*(5), 1029–1039.

Linden, S. C., Jackson, M. C., Subramanian, L., Wolf, C., Green, P., Healy, D., et al. (2010). Emotion-cognition interactions in schizophrenia: Implicit and explicit effects of facial expression. *Neuropsychologia, 48*(4), 997–1002.

Luck, P., & Dowrick, C. F. (2004). Don't look at me in that tone of voice! Disturbances in the perception of emotion in facial expression and vocal intonation by depressed patients. *Primary Care and Mental Health, 2*(2), 99–106.

Magnée, M. J. C. M., Oranje, B., Engeland, H. van, Kahn, R. S., & Kemner, C. (2009). Cross-sensory gating in schizophrenia and autism spectrum disorder: EEG evidence for impaired brain connectivity? *Neuropsychologia, 47*(7), 1728–1732.

Martin, F., Baudouin, J.-Y., Tiberghien, G., & Franck, N. (2005). Processing emotional expression and facial identity in schizophrenia. *Psychiatry Research, 134*(1), 43–53.

Mayes, A. K., Pipingas, A., Silberstein, R. B., & Johnston, P. (2009). Steady state visually evoked potential correlates of static and dynamic emotional face processing. *Brain Topography, 22*(3), 145–157.

Monk, C. S., Klein, R. G., Telzer, E. H., Schroth, E. A., Mannuzza, S., Moulton, J. L., III, et al. (2008). Amygdala and nucleus accumbens activation to emotional facial expressions in children and adolescents at risk for major depression. *American Journal of Psychiatry, 165*(1), 90–98.

Münte, T. F., Brack, M., Grootheer, O., Wieringa, B. M., Matzke, M., & Johannes, S. (1998). Brain potentials reveal the timing of face identity and expression judgments. *Neuroscience Research, 30*(1), 25–34.

Naranjo, C., Kornreich, C., Campanella, S., Noël, X., Vandriette, Y., Gillain, B., et al. (2010). Major depression is associated with impaired processing of emotion in music as well as in facial and vocal stimuli. *Journal of Affective Disorders, 128*(3), 243–251.

Neumeister, A., Drevets, W. C., Belfer, I., Luckenbaugh, D. A., Henry, S., Bonne, O., et al. (2006). Effects of a α_{2c}--adrenoreceptor gene polymorphism on neural responses to facial expressions in depression. *Neuropsychopharmacology, 31*(8), 1750–1756.

O'Doherty, J., Rolls, E., & Kringelbach, M. (2004). Neuroimaging studies of crossmodal integration for emotion. In G. Calvert, C. Spence, & B. Stein (Eds.), *The handbook of multisensory processes* (pp. 563–579). Cambridge, MA: MIT Press.

Paulmann, S., Jessen, S., & Kotz, S. A. (2009). Investigating the multimodal nature of human communication: Insights from ERPs. *Journal of Psychophysiology, 23*(2), 63–76.

Pomarol-Clotet, E., Hynes, F., Ashwin, C., Bullmore, E. T., McKenna, P. J., & Laws, K. R. (2010). Facial emotion processing in schizophrenia: A non-specific neuropsychological deficit? *Psychological Medicine, 40*(6), 911–919.

Quintana, J., Lee, J., Marcus, M., Kee, K., Wong, T., & Yerevanian, A. (2011). Brain dysfunctions during facial discrimination in schizophrenia: Selective association to affect decoding. *Psychiatry Research: Neuroimaging, 191*, 44–50.

Quintana, J., Wong, T., Ortiz-Portillo, E., Marder, S. R., & Mazziotta, J. C. (2003). Right lateral fusiform gyrus dysfunction during facial information processing in schizophrenia. *Biological Psychiatry, 53*(12), 1099–1112.

Rauch, A. V., Reker, M., Ohrmann, P., Pedersen, A., Bauer, J., Dannlowski, U., et al. (2010). Increased amygdala activation during automatic processing of facial emotion in schizophrenia. *Psychiatry Research, 182*(3), 200–206.

Rottenberg, J., & Gotlib, I. (2004). Socioemotional functioning in depression. In M. Power (Ed.), *Mood disorders: A handbook of science and practice* (pp. 61–77). Chichester, UK: Wiley.

Roux, P., Christophe, A., & Passerieux, C. (2010). The emotional paradox: Dissociation between explicit and implicit processing of emotional prosody in schizophrenia. *Neuropsychologia, 48*(12), 3642–3649.

Sachs, G., Steger-Wuchse, D., Kryspin-Exner, I., Gur, R. C., & Katschnig, H. (2004). Facial recognition deficits and cognition in schizophrenia. *Schizophrenia Research, 68*(1), 27–35.

Salgado-Pineda, P., Fakra, E., Delaveau, P., Hariri, A. R., & Blin, O. (2010). Differential patterns of initial and sustained responses in amygdala and cortical regions to emotional stimuli in schizophrenia patients and healthy participants. *Journal of Psychiatry & Neuroscience, 35*(1), 41–48.

Sato, W., Kochiyama, T., Yoshikawa, S., Naito, E., & Matsumura, M. (2004). Enhanced neural activity in response to dynamic facial expressions of emotion: An fMRI study. *Cognitive Brain Research, 20*(1), 81–91.

Schaefer, K. L., Baumann, J., Rich, B. A., Luckenbaugh, D. A., & Zarate, C. A., Jr. (2010). Perception of facial emotion in adults with bipolar or unipolar depression and controls. *Journal of Psychiatric Research, 44*(16), 1229–1235.

Schirmer, A., & Kotz, S. A. (2006). Beyond the right hemisphere: Brain mechanisms mediating vocal emotional processing. *Trends in Cognitive Sciences, 10*(1), 24–30.

Schirmer, A., Kotz, S. A., & Friederici, A. D. (2002). Sex differentiates the role of emotional prosody during word processing. *Cognitive Brain Research, 14*(2), 228–233.

Schirmer, A., Kotz, S. A., & Friederici, A. D. (2005). On the role of attention for the processing of emotions in speech: Sex differences revisited. *Cognitive Brain Research, 24*(3), 442–452.

Schneider, F., Weiss, U., Kessler, C., Salloum, J. B., Posse, S., Grodd, W., et al. (1998). Differential amygdala activation in schizophrenia during sadness. *Schizophrenia Research, 34*(3), 133–142.

Sheline, Y. I., Barch, D. M., Donnelly, J. M., Ollinger, J. M., Snyder, A. Z., & Mintun, M. A. (2001). Increased amygdala response to masked emotional faces in depressed subjects resolves with antidepressant treatment: An fMRI study. *Biological Psychiatry, 50*(9), 651–658.

Siegle, G. J., Thompson, W., Carter, C. S., Steinhauer, S. R., & Thase, M. E. (2007). Increased amygdala and decreased dorsolateral prefrontal BOLD responses in unipolar depression: Related and independent features. *Biological Psychiatry, 61*(2), 198–209.

Sloan, D. M., Bradley, M. M., Dimoulas, E., & Lang, P. J. (2002). Looking at facial expressions: Dysphoria and facial EMG. *Biological Psychology, 60*(2–3), 79–90.

Surguladze, S., Brammer, M. J., Keedwell, P., Giampietro, V., Young, A. W., Travis, M. J., et al. (2005). A differential pattern of neural response toward sad versus happy facial expressions in major depressive disorder. *Biological Psychiatry, 57*(3), 201–209.

Surguladze, S., Young, A. W., Senior, C., Brébion, G., Travis, M. J., & Phillips, M. L. (2004). Recognition accuracy and response bias to happy and sad facial expressions in patients with major depression. *Neuropsychology, 18*(2), 212–218.

Suslow, T., Droste, T., Roestel, C., & Arolt, V. (2005). Automatic processing of facial emotion in schizophrenia with and without affective negative symptoms. *Cognitive Neuropsychiatry, 10*(1), 35–56.

Suslow, T., Konrad, C., Kugel, H., Rumstadt, D., Zwitserlood, P., Schöning, S., et al. (2010). Automatic mood-congruent amygdala responses to masked facial expressions in major depression. *Biological Psychiatry, 67*(2), 155–160.

Tomlinson, E. K., Jones, C. A., Johnston, R. A., Meaden, A., & Wink, B. (2006). Facial emotion recognition from moving and static point-light images in schizophrenia. *Schizophrenia Research, 85*(1–3), 96–105.

Trautmann, S. A., Fehr, T., & Herrmann, M. (2009). Emotions in motion: Dynamic compared to static facial expressions of disgust and happiness reveal more widespread emotion-specific activations. *Brain Research, 1284*, 100–115.

Turetsky, B. I., Kohler, C. G., Indersmitten, T., Bhati, M. T., Charbonnier, D., & Gur, R. C. (2007). Facial emotion recognition in schizophrenia: When and why does it go awry? *Schizophrenia Research, 94*(1–3), 253–263.

Uekermann, J., Abdel-Hamid, M., Lehmkämper, C., Vollmoeller, W., & Daum, I. (2008). Perception of affective prosody in major depression: A link to executive functions? *Journal of the International Neuropsychological Society, 14*(4), 552–561.

van den Stock, J., de Jong, J. J., Hodiamont, P., & de Gelder, B. (in press). Perceiving emotions from bodily expressions and multisensory integration of emotion cues in schizophrenia. *Social Neuroscience.*

Vandoolaeghe, E., van Hunsel, F., Nuyten, D., & Maes, M. (1998). Auditory event related potentials in major depression: Prolonged P300 latency and increased P200 amplitude. *Journal of Affective Disorders, 48*(2–3), 105–113.

van Wingen, G. A., van Eijndhoven, P., Tendolkar, I., Buitelaar, J., Verkes, R. J., & Fernández, G. (2011). Neural basis of emotion recognition deficits in first-episode major depression. *Psychological Medicine, 41*(7), 1397–1405.

Victor, T. A., Furey, M. L., Fromm, S. J., Ohman, A., & Drevets, W. C. (2010). Relationship between amygdala responses to masked faces and mood state and treatment in major depressive disorder. *Archives of General Psychiatry, 67*(11), 1128–1138.

Vuilleumier, P., & Pourtois, G. (2007). Distributed and interactive brain mechanisms during emotion face perception: Evidence from functional neuroimaging. *Neuropsychologia, 45*(1), 174–194.

Wambacq, I. J., Shea-Miller, K. J., & Abubakr, A. (2004). Non-voluntary and voluntary processing of emotional prosody: An event–related potentials study. *Neuroreport, 15*(3), 555–559.

Werner, S., & Noppeney, U. (2010). Superadditive responses in superior temporal sulcus predict audiovisual benefits in object categorization. *Cerebral Cortex, 20*(8), 1829–1842.

Wildgruber, D., Ackermann, H., Kreifelts, B., & Ethofer, T. (2006). Cerebral processing of linguistic and emotional prosody: fMRI studies. *Progress in Brain Research, 156*, 249–268.

Wildgruber, D., Hertrich, I., Riecker, A., Erb, M., Anders, S., Grodd, W., et al. (2004). Distinct frontal regions subserve evaluation of linguistic and emotional aspects of speech intonation. *Cerebral Cortex, 14*(12), 1384–1389.

Wildgruber, D., Riecker, A., Hertrich, I., Erb, M., Grodd, W., Ethofer, T., et al. (2005). Identification of emotional intonation evaluated by fMRI. *NeuroImage, 24*(4), 1233–1241.

Williams, L. M., Das, P., Liddell, B. J., Olivieri, G., Peduto, A. S., David, A. S., et al. (2007). Fronto-limbic and autonomic disjunctions to negative emotion distinguish schizophrenia subtypes. *Psychiatry Research, 155*(1), 29–44.

Winston, J. S., O'Doherty, J., & Dolan, R. J. (2003). Common and distinct neural responses during direct and incidental processing of multiple facial emotions. *NeuroImage, 20*(1), 84–97.

Wynn, J. K., Lee, J., Horan, W. P., & Green, M. F. (2008). Using event related potentials to explore stages of facial affect recognition deficits in schizophrenia. *Schizophrenia Bulletin, 34*(4), 679–687.

Perceiving emotions from bodily expressions and multisensory integration of emotion cues in schizophrenia

Jan Van den Stock[1,2], Sjakko J. de Jong[1,3], Paul P. G. Hodiamont[1,4], and Beatrice de Gelder[1,2,5]

[1]Cognitive and Affective Neuroscience Laboratory, Tilburg University, Tilburg, The Netherlands
[2]Department of Neuroscience, KU Leuven, Leuven, Belgium
[3]GGZ, Breburg Groep, Tilburg, The Netherlands
[4]Department of Developmental, Clinical and Cross-cultural Psychology, Tilburg University, Tilburg, The Netherlands
[5]Martinos Center for Biomedical Imaging, Massachusetts General Hospital and Harvard Medical School, Boston, MA, USA

Most studies investigating emotion recognition in schizophrenia have focused on facial expressions and neglected bodily and vocal expressions. Furthermore, little is known about affective multisensory integration in schizophrenia. In the first experiment, the authors investigated recognition of static, face-blurred, whole-body expressions (instrumental, angry, fearful, and sad) with a two-alternative, forced-choice, simultaneous matching task in a sample of schizophrenia patients, nonschizophrenic psychotic patients, and matched controls. In the second experiment, dynamic, face-blurred, whole-body expressions (fearful and happy) were presented simultaneously with either congruent or incongruent human or animal vocalizations to schizophrenia patients and controls. Participants were instructed to categorize the emotion expressed by the body and to ignore the auditory information. The results of Experiment 1 show an emotion recognition impairment in the schizophrenia group and to a lesser extent in the nonschizophrenic psychosis group, and this for all four expressions. The findings of Experiment 2 show that schizophrenia patients are more influenced by the auditory information than controls, but only when the auditory information consists of human vocalizations. This shows that schizophrenia patients are impaired in recognizing whole-body expressions, and they show abnormal affective multisensory integration of bimodal stimuli originating from the same source.

Keywords: Schizophrenia; Body; Emotion; Audiovisual.

An important aspect of normal social functioning consists of recognizing intentions and emotions displayed by others. Emotion recognition in schizophrenia is hard to assess due to limited tools and studies that have predominantly focused on facial expressions (e.g., Borod, Martin, Alpert, Brozgold, & Welkowitz, 1993; Feinberg, Rifkin, Schaffer, & Walker, 1986; Heimberg, Gur, Erwin, Shtasel, & Gur, 1992; Kee, Horan, Wynn, Mintz, & Green, 2006; Kohler et al., 2003; Pomarol-Clotet et al., 2010; Wolwer, Streit, Polzer, & Gaebel,

Correspondence should be addressed to: Beatrice de Gelder, Martinos Center for Biomedical Imaging, Massachusetts General Hospital, Room 417, Building 36, First Street, Charlestown, MA 02129, USA. E-mail: degelder@nmr.mgh.harvard.edu

We are grateful to all participants, especially the patients for their willingness to volunteer for the study. Research was funded by a Human Frontiers Science Program grant (RGP0054/2004-C), a European Commission grant (FP6-2005-NEST-Path Imp 043403-COBOL) to B.d.G., and a NWO grant (Dutch Science Foundation).

1996) (for a review, see Mandal, Pandey, & Prasad, 1998). The findings point to a deficit in recognition of negative emotions (Mandal et al., 1998), and this has been linked to the social dysfunctions observed in schizophrenia patients (Pinkham, Hopfinger, Ruparel, & Penn, 2008b). From that perspective, a facial expression recognition deficit is not very surprising, but an important issue is whether one can generalize from a deficit in recognition of facial expressions to difficulties in recognizing emotional signals conveyed by other common channels like the voice and the body. With the exception of a few isolated reports (Argyle, 1988; Sprengelmeyer et al., 1999), the literature on how body expressions are processed has only taken off in the last decade. One of the first basic research questions concerned whether observers can easily recognize different emotional states from body expressions alone. The available data indicate that this is clearly the case (Van den Stock, Righart, & de Gelder, 2007), and this is not surprising, considering the high frequency of interactions with conspecifics. Repeated exposure to body language, be it emotional or neutral, leads to perceptual expertise and tuning of the visual system. When we investigate emotional body language, comparing the results with what is known from facial expression research is almost inevitable, considering the many similarities between both stimulus categories. Bodies and faces both provide information on diverse dimensions such as identity, emotion, gender, and age (de Gelder et al., 2010). An interesting approach in comparing findings from face and body research might be to focus on the differences rather than on the similarities. At face value, at least two significant differences between faces and bodies appear.

Firstly, faces provide significantly more information about identity than bodies. Headless bodies reveal little information about personal identity, whereas faces alone are sufficient for identification. The fact that bodies contain little identity information is related to the fact that bodies are usually clothed. Clothing may conceal bodily features that are sufficient for identification. On the other hand, it has been shown that people can recognize friends by dynamic information provided by the body alone (Cutting & Kozlowski, 1977), underscoring the importance of dynamic information conveyed by body expressions. Secondly, the emphasis on the function of facial expressions lies in communication, whereas whole-body expressions also serve adaptive behavioral functions, like fight or flight.

Only a few studies have investigated the perception of whole-body expressions in clinical populations that have been shown to display facial expression recognition deficits, as in Huntington's disease, autism, and prosopagnosia (de Gelder, Van den Stock, de Diego Balaguer, & Bachoud-Levi, 2008; Hadjikhani et al., 2009; Tamietto, Geminiani, Genero, & de Gelder, 2007; Van den Stock, van de Riet, Righart, & de Gelder, 2008b). The results point to similar mechanisms for both categories. However, no data have been reported so far regarding whole-body expression perception in schizophrenia.

Another understudied area in affective neuroscience concerns how emotional information conveyed by different sensory channels is integrated (de Gelder & Van den Stock, 2011). The few studies so far have focused on the combined perception of face–voice pairs. The ability to decode emotional cues in prosody and facial expressions may have a common processing and/or representational substrate in the human brain (Borod et al., 2000; de Gelder & Bertelson, 2003; Pourtois, Debatisse, Despland, & de Gelder, 2002), facilitating processing and integration of these distinct but often calibrated sources of information. Judging the emotional state of a speaker is possible via facial or vocal cues (Banse & Scherer, 1996; Scherer, Banse, Wallbott, & Goldbeck, 1991) alone, but both judgment accuracy and speed seem to benefit from combining the modalities; for example, response accuracy increases and response speed decreases when a face is paired with a voice expressing the same emotion. This improvement of performance occurs even when participants are instructed to ignore the voice and rate only the face, suggesting that extracting affective information from a voice may be automatic and/or mandatory (de Gelder & Vroomen, 2000). The fact that prosodic and facial expressions of emotion frequently correlate suggests that the underlying cognitive mechanisms are highly sensitive to shared associations activated by cues in each channel (de Gelder, Bocker, Tuomainen, Hensen, & Vroomen, 1999; Massaro & Egan, 1996).

To assess how emotional judgments of the face are biased by prosody, Massaro and Egan (1996) presented computer-generated faces expressing a *happy, angry,* or *neutral* emotion accompanied by a word spoken in one of the three emotional tones. De Gelder and Vroomen (2000) presented photographs taken from the Ekman and Friesen (1976) series with facial expressions rendered emotionally ambiguous by "morphing" the expressions between *happy* and *sad* as the two endpoints. The emotional prosody tended to facilitate how accurately and quickly subjects rate an emotionally *congruent as compared to an incongruent* face. These findings indicate that the emotional value of prosody–face events is registered and somehow integrated during perceptual tasks, affecting behavioral responses according to the emotion congruity of the combined events. Moreover, these cross-modal influences appear to be resistant to increased attentional demands induced by a dual task, implying that combining the two forms of input may be mandatory (Vroomen, Driver, & de Gelder, 2001). Recent studies have only begun to unravel the mechanisms behind cross-modal influences on perception of whole-body expressions, and the results are compatible with what has been previously reported for facial expressions (Van den Stock, Grèzes, & de Gelder, 2008a; Van den Stock, Peretz, Grèzes, & de Gelder, 2009; Van den Stock et al., 2007). However, no evidence has been reported so far on cross-modal influences on whole-body perception in schizophrenia. We recently used affective face–voice combinations that were either congruent (for example, a happy face presented simultaneously with a happy vocal expression) or incongruent (for example, a happy face paired with a fearful vocal expression), and we asked schizophrenia patients to rate one of the two modalities (de Gelder et al., 2005; de Jong, Hodiamont, Van den Stock, & de Gelder, 2009). The results showed anomalous, cross-modal bias effects in the patient group. For example, when schizophrenia patients were instructed to categorize the emotion expressed in the voice, they were less influenced than the controls by the facial expression (de Jong et al., 2009).

In the present study, we focus on the perception of whole-body expressions. In Experiment 1, we investigated the recognition of emotional body language in a group of schizophrenia patients, nonschizophrenic psychotic patients, and normal controls in order to determine whether the emotion-recognition deficit previously reported for faces (for a review, see Mandal et al., 1998) extends to whole-body expressions. So far, little is known about recognition of emotional body language in schizophrenia, but in view of the behavioral and neuroanatomical similarities between perception of faces and bodies in normals (for reviews, see de Gelder, 2006; de Gelder et al., 2010; Peelen & Downing, 2007), we hypothesized that the patients would be impaired in recognizing negative whole-body expressions.

In Experiment 2, we investigated how schizophrenia perceive multisensory emotional events, consisting of realistic body language combined with affective auditory utterances. We presented video clips of emotional body language, of people engaged in a common activity in an everyday situation. In addition to adding human vocal expressions, we also combined the video clips with animal vocalizations in order to investigate the role of environmental sounds. As reported previously, it is important to control for task variables, as attention may shift across conditions and trials from face to voice especially in clinical populations (Bertelson & de Gelder, 2004; de Gelder & Bertelson, 2003).

EXPERIMENT 1: RECOGNITION OF STATIC BODY LANGUAGE

Methods and materials

Participants

Thirty-one schizophrenia, 23 patients with non-schizophrenic psychosis, and a group of 21 normal controls matched for gender, age, and socioeconomic status participated in the study. The nonschizophrenic psychosis group consisted of patients with schizophreniform disorder ($n = 2$), schizoaffective disorder ($n = 5$), bipolar I disorder with psychosis ($n = 4$), depressive disorder with psychosis ($n = 1$), delusional disorder ($n = 1$), psychotic disorder not otherwise specified ($n = 9$), and dysthymic disorder ($n = 1$). All but three patients in the nonschizophrenic psychosis group and all but three patients in the schizophrenia group received antipsychotic medication. Demographic data are shown in Table 1. There was no significant difference in age, $F(2, 72) = 0.620$; $p < .541$, or gender ratio, $\chi^2 \leq 2.13$; $p < .14$, between the three groups. Only patients meeting the criteria

TABLE 1
Demographic data

Demographic data	Schizophrenia	Nonschizophrenic psychosis	Normal
Experiment 1			
N	31	23	21
Age (mean range)	33.7 (21-52)	35.7 (20-54)	32.4 (21-58)
Gender	23 M/8 F	14 M/9 F	13 M/8 F
Dexterity	28 R/3 L	20 R/3 L	20 R/1 L
Experiment 2			
N	16	/	16
Age (mean range)	36.8 (22-53)	/	38.0 (22-53)
Gender	15 M/1 F	/	9 M/7 F
Dexterity	15 R/1 L	/	13 R/3 L

for schizophrenia and nonschizophrenic psychosis set by the DSM-IV (APA, 2000) were included. All patients were under treatment at the local hospital. Diagnosis was established with the Schedules for Clinical Assessment in Neuropsychiatry (SCAN, version 2.1), a standardized interview for diagnosing axis I disorders, conducted by a trained psychiatrist. Exclusion criteria for patients consisted of: organic or substance-induced psychosis, current substance abuse, serious somatic illness, relevant neurological illness, auditory and /or visual handicap and language problems. Control subjects with a psychiatric disorder, a brain dysfunction, or a genetic predisposition to schizophrenia were excluded from participation. All participants were paid for participation (€22).

Materials and procedure

Materials consisted of pictures from our own database of body expressions and instrumental actions (for details on stimulus construction, see de Gelder, Snyder, Greve, Gerard, & Hadjikhani, 2004; Hadjikhani & de Gelder, 2003). The faces of the bodies were blurred. In a pilot study, the pictures were presented one by one for 4 s with an interstimulus interval of 7 s. Twenty participants were instructed to categorize the pictures according to expressed emotion by indicating one of three (anger, fear, sadness) response alternatives. For the instrumental actions, the participants were instructed to categorize the displayed action (combing hair, drinking, pouring water in glass, opening door, talking on telephone, or putting on trousers). Only pictures that were correctly recognized above 85% were selected for the experiment.

The experiment consisted of two blocks: one with bodily expressions and one with bodily actions. We included instrumental whole-body actions, because these displays elicit action representation (Johnson-Frey et al., 2003) and are thus appropriate to use as controls to investigate emotional body expressions. The procedure was identical in each block. Materials for the experiment consisted of 30 emotional bodies and 24 instrumental actions. A stimulus consisted of the presentation of a target at the top of the screen that had to be matched with one of two simultaneously presented probes underneath (Figure 1). The three pictures in a stimulus were always three different identities, but from the same gender. The instructions stated that the participant was to select the probe that matched the action or emotion of the target. The position of the correct probe was counterbalanced. Participants responded by pressing the corresponding button, indicating their choice for the left or right probe. The

stimulus was presented until the participant responded. During the 1000-ms intertrial interval, a blank screen was shown. The instrumental action block consisted of 48 trials (6 actions × 2 genders × 4 exemplars), and the bodily expression block consisted of 36 trials (3 expressions × 2 genders × 6 exemplars). For every emotion, 12 trials were presented, half with male images, half with female images. For example, the six male anger trials consisted of three trials where sadness was a distracter and three trials where fear was a distracter.

Results

Results are displayed in Figure 1. We calculated for every condition and participant the mean accuracy and median reaction times (RT) of the correct trials. Both RT and accuracy data were submitted to a repeated-measures ANOVA with Expression (four levels: angry, fearful, sad, and instrumental) as within-subjects factor and Group (three levels: schizophrenia, nonschizophrenic psychosis, and control) as between-subjects factor. This revealed for the accuracy data a significant main effect of Expression, $F(3, 210) = 13.269$; $MSE = 0.022$, $p < .001$, and Group, $F(2, 70) = 6.234$; $MSE = 0.033$, $p < .003$. The Expression × Group interaction was not significant. Tukey post hoc tests on the main effect of Group showed a significant difference between the control group and the schizophrenia group (mean difference $= 0.093$, $p < .002$), and a marginally significant difference between the control group and the nonschizophrenic psychosis group (mean difference $= 0.067$, $p < .054$). To follow up the main effect of Expression, we performed Bonferroni-corrected paired-sample t-tests between every combination of expressions ($n = 6$). This showed significant differences between angry and fearful, $t(74) = 5.911$; $p < .001$; between instrumental and fearful, $t(74) = 6.818$; $p < .001$; and between instrumental and sad expressions, $t(74) = 4.303$; $p < .001$. The difference between fearful and sad expressions was marginally significant, $t(74) = 2.483$; $p < .015$.

The analysis of RT showed a significant main effect of Expression, $F(3, 210) = 8.762$; $MSE = 116.649$, $p < .003$. The main effect of Group and the Expression × Group interaction were not significant. Bonferroni-corrected paired-sample post hoc t-tests showed a significant difference between angry and instrumental, $t(74) = 3.481$; $p < .001$; between fearful and sad, $t(74) = 3.326$; $p < .001$; between fearful and instrumental, $t(74) = 5.142$; $p < .001$; and between sad and instrumental, $t(74) = 4.009$; $p < .001$, expressions.

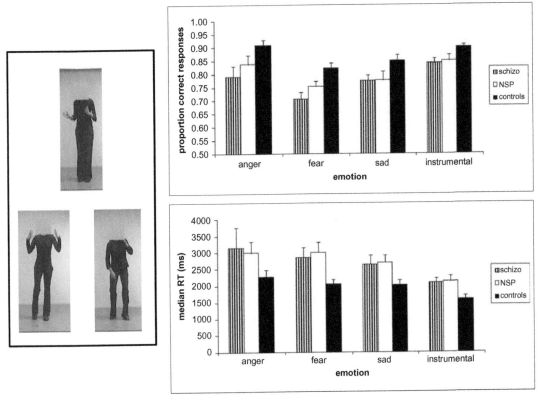

Figure 1. Left: example of stimulus in emotion block showing an angry body on top and bottom left and a fearful body on the bottom right. Right: Accuracy (top) and reaction time (bottom) of Experiment 1 as a function of expression and group (Schizo: schizophrenia group; NSP: nonschizophrenic psychosis group). Error bars represent 1 *SEM*.

Discussion

We presented patients with schizophrenia, patients with nonschizophrenic psychosis, and matched controls with a two-alternative, forced-choice, whole-body expression-matching task. The results show that, compared to the control group, the schizophrenia group exhibits a general impairment in recognizing emotional body language. The nonschizophrenic psychosis group occupies an intermediate position between the controls and schizophrenia group. It is unlikely that the observed effects can be explained by task difficulty, since the absence of an Expression × Group interaction reveals that the patients are not differentially impaired on recognition of specific emotions, while the main emotion effect indicates that not all emotions are equally well recognizable. Hence, the patients are not more impaired in recognizing the more difficult emotions.

The generalized whole-body emotion recognition deficit is consistent with findings from facial expression recognition studies (reviewed in Mandal et al., 1998). Schizophrenia patients are in general less able to make adequate emotional judgments of ambiguous facial expressions (Kee et al., 2006), and they attribute negative emotional valence to neutral face cues (Kohler et al., 2003). This may explain the nonspecific nature of emotion-recognition difficulties.

An important ability to adequately recognize facial expressions concerns the processing of the configuration of the face. It has been shown that this configurational perception mechanism is impaired in schizophrenia patients (Joshua & Rossell, 2009). Recent studies have reported similar configurational processing mechanisms for faces and bodies (Stekelenburg & de Gelder, 2004). The impaired recognition of whole-body expressions in schizophrenia may therefore have its roots in deficient configurational processing, and this may lead in social situations to inadequate interpretation of both facial and bodily expressions and ensuing social dysfunction.

EXPERIMENT 2: MULTISENSORY INTEGRATION OF DYNAMIC BODY LANGUAGE AND HUMAN AND ANIMAL VOCALIZATIONS

In everyday situations, fearful body language is usually accompanied by anxious screams. Recently, we showed that static whole-body expressions influence recognition of simultaneously presented vocal expressions (Van den Stock et al., 2007, experiment 3). In a follow-up study (Van den Stock et al., 2008a), we used dynamic stimuli in realistic situations to increase ecological validity, which may be an important factor in multisensory integration (Bertelson & de Gelder, 2004; de Gelder & Bertelson, 2003). We paired these visual stimuli with nonverbal vocalizations, and, more importantly, we also manipulated the nature of the bimodal combinations. We used auditory stimuli that were either produced by the same source as the visual stimuli (human vocalizations), or by a different source (animal vocalizations). The findings showed that both human and animal sounds influence recognition of dynamic body language. The second objective of the present study is to investigate the multisensory integration pattern of these everyday emotional events in schizophrenia.

Methods and materials

Participants

Sixteen schizophrenia meeting the criteria described in Experiment 1 and 16 matched controls participated in Experiment 2. All patients received antipsychotic medication. Demographic data are shown in Table 1. There was no significant difference of age between groups, $t(30) = 0.32$, $p < .751$. None of the participants of Experiment 2 participated in Experiment 1.

Materials and procedure

Visual stimuli. Video recordings were made of 12 semi-professional actors (6 women), coached by a professional director. They were instructed to approach a table, pick up a glass, drink from it, and put it back on the table. They performed this action once in a happy and once in a fearful manner. During the recording of the video clips, actors were provided with specific scenarios for every emotion; for example, the fearful scenario stated that the glass contained extremely hot water (see also Grèzes, Pichon, & de Gelder, 2007).

A fragment of 800 ms showing the actor grasping the glass was selected from each take. Facial expressions were blurred by motion-tracking software. In a pilot study, the 24 edited dynamic stimuli were presented 4 times to 14 participants. Participants were instructed to categorize as accurately and as fast as possible the emotion expressed by the actor (fear or happiness). The pilot session was preceded by eight familiarization trials. Sixteen stimuli were selected (2 genders × 4 actors × 2 emotions). Since we expected that recognition of the body language would improve when the body stimuli were combined with congruent auditory information, body stimuli that were recognized at ceiling were not selected. Mean recognition of the selected stimuli was 86.1% ($SD = 9.7$). A paired t-test between the fearful and happy body language showed no significant difference, $t(13) = 1.109$, $p < .287$.

Auditory stimuli. Audio recordings were made at a sampling rate of 44.1 kHz of 22 subjects (14 women), while they made nonverbal emotional vocalizations (fearful and happy). Specific scripts were provided for every target emotion. For example, for fear, the subjects were instructed to imagine that they were going to be attacked by a robber. Explicit instructions were given to refrain from pronouncing words. The most representative 800 ms from each recording was selected. In a pilot study, the sounds were presented 4 times to 15 participants in a randomized order. The participants were instructed to categorize as accurately and as fast as possible the emotion expressed by the voice (fear or happiness). The pilot session was preceded by three familiarization trials. From these results, eight fearful and eight happy sounds were selected. Mean recognition of the stimuli was 94.6% ($SD = 6.7$). A paired t-test between the fearful ($M = 96.1$) and happy ($M = 93.0$) vocalizations showed no significant difference, $t(14) = 0.474$, $p < .643$.

Environmental sounds of aggressive dog barking and joyful bird songs were downloaded from the Internet. Stimuli were selected on the basis of their emotion-inducing characteristics. The most representative 800-ms fragment of every sound was presented 4 times to 13 participants in a third pilot study. Instructions were to categorize as accurately and as fast as possible the emotion induced by the sound (fear or happiness). The session was preceded by three familiarization trials. Eight fear- and eight happiness-inducing sounds were selected. Mean recognition of the stimuli was 94.8% ($SD = 5.7$). A paired t-test between the fearful ($M = 97.5$) and happy ($M = 92.1$) vocalizations showed no significant difference, $t(12) = 1.469$, $p < .168$.

For each emotion, we compared the ratings of the animal vocalizations with those of the human vocalizations. Independent samples *t*-tests showed no differences between the pairs, $t(26) \leq 1.195$, $p < .243$. Experimental stimuli were then constructed with these visual and auditory materials. For this purpose, every video was paired once with a fearful and a happy human vocalization and once with a fearful and a happy animal vocalization, resulting in a total of 64 bimodal stimuli.

Procedure. The experiment consisted of an auditory (A), visual (V), and audiovisual (AV) block. In each block, all stimuli were presented twice in random order. The order of the blocks was counterbalanced. The AV block consisted of 128 trials, the V block of 32 trials, and the A block of 64 trials. A trial started with the presentation of a white fixation cross in the center of the screen against a dark background. The fixation cross had a variable duration to reduce temporal predictability (2000–3000 ms) and was followed by presentation of a stimulus (800 ms), after which a question mark appeared until the participant responded. In the AV and V blocks, participants were instructed to categorize the emotion expressed by the body in a two-alternative, forced-choice task by pressing the corresponding button (happy or fearful). Response buttons were counterbalanced across participants. Because we wanted to make sure participants saw the full length of the stimulus before they responded, they were instructed to respond only when the question mark appeared. In the A block, participants were presented with only auditory stimuli and instructed to categorize the emotion (happy or fearful).

Results

We excluded trials on which participants responded before the end of the stimulus (RT < 800 ms). On this basis, 64 trials (1.3%) were discarded. We computed the proportion of "happy" responses in the different conditions. Results are shown in Figure 2. The data with animal and human vocalizations were analyzed separately. Since the participants performed a delayed-RT task, no RT data were analyzed. A comparison between both groups on recognition of each of the four unimodal auditory conditions showed no significant difference, $t(31) < 1.850$, $p < .074$.

Human vocalizations

A repeated-measures ANOVA was performed on the proportion of "happy" responses, with Visual Emotion (two levels: fearful and happy) and (human) Auditory Emotion (three levels: fearful, happy, and no auditory stimulus) as within-subjects factors, and with Group (two levels: schizophrenia and control) as between-subjects factor. This revealed a main effect of Visual Emotion, $F(1, 31) = 124.154$, $MSE = 0.102$, $p < .001$; a main effect of Auditory Emotion, $F(1, 31) = 11.278$, $MSE = 0.035$, $p < .001$; and a significant two-way Auditory Emotion × Group interaction, $F(2, 62) = 3.310$, $MSE = 0.035$, $p < .043$. The Visual Emotion × Group interaction was marginally significant, $F(1, 31) = 3.937$, $MSE = 0.102$, $p < .056$.

The main effect of Visual Emotion indicates that the proportion of happy responses is higher for happy body language, as expected. The main effect of Auditory Emotion shows that the ratings of the bodily expressions are influenced by Auditory Emotion, while the Auditory Emotion × Group interaction indicates that this auditory influence is significantly different between the two groups. To follow up this interaction effect, we computed the influence of the auditory information for both groups separately, by calculating the ordinal difference between the unimodal and bimodal conditions: (fear video minus fear video paired with fear audio) + (fear video paired with happy audio minus fear video) + (happy video paired with happy audio minus happy video) + (happy video minus happy video paired with fearful audio). The resulting difference was higher for the schizophrenia group (0.46) than the control group (0.14), indicating that the schizophrenia are more influenced by the vocalizations than the controls.

We also compared the ratings of both unimodal conditions (fearful and happy body language) between both groups with independent samples *t*-tests. This showed no significant difference, $t(31) < 1.335$, $p < .192$, indicating that both patients and controls were equally able to recognize the unimodal whole-body expression videos.

Animal vocalizations

A repeated-measures ANOVA on the proportion of happy responses with Visual Emotion (fearful and happy) and (animal) Auditory Emotion (fearful, happy, and no auditory stimulus) as within-subjects factors, and Group (schizophrenia and control) as between-subjects factor, revealed a significant main effect of Visual Emotion, $F(1, 31) = 112.758$, $MSE = 0.115$, $p < .001$, and a significant Visual Emotion × Group interaction, $F(1, 31) = 4.456$, $MSE = 0.115$, $p < .043$. To follow up the interaction effect, we computed for both groups separately, the mean proportion of happy responses for the conditions with a happy

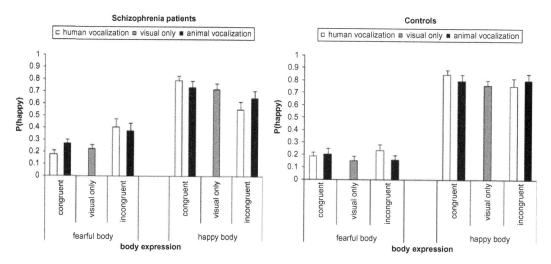

Figure 2. Proportion of "happy" responses in the bimodal and unimodal conditions, separated by group, emotion, auditory category, and congruence. Error bars represent 1 *SEM*.

video, regardless of (animal) auditory information, and we followed the same procedure for the fearful video conditions. The proportion of happy responses on the conditions with fear videos was significantly lower in the control group than in the schizophrenia group, $t(31) = 2.199$, $p < .035$), suggesting that the schizophrenia patients were less accurate in categorizing fearful videos. The difference between both groups in the conditions with happy videos was not significant.

Discussion

We presented schizophrenia patients and controls with videos of a person engaged in an everyday action (picking up a glass in a realistic situation). This action was performed in either a fearful or a happy manner. In the bimodal blocks, the videos were simultaneously presented with either a congruent or an incongruent vocal expression, which could be produced by a human or an animal. These stimuli were chosen to maximize ecological validity, which is an important factor in multisensory integration. It has been suggested that integrating information from multiple sensory channels leads to reduction of the ambiguity that is inherent in each single sensory channel (Bertelson & de Gelder, 2004; de Gelder & Bertelson, 2003). The results show that schizophrenia patients are more influenced by the task-irrelevant auditory information than the control group, but only for human and not for the animal vocalizations. The increased cross-modal bias of vocal expressions over body expressions may point to a greater impact of the auditory modality

under audiovisual perception conditions in schizophrenia patients. This explanation is also compatible with our previous study in which schizophrenia patients showed a reduced cross-modal bias of visual facial expression in the recognition of the emotion in a vocal expression (de Jong et al., 2009), and with a recent report from audiovisual speech perception in schizophrenia (Ross et al., 2007). But this does not explain why schizophrenia patients are more influenced by the human and not the animal vocalizations. The present results indicate that abnormal affective multisensory integration in schizophrenia is modulated by the association between the sources of different sensory channels.

Another possible explanation may be task difficulty. If schizophrenia patients have more difficulty in recognizing the visual stimulus, they might rely more on the information provided by the secondary stimulus. A direct test between both groups of the unimodal conditions reveals no significant difference, indicating that schizophrenia patients and controls perform equally in recognizing whole-body expressions as well as human and animal auditory vocalizations. This does not rule out the possibility, however, that they have more difficulty with audiovisual stimuli. The auditory information may be harder to ignore for the patients either because the focused-attention task requires them to shut out one input system, or because ignoring it is harder in the case of human sounds.

GENERAL DISCUSSION

In the first experiment, we tested recognition of static emotional body language in a group of schizophrenia patients, nonschizophrenic psychotic patients, and controls. The results show a general emotion recognition impairment in the schizophrenia group. The impairment is also present in the nonschizophrenic psychosis group, but to a lesser extent. The present study shows that the emotion-recognition difficulties in schizophrenia, which have been previously documented with studies using facial expressions (for a review, see Mandal et al., 1998), extend to the recognition of body language. This deficit is consistent with findings from neuroimaging studies, showing that brain structures involved in perceiving emotional body language (for reviews, see de Gelder, 2006; de Gelder et al., 2010) show abnormalities in schizophrenia. Perception of bodily expressions activates not only brain areas associated with emotion perception, but also areas involved in action representation (de Gelder et al., 2004; Grèzes et al., 2007; Pichon, de Gelder, & Grèzes, 2008), and both of these structures show abnormalities in schizophrenia (e.g., Bertrand et al., 2008; Gur et al., 2002; Michalopoulou et al., 2008; Phillips et al., 1999; Pinkham, Hopfinger, Pelphrey, Piven, & Penn, 2008a). Impaired recognition of body expressions in schizophrenia and nonschizophrenic psychosis might have its roots in a dysfunction of the brain network involved in emotion perception, but possibly also in a deficit of the brain areas involved in action representation.

Recently, it has been suggested that the motor abilities of the observer are an important aspect of body language recognition. A link has been suggested between disorders with movement deficits and anomalous recognition of bodily expressions (de Gelder, 2006; de Gelder et al., 2008). It is possible that the motor problems associated with schizophrenia, such as catatonia, play an important role in recognizing emotional body language. Static images contain less information than dynamic stimuli and require the brain to generate and fill in the movement information.

The second experiment focused on multisensory integration of dynamic emotional body language, on the one hand, and both human and animal vocalizations on the other hand. The data show an increased integration of both modalities in the schizophrenia group, but only when the auditory information consists of human voices. These findings are compatible with an auditory-dominance hypothesis in schizophrenia, as a previous study with face–voice combinations showed a reduced influence of the facial expression on recognition of the vocal expression in schizophrenia patients (de Jong et al., 2009; Ross et al., 2007). A critical evaluation of this hypothesis in future studies would include task manipulation and instructions to categorize the emotion expressed by the auditory stimulus.

At the neuroanatomical level, binding of emotional information in the face and voice has been associated with activity in the amygdala (Dolan, Morris, & de Gelder, 2001; Ethofer et al., 2006). Interestingly, abnormal amygdala activity has been reported in schizophrenia in response to facial expressions (e.g., Gur et al., 2002; Michalopoulou et al., 2008; Phillips et al., 1999), and it is therefore likely that the anomalous multisensory integration also results from abnormal amygdalar activity.

The patients show a general impairment in recognizing static emotional body expressions, whereas there is no significant difference between both groups in recognizing dynamic, ecologically valid, whole-body expressions. A possible explanation for this finding concerns the cognitive task demands. Recognizing isolated static expressions requires a higher flexibility in order to compensate for the lack of information, as in direction of movement and speed of movement. Future research is needed to identify the specific processes that are impaired in schizophrenia patients when recognizing affective stimuli, possibly in relation to ecological validity.

REFERENCES

APA (2000). *Diagnostic and statistical manual of mental disorders* (DSM-IV-TR). Washington, DC: American Psychiatric Association.

Argyle, M. (1988). *Bodily communication*. London: Methuen.

Banse, R., & Scherer, K. R. (1996). Acoustic profiles in vocal emotion expression. *Journal of Personality & Social Psychology*, *70*(3), 614–636.

Bertelson, P., & de Gelder, B. (2004). The psychology of multimodal perception. In C. Spence & J. Driver (Eds.), *Crossmodal space and crossmodal attention* (pp. 151–177). Oxford, UK: Oxford University Press.

Bertrand, M. C., Achim, A. M., Harvey, P. O., Sutton, H., Malla, A. K., & Lepage, M. (2008). Structural neural correlates of impairments in social cognition in first episode psychosis. *Social Neuroscience*, *3*(1), 79–88.

Borod, J. C., Martin, C. C., Alpert, M., Brozgold, A., & Welkowitz, J. (1993). Perception of facial emotion in schizophrenic and right brain-damaged patients. *Journal of Nervous and Mental Disease*, *181*(8), 494–502.

Borod, J. C., Pick, L. H., Hall, S., Sliwinski, M., Madigan, N., Obler, L. K., et al. (2000). Relationships among facial, prosodic, and lexical channels of emotional perceptual processing. *Cognition and Emotion*, *14*(2), 193–211.

Cutting, J. E., & Kozlowski, L. T. (1977). Recognizing friends by their walk: Gait perception without familiarity cues. *Bulletin of the Psychonomic Society*, *9*, 353–356.

de Gelder, B. (2006). Towards the neurobiology of emotional body language. *Nature Reviews Neuroscience, 7*(3), 242–249.

de Gelder, B., & Bertelson, P. (2003). Multisensory integration, perception and ecological validity. *Trends in Cognitive Sciences, 7*(10), 460–467.

de Gelder, B., Bocker, K. B., Tuomainen, J., Hensen, M., & Vroomen, J. (1999). The combined perception of emotion from voice and face: Early interaction revealed by human electric brain responses. *Neuroscience Letters, 260*(2), 133–136.

de Gelder, B., Snyder, J., Greve, D., Gerard, G., & Hadjikhani, N. (2004). Fear fosters flight: A mechanism for fear contagion when perceiving emotion expressed by a whole body. *Proceedings of the National Academy of Sciences of the United States of America, 101*(47), 16701–16706.

de Gelder, B., & Van den Stock, J. (2011). Real faces, real emotions: Perceiving facial expressions in naturalistic contexts of voices, bodies and scenes. In A. J. Calder, G. Rhodes, M. H. Johnson, & J. V. Haxby (Eds.), *The Oxford handbook of face perception* (pp. 535–550). New York, NY: Oxford University Press.

de Gelder, B., Van den Stock, J., de Diego Balaguer, R., & Bachoud-Levi, A. C. (2008). Huntington's disease impairs recognition of angry and instrumental body language. *Neuropsychologia, 46*(1), 369–373.

de Gelder, B., Van den Stock, J., Meeren, H. K., Sinke, C. B., Kret, M. E., & Tamietto, M. (2010). Standing up for the body. Recent progress in uncovering the networks involved in the perception of bodies and bodily expressions. *Neuroscience and Biobehavioral Reviews, 34*(4), 513–527.

de Gelder, B., & Vroomen, J. (2000). The perception of emotions by ear and by eye. *Cognition and Emotion, 14*(3), 289–311.

de Gelder, B., Vroomen, J., de Jong, S. J., Masthoff, E. D., Trompenaars, F. J., & Hodiamont, P. (2005). Multisensory integration of emotional faces and voices in schizophrenia. *Schizophrenics Research, 72*(2–3), 195–203.

de Jong, J. J., Hodiamont, P. P., Van den Stock, J., & de Gelder, B. (2009). Audiovisual emotion recognition in schizophrenia: Reduced integration of facial and vocal affect. *Schizophrenia Research, 107*(2–3), 286–293.

Dolan, R. J., Morris, J. S., & de Gelder, B. (2001). Crossmodal binding of fear in voice and face. *Proceedings of the National Academy of Sciences of the United States of America, 98*(17), 10006–10010.

Ekman, P., & Friesen, W. V. (1976). *Pictures of facial affect.* Palo Alto, CA: Consulting Psychologists Press.

Ethofer, T., Anders, S., Erb, M., Droll, C., Royen, L., Saur, R., et al. (2006). Impact of voice on emotional judgment of faces: An event-related fMRI study. *Human Brain Mapping, 27*(9), 707–714.

Feinberg, T. E., Rifkin, A., Schaffer, C., & Walker, E. (1986). Facial discrimination and emotional recognition in schizophrenia and affective disorders. *Archives of General Psychiatry, 43*(3), 276–279.

Grèzes, J., Pichon, S., & de Gelder, B. (2007). Perceiving fear in dynamic body expressions. *NeuroImage, 35*(2), 959–967.

Gur, R. E., McGrath, C., Chan, R. M., Schroeder, L., Turner, T., Turetsky, B. I., et al. (2002). An fMRI study of facial emotion processing in patients with schizophrenia. *American Journal of Psychiatry, 159*(12), 1992–1999.

Hadjikhani, N., & de Gelder, B. (2003). Seeing fearful body expressions activates the fusiform cortex and amygdala. *Current Biology, 13*(24), 2201–2205.

Hadjikhani, N., Joseph, R. M., Manoach, D. S., Naik, P., Snyder, J., Dominick, K., et al. (2009). Body expressions of emotion do not trigger fear contagion in autism spectrum disorder. *Social, Cognitive and Affective Neuroscience, 4*(1), 70–78.

Heimberg, C., Gur, R. E., Erwin, R. J., Shtasel, D. L., & Gur, R. C. (1992). Facial emotion discrimination. III. Behavioral findings in schizophrenia. *Psychiatry Research, 42*(3), 253–265.

Johnson-Frey, S. H., Maloof, F. R., Newman-Norlund, R., Farrer, C., Inati, S., & Grafton, S. T. (2003). Actions or hand–object interactions? Human inferior frontal cortex and action observation. *Neuron, 39*(6), 1053–1058.

Joshua, N., & Rossell, S. (2009). Configural face processing in schizophrenia. *Schizophrenia Research, 112*(1–3), 99–103.

Kee, K. S., Horan, W. P., Wynn, J. K., Mintz, J., & Green, M. F. (2006). An analysis of categorical perception of facial emotion in schizophrenia. *Schizophrenia Research, 87*(1–3), 228–237.

Kohler, C. G., Turner, T. H., Bilker, W. B., Brensinger, C. M., Siegel, S. J., Kanes, S. J., et al. (2003). Facial emotion recognition in schizophrenia: Intensity effects and error pattern. *American Journal of Psychiatry, 160*(10), 1768–1774.

Mandal, M. K., Pandey, R., & Prasad, A. B. (1998). Facial expressions of emotions and schizophrenia: A review. *Schizophrenia Bulletin, 24*(3), 399–412.

Massaro, D. W., & Egan, P. B. (1996). Perceiving affect from the voice and the face. *Psychonomic Bulletin and Review, 3*, 215–221.

Michalopoulou, P. G., Surguladze, S., Morley, L. A., Giampietro, V. P., Murray, R. M., & Shergill, S. S. (2008). Facial fear processing and psychotic symptoms in schizophrenia: Functional magnetic resonance imaging study. *British Journal of Psychiatry, 192*(3), 191–196.

Peelen, M. V., & Downing, P. E. (2007). The neural basis of visual body perception. *Nature Reviews Neuroscience, 8*(8), 636–648.

Phillips, M. L., Williams, L., Senior, C., Bullmore, E. T., Brammer, M. J., Andrew, C., et al. (1999). A differential neural response to threatening and non-threatening negative facial expressions in paranoid and non-paranoid schizophrenics. *Psychiatry Research, 92*(1), 11–31.

Pichon, S., de Gelder, B., & Grèzes, J. (2008). Emotional modulation of visual and motor areas by still and dynamic body expressions of anger. *Social Neuroscience, 3*(3–4), 199–212.

Pinkham, A. E., Hopfinger, J. B., Pelphrey, K. A., Piven, J., & Penn, D. L. (2008a). Neural bases for impaired social cognition in schizophrenia and autism spectrum disorders. *Schizophrenia Research, 99*(1–3), 164–175.

Pinkham, A. E., Hopfinger, J. B., Ruparel, K., & Penn, D. L. (2008b). An investigation of the relationship between activation of a social cognitive neural network and social functioning. *Schizophrenia Bulletin, 34*(4), 688–697.

Pomarol-Clotet, E., Hynes, F., Ashwin, C., Bullmore, E. T., McKenna, P. J., & Laws, K. R. (2010). Facial emotion processing in schizophrenia: A non-specific neuropsychological deficit? *Psychological Medicine*, *40*(6), 911–919.

Pourtois, G., Debatisse, D., Despland, P. A., & de Gelder, B. (2002). Facial expressions modulate the time course of long latency auditory brain potentials. *Cognitive Brain Research*, *14*(1), 99–105.

Ross, L. A., Saint-Amour, D., Leavitt, V. M., Molholm, S., Javitt, D. C., & Foxe, J. J. (2007). Impaired multisensory processing in schizophrenia: Deficits in the visual enhancement of speech comprehension under noisy environmental conditions. *Schizophrenia Research*, *97*(1–3), 173–183.

Scherer, K. R., Banse, R., Wallbott, H. G., & Goldbeck, T. (1991). Vocal cues in emotion encoding and decoding. *Motivation and Emotion*, *15*(2), 123–148.

Sprengelmeyer, R., Young, A. W., Schroeder, U., Grossenbacher, P. G., Federlein, J., Buttner, T., et al. (1999). Knowing no fear. *Proceedings of the Royal Society. Series B: Biological Sciences*, *266*(1437), 2451–2456.

Stekelenburg, J. J., & de Gelder, B. (2004). The neural correlates of perceiving human bodies: An ERP study on the body-inversion effect. *Neuroreport*, *15*(5), 777–780.

Tamietto, M., Geminiani, G., Genero, R., & de Gelder, B. (2007). Seeing fearful body language overcomes attentional deficits in patients with neglect. *Journal of Cognitive Neuroscience*, *19*(3), 445–454. doi: 10.1162/jocn.2007.19.3.445.

Van den Stock, J., Grèzes, J., & de Gelder, B. (2008a). Human and animal sounds influence recognition of body language. *Brain Research*, *1242*, 185–190.

Van den Stock, J., Peretz, I., Grèzes, J., & de Gelder, B. (2009). Instrumental music influences recognition of emotional body language. *Brain Topography*, *21*(3–4), 216–220.

Van den Stock, J., Righart, R., & de Gelder, B. (2007). Body expressions influence recognition of emotions in the face and voice. *Emotion*, *7*(3), 487–494.

Van den Stock, J., van de Riet, W. A., Righart, R., & de Gelder, B. (2008b). Neural correlates of perceiving emotional faces and bodies in developmental prosopagnosia: An event-related fMRI-study. *PLoS One*, *3*(9), e3195.

Vroomen, J., Driver, J., & de Gelder, B. (2001). Is crossmodal integration of emotional expressions independent of attentional resources? *Cognitive, Affective, & Behavioral Neuroscience*, *1*(4), 382–387.

Wolwer, W., Streit, M., Polzer, U., & Gaebel, W. (1996). Facial affect recognition in the course of schizophrenia. *European Archives of Psychiatry and Clinical Neuroscience*, *246*(3), 165–170.

Social impairment in schizophrenia revealed by Autism-Spectrum Quotient correlated with gray matter reduction

Akihiko Sasamoto[1], Jun Miyata[1], Kazuyuki Hirao[1,2], Hironobu Fujiwara[1], Ryosaku Kawada[1], Shinsuke Fujimoto[1], Yusuke Tanaka[1], Manabu Kubota[1], Nobukatsu Sawamoto[3], Hidenao Fukuyama[3], Hidehiko Takahashi[1], and Toshiya Murai[1]

[1]Department of Neuropsychiatry, Graduate School of Medicine, Kyoto University, Kyoto, Japan
[2]Department of Clinical Psychology, Kyoto Bunkyo University, Kyoto, Japan
[3]Human Brain Research Center, Graduate School of Medicine, Kyoto University, Kyoto, Japan

One of the difficulties facing schizophrenia patients is a failure to construct appropriate relationships with others in social situations. This impairment of social cognition is also found in autism-spectrum disorder (ASD). Considering such commonality between the two disorders, in this study we adopted the Autism-Spectrum Quotient (AQ) score to assess autistic traits, and explored the association between such traits and gray matter (GM) alterations of the brain in schizophrenia. Twenty schizophrenia patients and 25 healthy controls underwent structural magnetic resonance imaging (MRI), and AQ was assessed, comprising five subscales measuring different facets of autistic traits. Voxel-based morphometry (VBM) was applied to investigate the correlation between these AQ scores and regional GM alterations. Schizophrenia patients showed significantly higher scores in total AQ, and in four of the five subscales, compared to healthy controls. The total AQ score in schizophrenia showed significant negative correlation with GM volume reduction in the cortical area surrounding the left superior temporal sulcus (STS), which is considered to be important in social perception. Our findings suggest a possible neuroanatomical basis of autistic tendencies in schizophrenia.

Keywords: Schizophrenia; Social cognition; Autism-Spectrum Quotient; Voxel-based morphometry; Superior temporal sulcus.

Schizophrenia patients have difficulty in constructing appropriate mutual relationships with others in social situations. Individuals with schizophrenia are not skilled at "mentalizing" or "theory of mind" (ToM) (Premack & Woodruff, 1978), having decreased ability to infer the emotions, beliefs, or intentions of other individuals (Brüne, 2005; Frith & Corcoran, 1996; Hirao et al., 2008). Among various cognitive abilities, impairments in social cognition have been well correlated with poor social functioning in schizophrenia (Couture, Penn, & Roberts, 2006).

Impairments in social cognition and social functioning, such as difficulty with communication and inferring the feelings and thought of others, have also been revealed as key features of autism-spectrum disorder (ASD) (Attwood, 2006; Klin, Jones, Schultz, Volkmar, & Cohen, 2002; Pinkham, Hopfinger, Pelphrey, Piven, & Penn, 2008). Although schizophrenia and ASD have different symptoms and clinical courses and are usually classified as two distinct disorders, recent studies have indicated that these two disorders show similar patterns of social cognitive impairment, particularly in tasks requiring higher levels of social cognition (Couture et al., 2006; Hughes, Leboyer, & Bouyard 1997; Klin et al., 2002; Pinkham, Penn, Perkins, & Lieberman, 2003; Pinkham et al.,

Correspondence should be addressed to: Jun Miyata, Department of Neuropsychiatry, Graduate School of Medicine, Kyoto University, 54 Shogoin-Kawahara-cho, Kyoto 606-8507, Japan. E-mail: miyata10@kuhp.kyoto-u.ac.jp

http://dx.doi.org/10.1080/17470919.2011.575693

2008). Importantly, as mentioned above, dysfunction of empathy to infer or perceive others' emotions has repeatedly been emphasized in both schizophrenia (Bora, Gökçen, & Veznedaroglu, 2008; Haker & Rössler, 2009; Shamay-Tsoory, Shur, Harari, & Levkovitz, 2007) and ASD (Attwood, 2006; Baron-Cohen & Wheelwright, 2004). These findings suggest that social cognitive deficits, especially impaired empathic abilities, in both disorders may arise from dysfunctions in similar neural systems underlying social cognition, and that comparing these two disorders may help illuminate the general mechanisms of social cognition and its impairments (Pinkham et al., 2008).

A number of functional and lesion studies have shown that the amygdala plays a key role in perception of facial, emotional expression (Adolphs, Tranel, Damasio, & Damasio, 1994). By contrast, the prefrontal cortices are more associated with ToM (Baron-Cohen et al., 1994; Fletcher et al., 1995), and the superior temporal sulcus (STS) is related to both facial, emotional perception and ToM (Brunet-Gouet & Decety, 2006). Meanwhile, numerous neuroimaging studies have reported gray matter (GM) reduction of various cortical and subcortical regions in schizophrenia (Honea, Crow, Passingham, & Mackay, 2005; Shenton, Dickey, Frumin, & McCarley, 2001), which include a network critical for social cognition, such as the temporal cortex, amygdala, and prefrontal cortex (Lee, Farrow, Spence, & Woodruff, 2004). The structural brain alterations of ASD have also been reported in postmortem and structural magnetic resonance imaging (MRI) studies that highlighted the pathology of the frontal lobes, amygdala, and cerebellum (Amaral, Schumann, & Nordahl, 2008). The increased total brain volume and regional heterogeneous thickness of GM were also investigated (Hyde, Samson, Evans, & Mottron, 2010), although the results have been inconsistent (Amaral et al., 2008). Recent studies suggest that, in ASD, the developmental course of the brain is primarily disturbed, rather than the ultimate brain structure. In addition, the heterogeneity of symptoms in each individual with ASD is suggestive of underlying heterogeneous patterns of neuropathology in this disorder (Amaral et al., 2008).

In view of these findings, although schizophrenia and ASD have different overall patterns of structural brain abnormalities, they may share common neural underpinnings of social cognitive impairment. In this study, we aimed to focus on social cognitive impairments shared by schizophrenia and ASD to elucidate common and differential neural mechanisms for social cognition between these two disorders.

We therefore aimed to capture the "autistic tendency" in schizophrenia patients. We adopted the Autism-Spectrum Quotient (AQ) (Baron-Cohen, Wheelwright, Skinner, Martin, & Clubley, 2001; Wakabayashi, 2003) as a psychometric measure. The AQ was originally developed as the screening method for ASD among the general population. However, it has been gradually recognized that this scale is a useful tool to assess social skills in the daily life of neuropsychiatric populations other than ASD.

We hypothesized that schizophrenia patients would have impaired empathy and mentalizing, which are often common with ASD, and, by specifically evaluating AQ in this study, we tried to reveal that neuroanatomical underpinnings would be located within the structures involved in social cognition. Using a combination of AQ and voxel-based morphometry (VBM) (Ashburner & Friston, 2000), we explored a possible relationship between "autistic" social dysfunction and regional brain abnormalities in schizophrenia.

METHODS

Participants

The schizophrenia group consisted of 20 individuals (14 men and 6 women, all right-handed) referred to the Department of Neuropsychiatry, Kyoto University Hospital. Each patient fulfilled the criteria for schizophrenia based on the Structural Clinical Interview for DSM-IV axis I disorders, patient edition (SCDI-I/P, version 2.0; First, Spitzer, Gibbon, & Williams, 1996). Psychopathology was assessed by the Positive and Negative Syndrome Scale (PANSS) (Kay, Fiszbein, & Opler, 1987). All patients were receiving antipsychotic medication—typical ($n = 2$), atypical ($n = 12$), and typical and atypical ($n = 6$)— and were physically healthy at the time of scanning and cognitive tests. Exclusion criteria included a history of neurological injury or disease, and medical diseases or substance abuse that may affect brain function.

The comparison group comprised 25 healthy individuals (16 men and 9 women, all right-handed), who were matched with the schizophrenia group by age and gender. They were also assessed on the basis of SCID and had no history of neurological or psychiatric disease and no first-degree relative with psychotic episodes.

This study was approved by the Committee on Medical Ethics of Kyoto University and carried out in accordance with the Code of Ethics of the World

Medical Association. Written, informed consent was obtained from participants after presenting a complete description of the study.

Basic cognitive tasks

To evaluate each participant's basic cognitive ability, we administered the following cognitive tasks. Predicted IQ was assessed with the Japanese version of the National Adult Reading Test (JART) (Matsuoka, Uno, Kasai, Koyama, & Kim, 2006), which is considered to reflect the premorbid IQ of patients with schizophrenia. In addition, the estimated verbal and performance IQ scores were obtained from vocabulary and block design subtests in the Wechsler Adult Intelligence Scale–Revised (WAIS–R), respectively, by transforming the scores corrected for age into T scores.

AQ scale

We adopted the Japanese version of the AQ scale (Wakabayashi, 2003) to measure the autistic tendencies of individuals. This scale is a self-administered questionnaire originally developed to measure the degree of autistic traits of individuals with normal intelligence (Baron-Cohen et al., 2001). This questionnaire comprises 50 questions which are divided into five different subscales: social skill, attention switching, attention to detail, communication, and imagination. Each subscale has 10 questions. Each question is rated on four choices: definitely agree (1), slightly agree (2), slightly disagree (3), and definitely disagree (4). When the participant recorded abnormal or autistic-like behavior as either definitely or slightly agree/disagree, 1 point was scored for the item. Higher scores for total and for each subscale indicate higher autistic tendency.

MRI acquisition and preprocessing

All participants underwent MRI scans on a 3-T whole-body scanner equipped with an 8-channel, phased-array head coil (Trio, Siemens, Erlangen, Germany). The scanning parameters of the three-dimensional, magnetization-prepared, rapid gradient-echo (3D-MPRAGE) sequences were as follows: TE = 4.38 ms, TR = 2000 ms, TI = 990 ms, FOV = 240 mm, slice plane = axial, slice thickness = 1 mm, resolution = $0.94 \times 0.94 \times 1.0$ mm, and slice number = 208. The images were processed and analyzed by Statistical Parametric Mapping software (SPM5) (Wellcome Department of Imaging Neuroscience, London, UK)

running on Matlab 2007b (MathWorks, Natick, MA, USA). The VBM method was applied to investigate regional GM volume alterations, using the VBM 5.1 toolbox written by Gaser (http://dbm.neuro.uni-jena.de/vbm). The images were normalized and segmented into GM, white matter, and cerebrospinal fluid partitions in the unified segmentation step (Ashburner & Friston, 2005). The normalized and segmented images were resliced into $1 \times 1 \times 1$ mm³ voxels, and modulated by the Jacobian determinants for nonlinear warping only. Then the modulated GM images were smoothed with a Gaussian kernel of 12 mm full width at half maximum, on which all the analyses were performed.

Data analyses

Regional GM volume reductions in schizophrenia patients in comparison with controls

For the purpose of exploring the regional brain reduction in the schizophrenia group relative to controls, we performed analysis of covariance in SPM5, with the nuisance covariates of age and gender. The voxelwise statistical threshold was set at $p < .001$ uncorrected for multiple comparisons, with an extent threshold of 100 voxels. Montreal Neurological Institute (MNI) coordinates were transformed into Talairach coordinates (Talairach & Tournoux, 1988) by using the mni2tal.m Matlab script written by Matthew Brett (http://imaging.mrc-cbu.cam.ac.uk/imaging/MniTalairach).

Correlational analyses of AQ scores and GM volume

The correlational analyses of GM volume were carried out in the following manner. First, we separately performed multiple-regression analysis in SPM5 to explore the brain regions where GM volume was negatively correlated with total AQ in the whole brain, in the schizophrenia and healthy control groups. Age and gender were entered as nuisance covariates. The same statistical and extent thresholds used in the above-mentioned group comparison were applied.

Second, in addition, we analyzed GM volume correlations with each AQ subscale score in the whole brain in the schizophrenia group by the same analysis.

Third, using the Volume of Interest (VOI) function (sm_regions.m) in SPM5, we extracted the eigenvariates from the regions that correlated significantly with total AQ scores in schizophrenia

TABLE 1
Demographic, clinical and neuropsychological characteristics of participants

	Nc (N = 25)		Scz (N = 20)		
	Mean	SD	Mean	SD	p
Age (years)	34.5	9.4	34.5	8.8	.994
Gender (male/female)	16/9		14/6		
Handedness (left/right)	0/25		0/20		
Education (years)	15.2	2.8	13.2	2.0	.007*
Duration of illness (years)	–	–	11.8	8.9	
Drug (mg/day, haloperidol equivalent)	–	–	15.4	21.3	
PANSS positive	–	–	13.1	3.6	
PANSS negative	–	–	15.4	4.0	
JART	108.6	9.8	104.4	8.3	.127
VIQ	120.2	15.9	101.8	18.1	.001*
PIQ	120.0	12.4	111.3	13.2	.027*

Nc: normal controls; Scz: schizophrenia patients; PANSS: Positive and Negative Syndrome Scale; JART: Japanese version of the National Adult Reading Test; VIQ: Verbal Intelligence Quotient; PIQ: Performance Intelligence Quotient.

*Statistically significant ($p < .05$).

patients. Then, correlational analyses were performed in SPSS 15.0 (SPSS, Inc., Chicago, IL, USA) to investigate the association between GM volume and the five AQ subscales. In the schizophrenia group, clinical data (medication dose and PANSS scores; i.e., positive symptom, negative symptom, general psychopathology subscales and total scores) were also investigated as to whether they were associated with GM volume.

Finally, to confirm the correlation controlling for confounding factors, the partial correlations were analyzed between GM volumes and psychopathology and clinical variables in the schizophrenia group (age, gender, PANSS scores, JART, medication dose and duration of illness). The statistical significance level was set at $p < .05$.

RESULTS

Demographic and clinical data

Demographic data and clinical data are shown in Table 1. There were no significant differences in age, gender, handedness and predicted IQ (JART score) between the schizophrenia patients and the healthy controls. The estimated VIQ ($t = 3.638$, $p = .001$) and PIQ ($t = 2.287$, $p = .027$) were lower in the schizophrenia group than in the healthy controls, and the education years in patients were fewer than in healthy controls ($t = 2.827$, $p = .007$).

AQ scores

The results of the AQ questionnaire are shown in Table 2 and Figure 1. Unpaired t-tests were performed to investigate the relationships between schizophrenia patients and healthy controls. The schizophrenia group scored significantly higher (i.e., they had stronger autistic tendencies) in the total AQ score ($t = -5.355$, $p < .001$) and the four subscales; i.e., "social skill" ($t = -3.952$, $p < .001$), "attention switching" ($t = -2.990$, $p = .005$), "communication" ($t = -3.841$, $p < .001$), and "imagination" ($t = -4.295$, $p < .001$). There was no significant difference in "attention to detail" ($t = -1.702$, $p = .096$).

TABLE 2
Total AQ and subscale scores in the control (Nc) and schizophrenia (Scz) groups

	Nc (N = 25)		Scz (N = 20)			
	Mean	SD	Mean	SD	t(df = 43)	p
Total	14.48	6.89	25.35	6.61	−5.355	<.001*
Social skill	2.32	2.46	5.30	2.58	−3.952	<.001*
Attention switching	3.60	2.29	5.45	1.73	−2.990	.005*
Attention to detail	4.08	1.73	5.10	2.29	−1.702	.096
Communication	1.80	1.98	4.35	2.48	−3.841	<.001*
Imagination	2.68	1.73	5.15	2.13	−4.295	<.001*

*Statistically significant ($p < .05$).

total AQ score

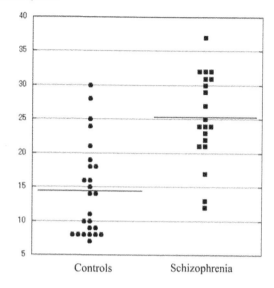

Figure 1. Scattered distributions of total AQ scores in the healthy control and schizophrenia groups. The two horizontal lines represent the mean score in each group.

Regional GM reductions in schizophrenia patients relative to controls

Schizophrenia patients exhibited significant GM volume reductions in several regions relative to healthy controls, including the bilateral dorsolateral prefrontal cortices, medial prefrontal cortices and orbitofrontal cortices, and right thalamus ($p < .001$ uncorrected; Figure 2). There were no regions in which schizophrenia patients exhibited larger GM volume than controls.

Correlational analysis between GM reductions, total AQ, and subscale scores

In the correlational analysis of healthy controls, the total AQ score was inversely correlated with the left paracentral gyrus, while in schizophrenia patients, total AQ score was inversely correlated with the volume of cortical area surrounding the left STS (Figure 3, Table 3).

Subsequently, on the subscales of AQ in the schizophrenia group, the "social skill" score was negatively correlated with a region in the left occipital

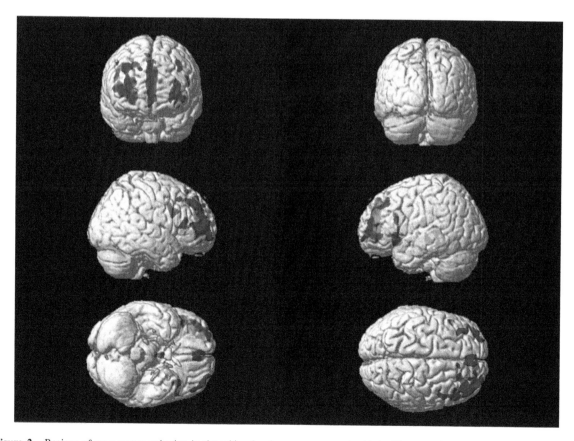

Figure 2. Regions of gray matter reduction in the schizophrenia group compared with healthy controls; uncorrected $p < .001$, extent threshold = 100 voxels.

(a)

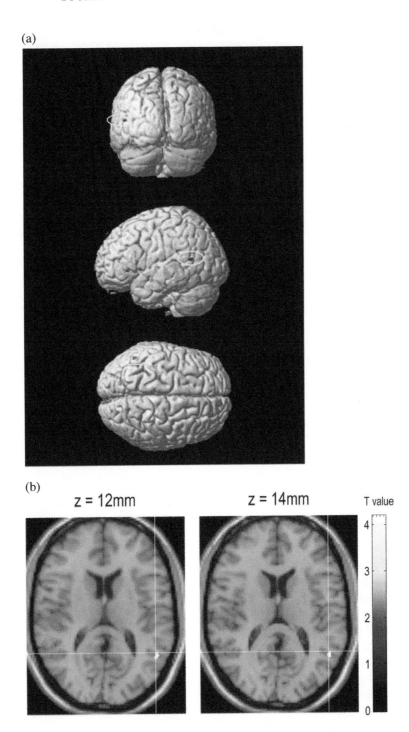

(b)

Figure 3. Regions of gray matter reduction were inversely correlated with total AQ score in the schizophrenia group; uncorrected $p < .001$, extent threshold $= 100$ voxels. (a) Significant cluster (encircled in yellow) in whole-brain display; (b) the significant cluster shown in an axial slice display (left–right orientation is according to the radiological convention).

lobe; the "attention switching" score with five regions located in the right superior temporal gyrus, left occipital lobe, and left superior temporal, left middle temporal, and left middle occipital gyri; the "communication" score with two regions in the left middle temporal and left superior temporal gyri; and the "imagination" score with one region in the left middle temporal gyrus; but the "attention to detail" score was not correlated with any regions (Table 4).

SOCIAL NEUROSCIENCE OF PSYCHIATRIC DISORDERS

TABLE 3
Regions where the GM volumes were negatively correlated with total AQ score in the control and schizophrenia group

| Regions | BA | Cluster centers[a] | | | Cluster size | Z |
		x	y	z		
Healthy controls						
Left paracentral lobule	6	−11	−22	46	371	3.79
Schizophrenia group						
Left superior temporal sulcus	22	−50	−55	16	341	3.38

[a]Coordinates according to the stereotaxic atlas of Talairach and Tournoux (1998).

TABLE 4
Regions where the GM volumes were negatively correlated with each AQ subscale score in the schizophrenia group

| Regions | BA | Cluster centers[a] | | | Cluster size | Z |
		x	y	z		
Subscale 1 (social skill)						
Left occipital lobe	30	18	−42	−2	280	3.28
Subscale 2 (attention switching)						
Right superior temporal gyrus	22	51	−56	16	5442	4.97
Left occipital lobe	18	−12	−85	25	363	3.51
Left superior temporal gyrus	22	−46	−17	−2	123	3.35
Left middle temporal gyrus	39	−42	−62	18	215	3.31
Left middle occipital gyrus	19	−35	−83	10	113	3.27
Subscale 3 (attention to detail)						
(no suprathreshold clusters)	−	−	−	−	−	−
Subscale 4 (communication)						
Left middle temporal gyrus	21	−56	3	−20	1014	3.69
Left superior temporal gyrus	22	−49	−54	16	201	3.30
Subscale 5 (imagination)						
Left middle temporal gyrus	39	−49	−56	13	525	3.63

[a]Coordinates according to the stereotaxic atlas of Talairach and Tournoux (1998).

Correlational analysis between left STS volumic eigenvariate values and AQ subscales in schizophrenia

Subsequently, we extracted the eigenvariates from the left STS region correlated with total AQ score in the schizophrenia group mentioned above and then investigated the relationship between the AQ scores and left STS volume (Figure 4). The eigenvariates of the left STS showed significant negative correlation with the total AQ score ($r = −.735$, $p < .001$). Among the subscales of the AQ, the eigenvariate of the left STS exhibited significant inverse correlation with scores of "social skill" ($r = −.576$, $p = .008$), "attention switching" ($r = −.592$, $p = .006$), "communication" ($r = −.614$, $p = .004$), and "imagination" ($r = −.605$, $p = .005$). However, the subscale "attention to detail" score was not significantly correlated with the eigenvariate of the left STS ($r = .202$, $p = .393$).

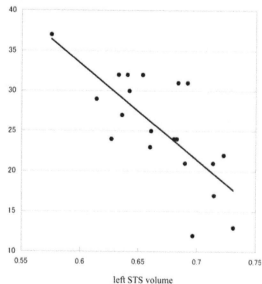

Figure 4. Correlation between left STS volume and total AQ score.

140

Partial correlational analysis

Among the clinical variables in the schizophrenia group, age and duration of illness were also significantly correlated with the left STS eigenvariate ($r = -.590$, $p = .006$ and $r = -.617$, $p = .004$). However, the correlation between the left STS eigenvariate and total AQ score was still significant ($r = -.761$, $p = .002$) after controlling for demographic and clinical variables (age, gender, PANSS scores, JART, medication dose, and duration of illness).

DISCUSSION

The major findings of this study are as follows:

(a) The total AQ score and four of the five subscale scores ("social skill," "attention switching," "communication," and "imagination") were significantly higher in the schizophrenia group than in healthy controls.

(b) In whole-brain analysis of VBM, schizophrenia patients, compared with healthy controls, exhibited basically bilateral and widespread GM volume reductions in the prefrontal cortices and thalamus.

(c) In schizophrenia subjects, the cortical volume of the left STS region was inversely correlated with the total AQ score, as well as four of the AQ subscales, excluding the item "attention to detail." VBM analysis on each AQ subscale in schizophrenia showed that "attention switching," "communication," and "imagination" were also negatively correlated with GM reduction of the middle to superior temporal regions.

The AQ was originally developed to screen for autism-spectrum conditions in adults of normal intelligence (Baron-Cohen et al., 2001). According to their results, 80% of ASD individuals, but only 2% of normal subjects, scored more than 32 for total AQ score. Moreover, in a large sample study applying the Japanese version of the AQ, 87.8% of ASD sufferers, but only 3% of normal subjects, scored more than 33 (Wakabayashi, 2003). In our current study, schizophrenia patients (25.35 ± 6.61) scored significantly higher than healthy controls (14.48 ± 6.89) in total AQ score (Table 2). Interestingly, the majority (19/20) of schizophrenia individuals in this study scored lower than 33, which is the cutoff point for ASD proposed by Wakabayashi (2003). Thus, the degree

of autistic traits in schizophrenia could be situated between those of healthy and ASD individuals.

In addition to the total score, schizophrenia individuals scored significantly higher than healthy controls in four of the five subscales; i.e., "social skill," "attention switching," "communication," and "imagination," but not in "attention to detail." In factor analyses of AQ in the general population and the autism-spectrum conditions in The Netherlands, a two-factor model was drawn; i.e., one factor was the four subscales ("social skill," "attention switching," "communication," and "imagination"), and the other was "attention to detail" (Hoekstra, Bartels, Cath, & Boomsma, 2008). The former four subscales were highly intercorrelated and termed "social interaction" factors, but poorly correlated with the "attention to detail" factor. Thus, our results may indicate that autistic tendency in schizophrenia is more closely related to the "social interaction" factors of AQ.

VBM analyses of the whole brain revealed that schizophrenia patients exhibited regional GM volume reductions in the bilateral dorsolateral prefrontal cortices, medial prefrontal cortices and orbitofrontal cortices, and right thalamus. Similar patterns of GM reduction were revealed in our previous studies (Hirao et al., 2008; Kawada et al., 2009; Miyata et al., 2009; Yamada et al., 2007) and were consistent with meta-analyses published recently (Ellison-Wright, Glahn, Laird, Thelen, & Bullmore, 2008; Glahn et al., 2008).

The total AQ score was revealed to be inversely correlated with the left STS volume in schizophrenia. In addition, among AQ subscales in schizophrenia (specifically "attention switching," "communication," and "imagination"), the most frequent GM negative correlation was recognized around the middle to superior temporal regions. These are the "social interaction" subscale factors in AQ. These results are compatible with our hypothesis that schizophrenia patients have autistic traits, such as emphatic impairments, and indicate that the temporal GM alteration near the STS region is the neuroanatomical underpinning for these impairments.

Originally, the specific role of the STS region in social perception was discovered by examination of the macaque monkey's responsiveness to gaze and head direction (Perrett et al., 1985). Subsequently, the corresponding functionality was investigated in the human brain, and it was revealed that this region is deeply associated with human biological perception. This STS region is known to be activated not only by actual physical movements, such as of the eyes, mouth, hands, and body, but also by static implied motion images and more general stimuli that signal the actions of another individual. Moreover, the STS region has been suggested to play a role in integrating information from the ventral object recognition system (the "what" system) and from the dorsal spatial location-movement system (the "where" system). This region is also thought to reciprocally interact with the amygdala and orbitofrontal cortices to attach emotional salience to sensory input (Allison, Puce, & McCarthy, 2000). Therefore, the STS region was considered to be a major player in "social perception" in the "social brain" (Zilbovicius et al., 2006).

The alterations of the STS region in psychiatric disorders have been intensively studied. STS abnormalities, such as decreased GM volume, rest hypoperfusion, and abnormal activation in social tasks, are highly implicated in ASD (Boddaert et al, 2004; Zilbovicius et al., 2000, 2006). These abnormalities were considered to be related to disorders of social perception (e.g., vocal cognition) and complex social cognition in ASD (e.g., interpretation and analysis of the intention of others; Zilbovicius et al., 2006). Volume reduction in the STS region in schizophrenia has also attracted attention (Antonova, Sharma, Morris, & Kumari, 2004). For example, reduction of the GM volume of the region surrounding the left STS, including the superior temporal gyrus (STG), was reported to be associated with auditory hallucinations (Barta, Pearlson, Powers, Richards, & Tune, 1990); poor psychomotor speed, as in verbal fluency and picture-naming accuracy (Sanfilipo et al., 2002; Vita et al., 1995); and the severity of formal thought disorder (Horn et al., 2010). Our study directly demonstrates an association between autistic traits and the STS region in schizophrenia, and indicates that these left STS-related disorders are associated, at least partly, with the impairments of social functioning in schizophrenia.

By contrast, the above-mentioned correlation was not demonstrated in normal subjects. This discrepancy of AQ–STS association between normal and schizophrenia subjects may be explained by the different distribution pattern of AQ scores between schizophrenia and normal subjects. As shown in Figure 1, only two control subjects scored more than the average AQ score for the schizophrenia subjects (25.35). Thus, a lack of association between STS volume and AQ in control subjects may be explained if one hypothesizes that the volume–AQ relationship exists in subjects with higher AQ scores, but not in those with lower AQ.

There are several limitations to this study. First, as the number of subjects was small, we could not analyze the differences among subtype and gender of schizophrenia individuals. For example, a previous

study reported that activation of the amygdala was different between paranoid and nonparanoid schizophrenia patients (Phillips et al., 1999; Williams et al., 2004). Thus, investigation of the effects of these clinical variables will be important in future studies. Secondly, only the left STS, and not the right STS, was significantly correlated with total AQ. This lateralized association could be due to the verbal nature of the questionnaire. This issue should be further investigated, using both verbal and nonverbal assessment of autistic traits and social cognition. Third, as the AQ questionnaire is self-completed, we could not objectively assess individual social and communication skills. If family and friends evaluated the autistic degree of the same individual, the result might be different. Given the above-mentioned limitations, the reported relationships between left STS volume reduction and autistic tendency in schizophrenia should be regarded as preliminary findings.

In conclusion, schizophrenia patients showed higher autistic traits than healthy people, especially in social interaction subscales of the AQ. This tendency was negatively associated with GM volume of the left STS area, and may imply the neuroanatomical basis of autistic tendencies of schizophrenia.

REFERENCES

Adolphs, R., Tranel, D., Damasio, H., & Damasio, A. (1994). Impaired recognition of emotion in facial expressions following bilateral damage to the human amygdala. *Nature, 372*, 669–672.

Allison, T., Puce, A., & McCarthy, G. (2000). Social perception from visual cues: Role of the STS region. *Trends in Cognitive Science, 4*, 267–278.

Amaral, D. G., Schumann, C. M., & Nordahl, C. W. (2008). Neuroanatomy of autism. *Trends in Neurosciences, 31*, 137–145.

Antonova, E., Sharma, T., Morris, R., & Kumari, V. (2004). The relationship between brain structure and neurocognition in schizophrenia: A selective review. *Schizophrenia Research, 70*, 117–145.

Ashburner, J., & Friston, K. J. (2000). Voxel-based morphometry – the methods. *NeuroImage, 11*, 805–821.

Attwood, T. (2006). Asperger's syndrome. *Learning Disability Review, 11*, 3–11.

Baron-Cohen, S., Ring, H., Moriarty, J., Schmitz, B., Costa, D., & Ell, P. (1994). Recognition of mental state terms. Clinical findings in children with autism and a functional neuroimaging study of normal adults. *British Journal of Psychiatry, 165*, 640–649.

Baron-Cohen, S., Wheelwright, S., Skinner, R., Martin, J., & Clubley, E. (2001). The Autism-Spectrum Quotient (AQ): Evidence from Asperger syndrome/high-functioning autism, males and females, scientists and mathematicians. *Journal of Autism and Developmental Disorders, 31*, 5–17.

Baron-Cohen, S., & Wheelwright, S. (2004). The Empathy Quotient: An investigation of adults with Asperger syndrome or high-functioning autism, and normal sex differences. *Journal of Autism and Developmental Disorders, 34*, 163–175.

Barta, P. E., Pearlson, G. D., Powers, R. E., Richards, S. S., & Tune, L. E. (1990). Auditory hallucinations and smaller superior temporal gyral volume in schizophrenia. *American Journal of Psychiatry, 147*, 1457–1462.

Boddaert, N., Chabane, N., Gervais, H., Good, C. D., Bourgeois, M., Plumet M. H., et al. (2004). Superior temporal sulcus anatomical abnormalities in childhood autism: A voxel-based morphometry MRI study. *NeuroImage, 23*, 364–369.

Bora, E., Gökçen, S., & Veznedaroglu, B. (2008). Empathic abilities in people with schizophrenia. *Psychiatry Research, 160*, 23–29.

Brüne, M. (2005). "Theory of mind" in schizophrenia: A review of the literature. *Schizophrenia Bulletin, 31*, 21–42.

Brunet-Gouet, E., & Decety, J. (2006). Social brain dysfunctions in schizophrenia: A review of neuroimaging studies. *Psychiatry Research, 148*, 75–92.

Couture, S. M., Penn, D. L., Losh, M., Adolphs, R., Hurley, R., & Piven, J. (2009). Comparison of social cognitive functioning in schizophrenia and high-functioning autism: More convergence than divergence. *Psychological Medicine, 12*, 1–11.

Couture, S. M., Penn, D. L., & Roberts, D. L. (2006). The functional significance of social cognition in schizophrenia: A review. *Schizophrenia Bulletin, 32*, 44–63.

Ellison-Wright, I., Glahn, D. C., Laird, A. R., Thelen, S. M., & Bullmore, E. (2008). The anatomy of first-episode and chronic schizophrenia: An anatomical likelihood estimation meta-analysis. *American Journal of Psychiatry, 165*, 1015–1023.

First, M. B., Spitzer, R. L., Gibbon, M., & Williams, J. B. W. (1996). *Structured clinical interview for DSM-IV axis disorders: Patient Edition* (SCID-I/P, version 2.0). New York, NY: Biometrics Research Department.

Fletcher, P. C., Frith, C. D., Grasby, P. M., Shallice, T., Frackowiak, R. S., & Dolan, R. J. (1995). Brain systems for encoding and retrieval of auditory-verbal memory. An in vivo study in humans. *Brain, 118*, 401–416.

Frith, C. D., & Corcoran, R. (1996). Exploring 'theory of mind' in people with schizophrenia. *Psychological Medicine, 26*, 521–530.

Glahn, D. C., Laird, A. R., Ellison-Wright, I., Thelen, S. M., Robinson, J. L., Lancaster, J., et al. (2008). Meta-analysis of gray matter anomalies in schizophrenia: Application of anatomic likelihood estimation and network analysis. *Biological Psychiatry, 64*, 774–781.

Haker, H., & Rössler, W. (2009). Empathy in schizophrenia: Impaired resonance. *European Archives of Psychiatry and Clinical Neuroscience, 259*, 352–361.

Hirao, K., Miyata, J., Fujiwara, H., Yamada, M., Namiki, C., Shimizu, M., et al. (2008). Theory of mind and frontal lobe pathology in schizophrenia: A voxel-based morphometry study. *Schizophrenia Research, 105*, 165–174.

Hoekstra, R. A., Bartels, M., Cath, D. C., & Boomsma, D. I. (2008). Factor structure, reliability and criterion validity of the Autism-Spectrum Quotient (AQ): A study in Dutch population and patient groups. *Journal of Autism and Developmental Disorders, 38*, 1555–1566.

Honea, R., Crow, T. J., Passingham, D., & Mackay, C. E. (2005). Regional deficits in brain volume in schizophrenia: A meta-analysis of voxel-based morphometry studies. *American Journal of Psychiatry, 162*, 2233–2245.

Horn, H., Federspiel, A., Wirth, M., Müller, T. J., Wiest, R., Walther, S., et al. (2010). Gray matter volume differences specific to formal thought disorder in schizophrenia. *Psychiatry Research, 182*, 183–186.

Hughes, C., Leboyer, M., & Bouvard, M. (1997). Executive function in parents of children with autism. *Psychological Medicine, 27*, 209–220.

Hyde, K. L., Samson, F., Evans, A. C., & Mottron, L. (2010). Neuroanatomical differences in brain areas implicated in perceptual and other core features of autism revealed by cortical thickness analysis and voxel-based morphometry. *Human Brain Mapping, 31*, 556–566.

Kawada, R., Yoshizumi, M., Hirao, K., Fujiwara, H., Miyata, J., Shimizu, M., et al. (2009). Brain volume and dysexecutive behavior in schizophrenia. *Progress in Neuro-Psychopharmacology and Biological Psychiatry, 33*, 1255–1260.

Kay, S. R., Fiszbein, A., & Opler, L. A. (1987). The Positive and Negative Syndrome Scale (PANSS) for schizophrenia. *Schizophrenia Bulletin, 13*, 261–276.

Klin, A., Jones, W., Schultz, R., Volkmar, F., & Cohen, D. (2002). Visual fixation patterns during viewing of naturalistic social situations as predictors of social competence in individuals with autism. *Archives of General Psychiatry, 59*, 809–816.

Lee, K. H., Farrow, F. D., Spence, S. A., & Woodruff, P. W. (2004). Social cognition, brain networks and schizophrenia. *Psychological Medicine, 34*, 391–400.

Matsuoka, K., Uno, M., Kasai, K., Koyama, K., & Kim, Y. (2006). Estimation of premorbid IQ in individuals with Alzheimer's disease using Japanese ideographic script (Kanji) compound words: Japanese version of National Adult Reading Test. *Psychiatry and Clinical Neurosciences, 60*, 332–339.

Miyata, J., Hirao, K., Namiki, C., Fujiwara, H., Shimizu, M., Fukuyama, H., et al. (2009). Reduced white matter integrity correlated with cortico-subcortical gray matter deficits in schizophrenia. *Schizophrenia Research, 111*, 78–85.

Perrett, D. I., Smith, P. A., Potter, D. D., Mistlin, A. J., Head, A. S., Milner, A. D., et al. (1985). Visual cells in the temporal cortex sensitive to face view and gaze direction. *Proceedings of the Royal Society. Series B: Biological Sciences, 223*, 293–317.

Phillips, M. L., Williams, L., Senior, C., Bullmore, B. T., Brammer, M. J., Andrew, C., et al. (1999). A differential neural response to threatening and non-threatening negative facial expressions in paranoid and non-paranoid schizophrenics. *Psychiatry Research, 92*, 11–31.

Pinkham, A. E., Hopfinger, J. B., Pelphrey, K. A., Piven, J., & Penn, D. L. (2008). Neural bases for impaired social cognition in schizophrenia and autism spectrum disorders. *Schizophrenia Research, 99*, 164–175.

Pinkham, A. E., Penn, D. L., Perkins, D. O., & Lieberman, J. (2003). Implications for the neural basis of social cognition for the study of schizophrenia. *American Journal of Psychiatry, 160*, 815–824.

Premack, D., & Woodruff, G. (1978). Does the chimpanzee have a theory of mind? *Behavioral and Brain Sciences, 1*, 515–526.

Sanfilipo, M., Lafargue, T., Rusinek, H., Arena, L., Loneragan, C., Lautin, A., et al. (2002). Cognitive performance in schizophrenia: Relationship to regional brain volumes and psychiatric symptoms. *Psychiatry Research, 116,* 1–23.

Shamay-Tsoory, S. G., Shur, S., Harari, H., & Levkovitz, Y. (2007). Neurocognitive basis of impaired empathy in schizophrenia. *Neuropsychology, 21,* 431–438.

Shenton, M. E., Dickey, C. C., Frumin, M., & McCarley, R. W. (2001). A review of MRI findings in schizophrenia. *Schizophrenia Research, 49,* 1–52.

Talairach, J., & Tournoux, P. (1988). *Co-planar stereotaxic atlas of the human brain.* New York, NY: Thieme Medical Publishers.

Vita, A., Dieci, M., Giobbio, G. M., Caputo, A., Ghiringhelli, L., Comazzi, M., et al. (1995). Language and thought disorder in schizophrenia: Brain morphological correlates. *Schizophrenia Research, 15,* 243–251.

Wakabayashi, A. (2003). On Autism-Spectrum Quotient Japanese version. In Y. Tojo (Ed.), *Educational supports and assessments for children in autism and ADHD*

(pp. 47–52). Kanagawa, Japan: National Institute for Special Needs in Education (in Japanese).

Williams, L. M., Das, P., Harris, A. W. F., Liddell, B. B., Brammer, M. J., Olivieri, G., et al. (2004). Dysregulation of arousal and amygdala-prefrontal systems in paranoid schizophrenia. *American Journal of Psychiatry, 161,* 480–489.

Yamada, M., Hirao, K., Namiki, C., Hanakawa, T., Fukuyama, H., Hayashi, T., et al. (2007). Social cognition and frontal lobe pathology in schizophrenia: A voxel-based morphometric study. *NeuroImage, 35,* 292–298.

Zilbovicius, M., Boddaert, N., Belin, P., Poline, J. B., Remy, P., Mangin, J. F., et al. (2000). Temporal lobe dysfunction in childhood autism: A PET study. Positron emission tomography. *American Journal of Psychiatry, 157,* 1988–1993.

Zilbovicius, M., Meresse, I., Chabane, N., Brunelle, F., Samson, Y., & Boddaert, N. (2006). Autism, the superior temporal sulcus and social perception. *Trends in Neurosciences, 29,* 359–366.

Event-related potential correlates of suspicious thoughts in individuals with schizotypal personality features

Xue-bing Li[1], Jia Huang[1], Eric F. C. Cheung[2], Qi-yong Gong[3], and Raymond C. K. Chan[1]

[1]Neuropsychology and Applied Cognitive Neuroscience Laboratory, Key Laboratory of Mental Health, Institute of Psychology, Chinese Academy of Sciences, Beijing, China
[2]Castle Peak Hospital, Hong Kong Special Administrative Region, Hong Kong, China
[3]Huaxi MR Research Centre, Department of Radiology, West China Hospital/West China School of Medicine, Sichuan University, Chengdu, China

Suspiciousness is a common feature of schizophrenia. However, suspicious thoughts are also commonly experienced by the general population. This study aimed to examine the underlying neural mechanism of suspicious thoughts in individuals with and without schizotypal personality disorder (SPD)-proneness, using an event-related potential (ERP) paradigm. Electroencephalography (EEG) was recorded when the "feeling of being seen through" was evoked in the participants. The findings showed a prominent positive deflection of the difference wave within the time window 250–400 ms after stimuli presentation in both SPD-prone and non-SPD-prone groups. Furthermore, the P3 amplitude was significantly reduced in the SPD-prone group compared to the non-SPD-prone group. The current density analysis also indicated hypoactivity in both frontal and temporal regions in the SPD-prone group, suggesting that the frontotemporal cortical network may play a role in the onset of suspicious thoughts. The P3 of difference wave was inversely correlated with the cognitive-perception factor and the suspiciousness/paranoid ideation trait, which provided preliminary electrophysiological evidence for the association of suspiciousness with SPD features.

Keywords: Schizotypal personality; Suspicious thoughts; Event-related potentials.

Delusion is a cardinal symptom of schizophrenia. On the basis of findings from a World Health Organization prospective study in 10 countries of individuals with signs of schizophrenia, delusions of reference and persecutory delusions were rated as the two most common symptoms of psychosis, occurring in more than 50% of cases (Sartorius et al., 1986). However, many researchers have argued that psychotic symptoms such as delusions may be better understood as a continuum with normal experience (Freeman, 2007; van Os,

Hanssen, Bijl, & Ravelli, 2000). A number of surveys on the prevalence of psychotic symptoms in nonclinical samples have found that 12–28% of individuals in the general population reported at least one delusional experience unrelated to drug use or physical illness (Bijl, van Zessen, Ravelli, de Rijk, & Langendoen, 1998; Claridge, 1994; Freeman et al., 2005b; Johns et al., 2004; Kendler, Gallagher, Abelson, & Kessler, 1996; Scott, Chant, Andrews, & McGrath, 2006; van Os et al., 2000), and 10–15% of the general

Correspondence should be addressed to: Raymond C. K. Chan, 4A Datun Road, Chaoyang District, Beijing 100101, China. E-mail: rckchan@psych.ac.cn

This study was supported by the Project-Oriented Hundred Talents Programme (O7CX031003), the Research Initiation Fund (O9CX083008), the Knowledge Innovation Project of the Chinese Academy of Sciences (KSCX2-YW-R-131), the National Basic Research Programme (973 Programme No. 2007CB512302/5), the National Nature Science Foundation of China (NSFC 30900441), and the Outstanding Young Investigator Award from the National Science Foundation China (81088001) to Raymond Chan.

population experienced paranoid thoughts (Freeman, 2007). Delusions in psychosis therefore represent the severe end of a continuum. People in the general population may also have a similar but attenuated form of experience in everyday life. For example, the phenomena of "distant intentionality" (Braud, 1994), "remote staring" (Braud, Shafer, & Andrews, 1993a, 1993b), or "feeling of being seen through" (Chan, Huang, Wang, & Gong, 2009), i.e., one's own thought has been read or uncovered, are not uncommon in the general population.

Schizotypal personality disorder (SPD) traits or "psychosis proneness" may represent the distribution of the schizophrenia phenotype in the general population and has been defined within the framework of a schizotaxic, neurophysiological defect (Meehl, 1962, 1990). SPD is characterized by cognitive-perceptual and interpersonal disturbances, as well as disorganized behavior and speech (Raine, 2006). Behavioral genetic studies have shown that individuals with schizotypal personality traits appear to share liability genes for schizophrenia (Baron & Risch, 1987; Cadenhead & Braff, 2002). Neuroimaging studies also found that these at-risk individuals have structural or functional brain abnormalities similar to those of schizophrenia (Lagioia, Van De Ville, Debbane, Lazeyras, & Eliez, 2010; Moorhead et al., 2009; Peters et al., 2010; Raine, Sheard, Reynolds, & Lencz, 1992; Volpe et al., 2008), and proneness to SPD is significantly associated with reduced volume in the prefrontal and left medial temporal lobes in the general population (Moorhead et al., 2009; Raine et al., 1992). In nonclinical adolescent samples, a positive correlation between Schizotypal Personality Questionnaire (SPQ) scores and a functional connectivity measure has been reported for the visual network involving the occipital and bilateral temporal regions (Lagioia et al., 2010). Diffusion tensor fiber tracking studies also revealed abnormalities in white matter fiber tracts in individuals with high psychotic traits (Peters et al., 2010; Volpe et al., 2008).These lines of evidence suggest that schizotypal personality traits may be related to altered connectivity and brain structural/functional deficits.

Previous studies have demonstrated a link between paranoid belief and schizotypal personality as well as other features of schizophrenia (Chan et al., 2010; Thalbourne, 1994; Webb & Levinson, 1993). For example, Chan et al. (2010) examined the relationship between paranoid ideation and SPD in Chinese nonclinical samples. In this series of studies, more than 4000 undergraduates completed a checklist for paranoid ideation and the SPQ. Participants were classified into individuals with and without SPD features by their SPQ scores. They found the prevalence of paranoid ideation to be higher in individuals with SPD features. These studies provided evidence to support

the relationship between SPD personality and paranoid ideation in nonclinical samples. However, most of the studies exploring unfounded suspicion and paranoid ideation in nonclinical samples used clinical ratings or self-reported questionnaires (e.g., Fenigstein & Vanable, 1992; Freeman et al., 2005a), and the clinical validity and reliability of these rating scales were rarely examined.

Event-related potentials (ERPs) have been extensively applied in psychosis research and have provided important insights into the origins of paranoid thoughts. For example, preliminary clinical studies using ERP paradigms to investigate the neural activity underlying delusion have suggested that the N400 was associated with the severity of delusion (Debruille et al., 2007), whereas the N200 and the N300 were associated with reality distortion (Guillem et al., 2003; Williams, Gordon, Wright, & Bahramali, 2000) in schizophrenia. In a healthy population, Sumich, Kumari, Gordon, Tunstall, and Brammer (2008) used multiple linear regression to examine the relationships between SPD features and ERP, and found that paranoid ideation was inversely associated with the P300 amplitude at left-central scalp sites. Additionally, a leftward shift of P3a topographic descriptors and an increased cortical activation in left frontotemporal areas were observed in individuals with schizotypal and paranoid features (Mucci et al., 2005). However, all of these studies employed a traditional paradigm to evoke specific ERP components (e.g., oddball paradigm for P300, speech task for N400) and examined their relationships with psychotic symptom severity. More recently, Chan et al. (2009) designed a novel, digit-guessing task to induce a feeling of being seen through by others, which facilitates the examination of suspicious or paranoid thoughts directly with ERP. When the "feeling of being seen through" was induced, a prominent positive deflection of the ERP difference wave was observed around 250–350 ms after stimulus presentation.

In this study, we aimed to replicate and extend Chan et al.'s (2009) study in individuals with SPD features. We hypothesized that there would be a similar positive deflection elicited in individuals with SPD features, when the "feeling of being seen through" occurred. Moreover, given the nature and characteristics of SPD traits and their relationship to schizophrenia, we further hypothesized that there would be a distinct difference in the positive deflection between individuals with and without SPD features. In view of the association between delusion-like experiences and positive schizotypal personality, we hypothesized that the P3 amplitude would be more relevant to cognitive perceptual disturbance in SPD than to interpersonal disturbance and disorganized behavior or speech.

146

METHODS

Participants

A total of 501 university undergraduates completed a full version of the SPQ (Raine, 1991), Chinese version (Chen, Hsiao, & Lin, 1997). The SPQ measures schizotypal personality traits according to the nine features of SPD in the DSM-III-R (American Psychiatric Association, 1987). It has been shown that 55% of participants scoring in the top 10% of the SPQ had a clinical diagnosis of SPD (Raine, 1991). It is a 74-item questionnaire requiring simple "yes" or "no" answers. It captures specifically the nine traits of SPD; namely, idea of reference, excessive social anxiety, odd beliefs or magical thinking, unusual perceptual experiences, odd or eccentric behavior, lack of close friends, odd speech, constricted affect, and suspiciousness/paranoid ideation, and three factors of schizotypal personality (cognitive-perceptual, interpersonal, and disorganization). Students with a raw SPQ score in the top 10th percentile in the current sample were considered to be psychometrically defined SPD-prone individuals, while those with a SPQ score below the cutoff were randomly selected as comparison controls. All the participants (both SPD-prone and non-SPD-prone) were screened by a trained research assistant to ascertain the absence of a history of psychiatric and neurological disorders. None had a history of psychosis among their first-degree relatives. Finally, 33 volunteers completed the experiment, with 17 in the SPD-prone group (7 male and 10 female) and 16 in the non-SPD-prone group (8 male and 8 female). Overall, the participants had a mean age of 23 years (range = 18–30 years). All were right-handed and had normal or corrected to normal eyesight. All participants involved in the ERP experiment received monetary compensation and gave informed consent in accordance with the ethics committee of the Institute of Psychology at the Chinese Academy of Sciences.

Stimuli and procedure

In the ERP experiment, each participant was asked to come with a friend. The participants were instructed to input a digit from 1 to 9 at random on the computer and were told that their friend would guess which number they had input and the number guessed by their friend would be shown on the screen as a feedback. Participants were required to pay attention to the screen's feedback. However, the number of "correct guesses" or "incorrect guesses" was in fact controlled by the computer, and the participants were not told that their friend had not done anything until the experiment was completed. The stimulus procedure is shown in Figure 1.

Each trial began with an input box appearing at the center of the screen, which asked the participant to input a digit from 1 to 9 at random. If the participant input the digit out of this range or accidentally pressed the "enter" key without inputting, the program would ask the participant to input a proper digit again. If he or she input an appropriate digit, a text box would appear to inform the participant to wait for their friend to "guess" their inputs. This text box would appear for 1000–2000 ms at random. Then a "feedback" would be given to the participants for 1000 ms, and at the same time a trigger code would be sent to the electroencephalography (EEG) recording computer. Different trigger codes would be sent depending on whether the feedback was right or wrong. Afterwards, a blank screen would appear for 1000 ms to provide sufficient time for participants to ponder on the feedback.

The experiment contained four blocks and 40 trials in each block. In order to prevent arousing suspicion about the feedback digit in participants, the "correct guesses" number was kept at a low frequency (about 30%).

ERP recordings and data analysis

EEG recording was made from 64 scalp sites, using Ag/AgCl electrodes mounted in an elastic cap according to the modified expanded 10–20 system (NeuroScan, Inc. Charlotte, NC, USA). Continuous data were recorded relative to linked mastoids with a forehead ground. Vertical electrooculographic (EOG) recording electrodes were positioned above and below the left eye, and horizontal EOG recording electrodes

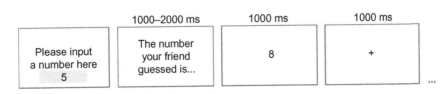

Figure 1. Schematic diagram of the task trials. The example shows that the participant inputs a number "5," and an incorrect feedback "8" is given by the computer.

were positioned at the outer canthi of both eyes. All electrode impedances were kept below 5 kΩ. The EEG and EOG were sampled at a digitization rate of 500 Hz and were bandpass filtered between 0.05 and 100 Hz.

The EEG data were analyzed by the Scan 4.3 software (NeuroScan, Inc.). Ocular artifacts were identified and corrected with an eye-movement correction algorithm. Any trials in which EEG voltages exceeded a threshold of ±100 μV during the recording epoch were excluded from the analysis. The data were corrected by subtracting from each sample the average activity of that channel during the baseline period. For each feedback stimulus, a 1200-ms epoch of data was extracted from the continuous data files for analysis (from 200 ms before the feedback onset to 1000 ms after the feedback onset). As a result, based on the feedback type, each participant had two average ERP waves (correct guess vs. incorrect guess). By a brief view from the scalp topography, a positive component during 250–450 ms in all conditions appeared greatest in the posterior scalp, and we labeled it as P3b.

Using the method of Chan et al. (2009), we computed the difference wave by subtracting the ERP of "incorrect guess" trials from the ERP of "correct guess" trials, and a positive deflection was observed during the 250–450-ms window time in both groups. To maintain consistency, we also called this component P3 as in Chan et al.'s 2009 study. To minimize overlap between "P3" and other ERP components, the

peak amplitude of P3 was measured as the maximum difference wave between "correct guess" and "incorrect guess" conditions in a window of 250–450 ms after the feedback presentation, relative to the baseline.

The following 18 main electrode sites were chosen for statistical analysis: FPz, Fz, FCz, Cz, Pz, Oz, FP1, FP2, F3, F4, FC3, FC4, C3, C4, P3, P4, O1, and O2 (Figure 2). The mean amplitude of P3b (250–450-ms time windows) of the original ERP wave was entered into a 2 (Group: SPD-proneness vs. non-SPD-proneness) × 2 (Feedback type: correct vs. incorrect) × 3 (Electrode laterality: left, middle, vs. right) × 6 (anterior-posterior scalp location: FP, F, FC, C, P, vs. O). Then the peak amplitude and latency of P3 of the difference wave were entered into a 2 (Group) × 3 (Electrode) × 6 (anterior-posterior scalp location) ANOVA, using the group as the between-subject factor. The Greenhouse–Geisser correction was used to compensate for sphericity violations.

Source analysis was performed on the "correct guess" minus "incorrect guess" difference wave, using the Curry V6.0 software (NeuroScan, Inc.). After the grand-averaged ERPs were obtained, the standard magnetic resonance imaging (MRI) head image supplied by the Curry software and the fiducial landmarks (the left and right preauricular points and the nasion) identified on it were fed into the software. A computer algorithm was automatically performed to calculate the best-fit sphere encompassed by the array

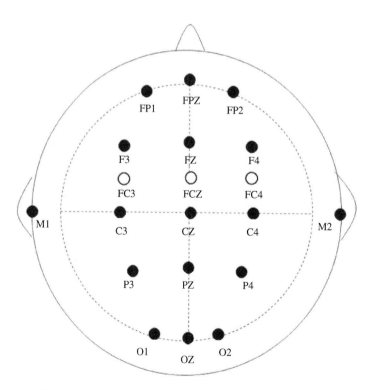

Figure 2. Position of the 18 channels of interest selected for analysis.

of electrode sites and to determine their spherical coordinates. The spherical coordinates for each recording site were used for the ERP current density analysis.

In this study, the low-resolution electromagnetic tomography method (LORETA) was used to reconstruct the distributed sources over the time range of 250–450 ms for the difference waves in a three-shell head model in the SPD-prone and non-SPD-prone groups separately.

RESULTS

ANOVA analysis

A four-way ANOVA of the mean amplitude of P3b revealed significant main effects of group, $F(1, 31) = 17.636$, $p < .001$; feedback type, $F(1, 31) = 19.719$, $p < .001$; and anterior-posterior electrodes, $F(5, 155) = 7.959$, $p = .005$. There was also a significant interaction effect of feedback type × group, $F(1, 31) = 9.874$, $p = .004$. The feedback type × group interaction effect was examined by comparing the effects of feedback type in each group and then comparing the group effect at each level of feedback type. At each level of feedback type, the non-SPD-prone group elicited a larger P3b than the SPD-prone group. However, the "correct" feedback stimuli elicited larger P3b amplitudes than "incorrect" feedback stimuli in the non-SPD-prone group, $F(1, 15) = 43.64$, $p < .001$, but not in the SPD-prone group, $F(1, 15) = 0.621$,

$p = .442$. The anterior-posterior electrodes main effect revealed P3b amplitude to be largest at the Pz electrode in both feedback type and groups.

ANOVA on P3 peak amplitudes of difference wave revealed significant main effects of group, $F(1, 31) = 13.888$, $p = .001$; electrode laterality, $F(2, 62) = 8.747$, $p = .01$; and anterior-posterior electrodes, $F(5, 155) = 14.983$, $p < .001$. There was no significant interaction between electrode site and group. The main effect of group on the P3 component of the "correct guess–incorrect guess" different wave indicated that the P3 amplitude was significantly larger in the non-SPD-prone group (7.93 μV at FCz) than the SPD-prone group (4.52 μV at FCz) (Figure 3). The laterality main effect showed that the P3 amplitudes in the midline were larger than those in the lateral scalp sites. The anterior-posterior electrodes main effect showed that the frontal P3 amplitudes were generally larger than the parietal-occipital P3 amplitudes. Overall, the P3 component was largest at the FCz electrode.

The same ANOVA analysis was conducted for P3 latency. Neither group effect nor group × electrode sites interaction effects were observed.

Current density analysis

The current density distribution of the difference waves was reconstructed in the time range 250–450 ms. The reconstructed results showed strong current density in the right inferior frontal gyrus (BA47),

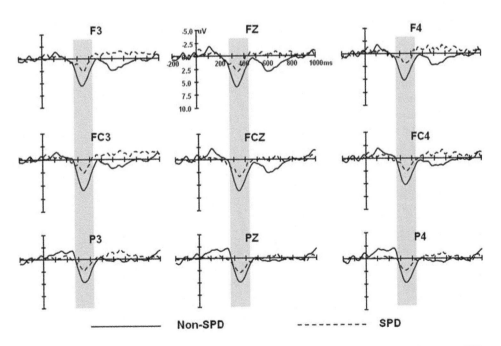

Figure 3. Grand-averaged different waveforms to feedback stimuli ("correct guess—incorrect guess") in SPD and non-SPD groups at nine electrode sites.

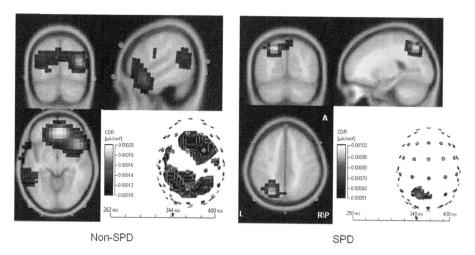

Non-SPD SPD

Figure 4. The averaged current density distribution of the difference waves in non-SPD (left) and SPD (right) groups at the time of difference wave peak (344 ms/non-SPD, 348 ms/SPD, respectively), reconstructed by the LORETA method, overlaid onto the cortex.

the anterior cingulate (BA32), the left lateral frontal gyrus (BA9), the left superior temporal gyrus (BA38), the bilateral middle temporal gyrus (BA21), and the parietal-occipital gyrus (BA18/19) in the non-SPD-prone group (Figure 4, left). In the SPD-prone group, only the superior parietal-occipital lobule (BA7/19) was activated in the current density mapping (Figure 4, right).

Correlation analysis

In order to examine the potential relationship between SPD features and brain activities, nonparametric Spearman correlation was conducted on P3 amplitude and each subscale score of SPQ in individuals with SPD features ($n = 17$). Three scalp sites of interest in the midline (Fz, FCz, Cz) were examined. In the SPD-prone group, significant correlations were found between the P3 amplitude and the suspiciousness/paranoid ideation trait at Fz ($r = -.476$, $p < .05$) and FCz ($r = -.523$, $p < .05$), and between P3 peak amplitude and the cognitive-perceptual factor at FCz ($r = -0.5$, $p < .05$) and Cz ($r = -.524$, $p < .05$) (Figure 5).

DISCUSSION

The first major finding of the current study was that there was a distinct difference in P3 amplitude between individuals with and without SPD features, and the positive component for the SPD-prone group was smaller than for the non-SPD-prone group. In our previous ERP study, the P3 component was induced when participants had the "feeling of being

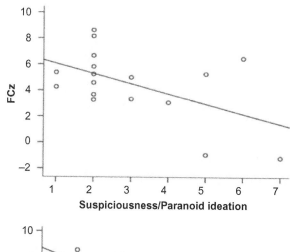

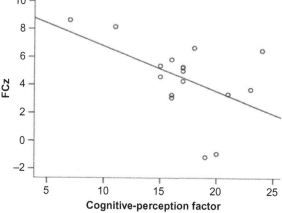

Figure 5. Correlation between P3 peak amplitude (in μV) of FCz site and suspiciousness/paranoid ideation subscale scores (top panel); and cognitive-perception factor of SPQ (bottom panel) for the individuals with SPD features ($n = 17$).

seen through," and the peak amplitude was inversely correlated with the participants' paranoid ideation scores, suggesting that people endorsing a higher level of suspicious thoughts and paranoid ideation would exhibit a smaller amplitude in this specific ERP component. One possible explanation is that for nonclinical participants, this may be a "usual" neural response when they recognized that their digits inputs were guessed correctly by others. For those with a higher level of paranoid ideation, this "usual" reaction might be inhibited, resulting in lower amplitude in the ERP component (Chan et al., 2009). The same explanation may be applicable in this study. The SPD-prone participants in this study were not surprised by the correct guesses by their friends, because they might have more suspicious or paranoid thoughts and believed that they were regularly being seen through by others all the time.

The P3 component in this study could be confounded by other ERP components, such as P3b, P3a, and feedback negativity (FN). The P3b is usually evoked in the oddball paradigm—the probability difference between target stimuli (infrequent) and nontarget stimuli (frequent) could generate a P3b component. In this study, we had kept the number of "correct guesses" to a low frequency (about 30%) to prevent participants from suspecting that the feedback digit was not guessed by their friends. However, our previous study had provided evidence that the "P3" was different from the P3b by exploring the role of feedback stimuli frequency in the generation of the "P3" component. In Chan's 2009 study, three frequency conditions of "correct guess" were involved; namely, high frequency (about 70%), middle frequency (about 50%), and low frequency (about 30%). The "P3" also existed in most of the scalp sites in the middle- and high-frequency conditions where the frequency of the "correct guess" was about the same as or greater than the "incorrect guess." As a result, the "P3" here might carry a different psychological meaning than simply probability difference. In addition, a discussion about the functional significance of the P3 component has been largely attributed to expectancy (see review by Kotchoubey et al., 1997). Researchers have found that the P3 reflects a mechanism of actualization of experience in the working memory related to receiving unexpected information (Donchin & Coles, 1988). This theory could explain the P3 component in our study. "Being seen through" might be unexpected information for the participants, leading to the P3 component being evoked.

Apart from the P3b, a smaller, frontocentrally distributed "P3a" component that peaks about 60–80 ms earlier than the P3b is elicited strongly in an auditory oddball paradigm that also incorporates an infrequent distracter stimulus or nonrepeating novel stimuli that require no behavioral response (Courchesne, Hillyard, & Galambos, 1975). These stimuli evoke a larger P3a

response than do targets, and manifest the involuntary capture of attention away from the central, ongoing task of detecting and responding to the task-relevant target stimuli (Debener, Kranczioch, Herrmann, & Engel, 2002; Friedman, Cycowicz, & Gaeta, 2001; Polich, 1998). It is obvious that there are no novel stimuli as distracters in our study, and the positive component of ERP evoked by feedback stimuli in original waveform was largest at posterior electrodes that are different from P3a largest at frontal electrodes.

Moreover, FN is most easily confused with "P3." Excluding the waveform direction feature, they have similar time windows (around 300 ms after the onset of feedback stimuli) and distribution sites over the scalp (largest at FCz site), and they both were computed as a difference wave by subtracting the ERP on right feedback trials from the ERP on wrong feedback trials. It is not clear whether the P3 in our study was in fact the same as FN. Previous studies have suggested that FN is influenced by the valence of the outcome in gambling tasks and outcome expectedness in time-estimation tasks. In gambling tasks, the amplitude of FN appeared to be greater when it followed losses rather than gains (Gehring & Willoughby, 2002), whereas, in time-estimation tasks, the amplitude of FN varied with the magnitude of the participants' prediction errors (Miltner, Braun, & Coles, 1997). Both explanations of FN did not fit the ERP paradigm in this study, as the feedback digits were unrelated to monetary reward, and our participants did not have error prediction to these feedbacks. Thus, the "P3" here may present specific mental processing (e.g., "feeling of being seen through") different from FN.

The current density analysis indicated that individuals with SPD proneness demonstrated extensive hypoactivation in their frontal lobes, anterior cingulate gyri, and temporal lobes. These findings are consistent with the preliminary findings on neuroanatomical features of SPD. For instance, Buchsbaum et al. (1997) reported that individuals with SPD showed decreased left and increased right prefrontal activations during the Wisconsin Card Sorting Task. High schizotypy scorers were found to exhibit both reduced dorsal and ventromedial prefrontal functioning during a self–other processing task (Platek et al., 2005), and increased right but decreased left activation in the dorsolateral prefrontal cortex during an emotional Stroop task (Mohanty et al., 2005). Moreover, reduced left temporal neural activity has been observed in those with SPD (Fukuzako, Kodama, & Fukuzako, 2002). However, fewer studies have examined brain functions of individuals with SPD. The hypoactivation regions, especially prefrontal and temporoparietal association cortices, in SPD-prone individuals in this study have been reported as a parallel network that facilitates arousal, attention, and motivation functions (Sackeim et al., 1990). Besides, some studies have suggested that these regions are associated with a number of higher cognitive functions including planning, decision-making, judgment, and impulse control (Stuss & Anderson, 2004). In our study, as the source of the P3 component, these hypoactive regions in SPD-prone participants may have a role in the generation of suspicious thoughts.

Correlation analysis also found a statistically meaningful relationship between ERP components and SPD proneness; that is, frontocentral brain activity was inversely associated with suspiciousness/paranoid ideation. Sumich et al. (2008) reported paranoid ideation to be negatively associated with the left-central P300 amplitude during an oddball paradigm in a nonclinical sample. Debruille et al. (2007) found a negative correlation between the N400 amplitude and delusion severity in a semantic processing task in schizophrenia. Thus, our findings are consistent with previous studies, and the digit-guessing feedback used in our paradigm provides another method of investigating subjective anomalous experiences associated with delusion formation in a nonclinical sample.

In addition, a negative correlation was seen between the "P3" component and the cognitive-perceptual factor of the SPQ from anterior to posterior central regions, suggesting that SPD-prone individuals with high positive symptoms would exhibit a lower amplitude of the P3 component than SPD-prone individuals with nonpositive symptoms. These findings reflect a close connection between paranoid ideation and positive SPD traits.

Our study has several limitations. First, "the feeling of being seen through" induced by the digit-guessing task might have been confounded by something more than the results of suspiciousness, such as odd beliefs or magical thinking (e.g., belief in paranormal/supernatural phenomena such as telepathy, clairvoyance, and precognition). It was possible that SPD-prone individuals were more likely to believe that their friends had the ability of telepathy or mind reading than non-SPD-prone individuals, such that they were not surprised to see "correct guesses." However, the significant inverse correlation between suspiciousness/paranoid ideation trait scores on the SPQ and the P3 component does not support the idea that our findings were confounded by odd beliefs/magical thinking traits.

Secondly, the signal-processing technique we used had low spatial resolution. Although source localization techniques have provided some cerebral current activations, the specific cerebral regions involved in the processing of suspicious thoughts were not definitively localized. Other neuroimaging methods could be used to more precisely elucidate the involved brain regions. Further studies should investigate the relationship between cerebral activation in SPD-prone and non-SPD-prone individuals and other indices of mechanisms implicated in paranoid ideation (e.g., measured by questionnaire, or behavioral performance).

Thirdly, the correlation analysis was conducted on the data of the SPD-prone group only; as a result, the sample size was small ($n = 17$). However, putting both the SPD-prone and the non-SPD-prone groups into the correlation analysis would result in correlations between all nine SPD traits and the ERP component, because these two groups have significant differences on the nine SPD traits. Further studies should examine the positive or negative features of schizotypal personality.

Notwithstanding these limitations, our results provide electrophysiological evidence that SPD-prone individuals have more suspicious thoughts than non-SPD-prone individuals. The current density analysis suggests that frontal and temporal regions may play an important role in the generation of suspicious thoughts. In addition, the P3 of the difference wave was negatively correlated with the SPQ cognitive-perception factor and the suspiciousness/paranoid ideation trait, providing further evidence for the association between paranoid ideation and positive features of SPD.

REFERENCES

American Psychiatric Association. (1987). DSM-IIII-R: Diagnostic and Statistical Manual of Mental Disorders. 3rd ed., revised. Washington, DC: The Association.

Baron, M., & Risch, N. (1987). The spectrum concept of schizophrenia: Evidence for a genetic-environmental continuum. *Journal of Psychiatric Research, 21*(3), 257–267.

Bijl, R. V., van Zessen, G., Ravelli, A., de Rijk, C., & Langendoen, Y. (1998). The Netherlands Mental Health Survey and Incidence Study (NEMESIS): Objectives and design. *Social Psychiatry and Psychiatric Epidemiology, 33*(12), 581–586.

Braud, W. G. (1994). Empirical explorations of prayer, distant healing, and remote mental influence. *Journal of Religion and Psychical Research, 17*(2), 62–73.

Braud, W. G., Shafer, D., & Andrews, S. (1993a). Further studies of autonomic detection of remote staring: Replication, new control procedures, and personality correlates. *Journal of Parapsychology, 57*(4), 391–409.

Braud, W. G., Shafer, D., & Andrews, S. (1993b). Reactions to an unseen gaze (remote attention): A review, with new data on autonomic staring detection. *Journal of Parapsychology, 57*(4), 373–390.

Buchsbaum, M. S., Trestman, R. L., Hazlett, E., Siegel, B. V., Jr., Schaefer, C. H., Luu-Hsia, C., et al. (1997). Regional cerebral blood flow during the Wisconsin Card Sort Test in schizotypal personality disorder. *Schizophrenia Research, 27*(1), 21–28.

Cadenhead, K. S., & Braff, D. L. (2002). Endophenotyping schizotypy: A prelude to genetic studies within the schizophrenia spectrum. *Schizophrenia Research, 54* (1–2), 47–57.

Chan, R. C., Huang, J., Wang, Y., & Gong, Q. Y. (2009). Psychological investigation of the "feeling of being seen through" in a non-clinical sample using an ERP paradigm. *Brain Research, 1254,* 63–73.

Chan, R. C., Li, X., Lai, M. K., Li, H., Wang, Y., Cui, J., et al. (2010). Exploratory study on the base-rate of paranoid ideation in a non-clinical Chinese sample. *Psychiatry Research, 185,* 254–260.

Chen, W. J., Hsiao, C. K., & Lin, C. C. (1997). Schizotypy in community samples: The three-factor structure and correlation with sustained attention. *Journal of Abnormal Psychology, 106*(4), 649–654.

Claridge, G. (1994). Single indicator of risk for schizophrenia: Probable fact or likely myth? *Schizophrenia Bulletin, 20*(1), 151–168.

Courchesne, E., Hillyard, S. A., & Galambos, R. (1975). Stimulus novelty, task relevance and the visual evoked potential in man. *Electroencephalography and Clinical Neurophysiology, 39*(2), 131–143.

Debener, S., Kranczioch, C., Herrmann, C. S., & Engel, A. K. (2002). Auditory novelty oddball allows reliable distinction of top-down and bottom-up processes of attention. *International Journal of Psychophysiology, 46*(1), 77–84.

Debruille, J. B., Kumar, N., Saheb, D., Chintoh, A., Gharghi, D., Lionnet, C., et al. (2007). Delusions and processing of discrepant information: An event-related brain potential study. *Schizophrenia Research, 89*(1–3), 261–277.

Donchin, E., & Coles, M. G. H. (1988). Is the P300 component a manifestation of context updating? *Behavioral and Brain Sciences, 11,* 357–374.

Fenigstein, A., & Vanable, P. A. (1992). Paranoia and self-consciousness. *Journal of Personality and Social Psychology, 62*(1), 129–138.

Freeman, D. (2007). Suspicious minds: The psychology of persecutory delusions. *Clinical Psychology Review, 27*(4), 425–457.

Freeman, D., Dunn, G., Garety, P. A., Bebbington, P., Slater, M., Kuipers, E., et al. (2005a). The psychology of persecutory ideation. I. A questionnaire survey. *Journal of Nervous and Mental Disease, 193*(5), 302–308.

Freeman, D., Garety, P. A., Bebbington, P., Slater, M., Kuipers, E., Fowler, D., et al. (2005b). The psychology of persecutory ideation. II. A virtual reality experimental study. *Journal of Nervous and Mental Disease, 193*(5), 309–315.

Friedman, D., Cycowicz, Y. M., & Gaeta, H. (2001). The novelty P3: An event-related brain potential (ERP) sign of the brain's evaluation of novelty. *Neuroscience and Biobehavioral Review, 25*(4), 355–373.

Fukuzako, H., Kodama, S., & Fukuzako, T. (2002). Phosphorus metabolite changes in temporal lobes of subjects with schizotypal personality disorder. *Schizophrenia Research, 58*(2–3), 201–203.

Gehring, W. J., & Willoughby, A. R. (2002). The medial frontal cortex and the rapid processing of monetary gains and losses. *Science, 295*(5563), 2279–2282.

Guillem, F., Bicu, M., Pampoulova, T., Hooper, R., Bloom, D., Wolf, M. A., et al. (2003). The cognitive and anatomo-functional basis of reality distortion in schizophrenia: A view from memory event-related potentials. *Psychiatry Research, 117*(2), 137–158.

Johns, L. C., Cannon, M., Singleton, N., Murray, R. M., Farrell, M., Brugha, T., et al. (2004). Prevalence and correlates of self-reported psychotic symptoms in the British population. *British Journal of Psychiatry, 185,* 298–305.

Kendler, K. S., Gallagher, T. J., Abelson, J. M., & Kessler, R. C. (1996). Lifetime prevalence, demographic risk factors, and diagnostic validity of nonaffective psychosis as assessed in a US community sample. The National Comorbidity Survey. *Archives of General Psychiatry, 53*(11), 1022–1031.

Kotchoubey, B., Grozinger, B., Kornhuber, A.W. & Kornhuber, H.H. (1997). Electrophysiological analysis of expectancy: P3 in informed guessing. *International Journal of Neuroscience, 91,* 105–122.

Lagioia, A., Van De Ville, D., Debbane, M., Lazeyras, F., & Eliez, S. (2010). Adolescent resting state networks and their associations with schizotypal trait expression. *Frontiers in Systems Neuroscience, 4,* pii: 35.

Meehl, P. E. (1962). Schizotaxia, schizotypy, schizophrenia. *American Psychologist, 17,* 827–838.

Meehl, P. E. (1990). Toward an integrated theory of schizotaxia, schizotypy and schizophrenia. *Journal of Personality Disorders, 4,* 1–99.

Miltner, W. H. R., Braun, C. H., & Coles, M. G. H. (1997). Event-related brain potentials following incorrect feedback in a time-estimation task: Evidence for a "generic" neural system for error detection. *Journal of Cognitive Neuroscience, 9*(6), 788–798.

Mohanty, A., Herrington, J. D., Koven, N. S., Fisher, J. E., Wenzel, E. A., Webb, A. G., et al. (2005). Neural mechanisms of affective interference in schizotypy. *Journal of Abnormal Psychology, 114*(1), 16–27.

Moorhead, T. W., Stanfield, A., Spencer, M., Hall, J., McIntosh, A., Owens, D. C., et al. (2009). Progressive temporal lobe grey matter loss in adolescents with schizotypal traits and mild intellectual impairment. *Psychiatry Research, 174*(2), 105–109.

Mucci, A., Galderisi, S., Bucci, P., Tresca, E., Forte, A., Koenig, T., et al. (2005). Hemispheric lateralization patterns and psychotic experiences in healthy subjects. *Psychiatry Research, 139*(2), 141–154.

Peters, B. D., Dingemans, P. M., Dekker, N., Blaas, J., Akkerman, E., van Amelsvoort, T. A., et al. (2010). White matter connectivity and psychosis in ultra-high-risk subjects: A diffusion tensor fiber tracking study. *Psychiatry Research, 181*(1), 44–50.

Platek, S. M., Fonteyn, L. C., Izzetoglu, M., Myers, T. E., Ayaz, H., Li, C., et al. (2005). Functional near infrared spectroscopy reveals differences in self–other processing as a function of schizotypal personality traits. *Schizophrenia Research, 73*(1), 125–127.

Polich, J. (1998). P300 clinical utility and control of variability. *Journal of Clinical Neurophysiology, 15*(1), 14–33.

Raine, A. (1991). The SPQ: A scale for the assessment of schizotypal personality based on DSM-III-R criteria. *Schizophrenia Bulletin, 17*(4), 555–564.

Raine, A. (2006). Schizotypal personality: Neuro-developmental and psychosocial trajectories. *Annual Review of Clinical Psychology, 2*, 291–326.

Raine, A., Sheard, C., Reynolds, G. P., & Lencz, T. (1992). Pre-frontal structural and functional deficits associated with individual differences in schizotypal personality. *Schizophrenia Research, 7*(3), 237–247.

Sackeim, H. A., Prohovnik, I., Moeller, J. R., Brown, R. P., Apter, S., Prudic, J., et al. (1990). Regional cerebral blood flow in mood disorders. I. Comparison of major depressives and normal controls at rest. *Archives of General Psychiatry, 47*(1), 60–70.

Sartorius, N., Jablensky, A., Korten, A., Ernberg, G., Anker, M., Cooper, J. E., et al. (1986). Early manifestations and first-contact incidence of schizophrenia in different cultures. A preliminary report on the initial evaluation phase of the WHO Collaborative Study on determinants of outcome of severe mental disorders. *Psychological Medicine, 16*(4), 909–928.

Scott, J., Chant, D., Andrews, G., & McGrath, J. (2006). Psychotic-like experiences in the general community: The correlates of CIDI psychosis screen items in an Australian sample. *Psychological Medicine, 36*(2), 231–238.

Stuss, D. T., & Anderson, V. (2004). The frontal lobes and theory of mind: Developmental concepts from adult focal lesion research. *Brain and Cognition, 55*(1), 69–83.

Sumich, A., Kumari, V., Gordon, E., Tunstall, N., & Brammer, M. (2008). Event-related potential correlates of paranormal ideation and unusual experiences. *Cortex, 44*(10), 1342–1352.

Thalbourne, M. A. (1994). Belief in the paranormal and its relationship to schizophrenia-relevant measures: A confirmatory study. *British Journal of Clinical Psychology, 33*(Pt 1), 78–80.

van Os, J., Hanssen, M., Bijl, R. V., & Ravelli, A. (2000). Strauss (1969) revisited: A psychosis continuum in the general population? *Schizophrenia Research, 45*(1–2), 11–20.

Volpe, U., Federspiel, A., Mucci, A., Dierks, T., Frank, A., Wahlund, L. O., et al. (2008). Cerebral connectivity and psychotic personality traits. A diffusion tensor imaging study. *European Archives of Psychiatry and Clinical Neuroscience, 258*(5), 292–299.

Webb, C. T., & Levinson, D. F. (1993). Schizotypal and paranoid personality disorder in the relatives of patients with schizophrenia and affective disorders: A review. *Schizophrenia Research, 11*(1), 81–92.

Williams, L. M., Gordon, E., Wright, J., & Bahramali, H. (2000). Late component ERPs are associated with three syndromes in schizophrenia. *International Journal of Neuroscience, 105*(1–4), 37–52.

Theory of mind in schizophrenia: Exploring neural mechanisms of belief attribution

Junghee Lee[1,2], Javier Quintana[1,2], Poorang Nori[1,2], and Michael F. Green[1,2]

[1] Semel Institute for Neuroscience and Human Behavior, University of California (UCLA), Los Angeles, CA, USA

[2] Veterans Affairs Greater Los Angeles Healthcare System, Los Angeles, CA, USA

Background: Although previous behavioral studies have shown that schizophrenia patients have impaired theory of mind (ToM), the neural mechanisms associated with this impairment are poorly understood. This study aimed to identify the neural mechanisms of ToM in schizophrenia, using functional magnetic resonance imaging (fMRI) with a belief attribution task.

Methods: In the scanner, 12 schizophrenia patients and 13 healthy control subjects performed the belief attribution task with three conditions: a false belief condition, a false photograph condition, and a simple reading condition.

Results: For the false belief versus simple reading conditions, schizophrenia patients showed reduced neural activation in areas including the temporoparietal junction (TPJ) and medial prefrontal cortex (MPFC) compared with controls. Further, during the false belief versus false photograph conditions, we observed increased activations in the TPJ and the MPFC in healthy controls, but not in schizophrenia patients. For the false photograph versus simple reading condition, both groups showed comparable neural activations.

Conclusions: Schizophrenia patients showed reduced task-related activation in the TPJ and the MPFC during the false belief condition compared with controls, but not for the false photograph condition. This pattern suggests that reduced activation in these regions is associated with, and specific to, impaired ToM in schizophrenia.

Keyword: Theory of mind; Belief attribution; Schizophrenia; fMRI; Social cognition.

Schizophrenia is a complex and severe mental disorder that affects approximately 1% of the population worldwide. It is associated with poor functioning that is observed in the form of poor family relationships, difficulty in maintaining employment, and social withdrawal. Impaired functioning in daily life is present before the onset of psychosis (Davidson et al., 1999) and tends to persist throughout the course of illness (Murray & Lopez, 1997; WHO, 2008). Psychotic symptoms are not generally strongly associated with functional outcome, but other factors, such as cognition and negative symptoms (e.g., lack of drive, flat affect), appear to be consistent determinants (Green, 1996; Green, Kern, Braff, & Mintz, 2000; Milev, Ho, Arndt, & Andreasen, 2005). Within the general area of cognition, social cognition has been shown to be a key determinant of poor functioning in schizophrenia (Kerr & Neale, 1993; Penn, Corrigan, Bentall, Racenstein, & Newman, 1997; Penn, Ritchie, Francis, Combs, & Martin, 2002). This study aims to investigate the neural

Correspondence should be addressed to: Junghee Lee, 300 Medical Plaza, Room 2261, Semel Institute for Neuroscience and Human Behavior, UCLA, Los Angeles, CA 90095-6968, USA. E-mail: jungheelee@ucla.edu

Support for this study came from the Veterans Integrated Service Network 22 Mental Illness Research, Education and Clinical Center (to Lee); a Merit Review Award from the Veterans Administration Medical Research Service (to Quintana); and NIMH grants MH043292 and MH065707 (to Green). We wish to thank Dr. Rebecca Saxe for providing the vignettes for the task. For generous support of the UCLA Brain Mapping Center, we also thank the Brain Mapping Medical Research Organization, Brain Mapping Support Foundation, Pierson-Lovelace Foundation, Ahmanson Foundation, William M. and Linda R. Dietel Philanthropic Fund at the Northern Piedmont Community Foundation, Tamkin Foundation, Jennifer Jones-Simon Foundation, Capital Group Companies Charitable Foundation, Robson Family, and Northstar Fund. Financial disclosure: The authors have nothing to disclose.

mechanisms in schizophrenia of one area of social cognition, theory of mind, using functional magnetic resonance imaging (fMRI).

Theory of mind (ToM; also called mental state attribution) is a social cognitive construct that refers to the ability to make high-level inferences about one's own and other persons' mental states such as thoughts, beliefs, desires, and feelings (Premack & Woodruff, 1978). Such an ability starts at a very early stage of development (around ages 3–4) (Dennett, 1978; Leslie, 1987). From an operational perspective, ToM makes it possible to assess or understand someone else's mental states in specific situations, and thus to interpret and anticipate his behaviors (e.g., He is missing his wallet. He is going back to his office because he *thinks* he left his wallet there). The ability to understand and predict others' behavior has obvious relevance for successful social interactions.

Previous studies using a variety of behavioral paradigms have found that schizophrenia patients show impairment in their ability to attribute mental states to others. Schizophrenia patients showed poor performance on picture-sequencing tasks (Sarfati, Hardy-Bayle, Besche, & Widlocher, 1997), on tasks detecting irony or sarcasm (Kern et al., 2009; Mitchley, Barber, Gray, Brooks, & Livingston, 1998), on tasks attributing spontaneous mental states to non-human objects (Horan et al., 2009), and on tasks of false-belief stories (Frith & Corcoran, 1996; Pickup & Frith, 2001). A recent meta-analysis (Bora, Yucel, & Pantelis, 2009) found a large effect size (1.10) for the patient–control difference, and the effect size remained large when the comparison was focused on remitted patients only (0.80). Further, the patient–group difference appears to be similar across all phases of the illness. A recent study from our group (Green et al., in press) compared the patient–group differences in ToM among individuals considered to be in a prodromal phase, first-episode schizophrenia patients, and chronic schizophrenia patients, and found that the group differences were relatively constant across all phases of illness, indicating good stability of the social cognitive impairment.

One might wonder whether impaired ToM in schizophrenia is due to general cognitive impairments; however, studies have shown that impaired ToM in schizophrenia cannot be explained by nonsocial cognitive impairments (Langdon, Coltheart, Ward, & Catts, 2001; Mazza, De Risio, Surian, Roncone, & Casacchia, 2001; Schenkel, Spaulding, & Silverstein, 2005). For example, one study (Schenkel et al., 2005) showed that ToM in schizophrenia was not significantly related to verbal fluency or verbal intelligence. Another study (2001) failed to find a significant correlation between ToM and executive function (measured by the Wisconsin Card Sorting Test) or general intelligence. In addition, Langdon et al. (2001) showed that schizophrenia patients had performance comparable to controls when making inferences about nonsocial cause-and-effect relations or social knowledge in general (without a ToM component), but they showed impaired ToM performance. These findings suggest that, even though schizophrenia patients have impaired cognitive functions, impaired ToM in schizophrenia cannot be fully explained by deficits in other cognitive processes.

Most of the studies of neural mechanisms associated with ToM in healthy individuals employed either a false belief task or a cartoon task. Commonly activated areas include the medial prefrontal cortex, the temporal-parietal junction (TPJ), and the precuneus/posterior cingulate cortex (Frith & Frith, 2006; Gallagher & Frith, 2003; Saxe & Kanwisher, 2003; Saxe & Powell, 2006; Siegal & Varley, 2002), known as the ToM network. A general question is whether activation of these areas is due to the general reasoning demands of the task or is specific to attributing mental states to other people. Kanwisher, Saxe, and colleagues attempted to address this issue by conducting a series of fMRI studies that compared regional brain activation during stories that involved belief attribution with stories that required similar reasoning ability, but without belief attribution (Saxe & Kanwisher, 2003; Saxe & Powell, 2006). For example, they compared the neural activations of belief attribution with nonsocial reasoning ability. To measure belief attribution, false belief stories were used in which subjects were asked to make inferences about the belief of another person that differs from the current state of the world. To measure nonsocial reasoning ability, a false photograph task was used that required subjects to make inferences about the physical world that differ from the current state of the world. In other words, both the false belief task and the false photograph task asked subjects to inhibit the current or "true" representation and make a decision based on the "false" representation of the belief state (as in the false belief task) or the physical world (as in the false photograph task). While both the false belief task and the false photograph task require inhibitory control process, only the false belief condition involves belief attribution, one component of ToM. They found that subjects showed greater activation in the medial prefrontal cortex, the bilateral TPJ, and the posterior cingulate cortex when they read stories about belief attribution than when they read the false photograph stories. In a subsequent study, Kanwisher and Saxe added a human nonbelief condition and showed that activations in those brain regions was not due to the fact that stories involved human beings; instead, they were specific to the belief attribution component (Saxe & Kanwisher, 2003; Saxe & Powell, 2006). It is possible that these brain regions are related only to the belief attribution component of ToM, but not to ToM in general (e.g., attributing emotion, desire, or other mental states). Several studies showed that the bilateral TPJ and the posterior cingulate cortex were activated only during the belief attribution, whereas the medial prefrontal cortex was activated for belief attribution and also other mental state attributions (Jenkinson & Mitchell, 2010; Saxe & Powell, 2006; van Overwalle, 2009). These findings suggest that the TPJ and posterior cingulate cortex may be specifically involved in belief attribution, but the medial prefrontal cortex is related to ToM more broadly.

Only a few studies have examined ToM in schizophrenia with functional neuroimaging. One fMRI study (Russell et al., 2000) reported reduced activation in the left prefrontal cortex when schizophrenia patients were asked to describe the mental state reflected in photographs of eyes. Another study used a cartoon task to evaluate how people infer social intention and found decreased activations in the TPJ and the medial prefrontal cortex in schizophrenia patients (Walter et al., 2009). In contrast, another study that used a similar cartoon task found increased activations in the medial prefrontal cortex and TPJ in schizophrenia patients (Brune et al., 2008). Although these studies suggest that abnormal brain activation is associated with impaired ToM in schizophrenia, the directions of the findings are inconsistent across studies. Further, the methods are limited to visual assessment of ToM, and they did not fully control for the extent of nonspecific task-related demands.

In this study, we investigated the neural mechanisms of ToM in schizophrenia, using a well-validated fMRI paradigm in social neuroscience: the belief attribution task (Saxe & Kanwisher, 2003). This task, as adapted for use in schizophrenia, consists of three conditions: a false belief condition, a false photograph condition, and a simple reading condition. In the false belief condition, a character's belief is false; that is, it is different from the actual situation (Dennett, 1978). To perform this condition well, subjects must predict the character's behavior based on the character's (inaccurate) belief, and not based on the actual situation. The false photograph condition was intended to control for the general problem-solving structure of the false belief vignette (Zaitchik, 1990). These vignettes have the same story structure but differ in that the false photograph condition requires similar general reasoning ability without the belief attribution. To adapt this task

for use in schizophrenia, we added a reading condition as a control for the reading ability and associated regional activity required by the false belief condition. Thus, having three conditions allowed us to determine neural substrates of ToM deficits in schizophrenia while controlling for other nonspecific task-related demands (i.e., general reasoning ability, ability to read simple vignettes). With the belief attribution task, we hypothesized that schizophrenia patients would show reduced task-related activations in brain regions associated with ToM during the false belief condition, but not during the false photograph condition, compared to healthy controls.

METHODS

Participants

Fourteen (3 female) patients with schizophrenia and 14 (3 female) healthy controls participated in this study. Schizophrenia patients were recruited from outpatient clinics at the Veterans Affairs Greater Los Angeles Healthcare System (VAGLAHS) and from local board and care facilities. Patients met diagnostic criteria for schizophrenia according to the Structured Clinical Interview for DSM-IV Axis I Disorders (SCID) . Exclusion criteria for patients included the following: (1) substance abuse or dependence in the last 6 months; (2) mental retardation based on review of medical records; (3) history of loss of consciousness for more than 1 h; (4) an identifiable neurological disorder; or (5) insufficient fluency in English to understand testing procedures. All patients were medicated and clinically stable at the time of testing.

Healthy control participants were recruited through flyers distributed in the local community and website postings. Exclusion criteria for control participants included the following: (1) history of schizophrenia or other psychotic disorder, bipolar disorder, recurrent depression, history of substance dependence, or any substance abuse in the last 6 months based on the SCID (First et al., 1997); (2) any of the following Axis II disorders: avoidant, paranoid, schizoid, or schizotypal, based on the SCID for Axis II disorders (First, Gibbon, Spitzer, Williams, & Benjamin, 1996); (3) schizophrenia or other psychotic disorder in a first-degree relative; (4) any significant neurological disorder or head injury with loss of consciousness for more than 1 h; or (5) insufficient fluency in English to understand testing procedures.

All SCID interviewers were trained to a minimum kappa of .75 for key psychotic and mood items through the Treatment Unit of the VA VISN 22 Mental Illness

157

Research, Education, and Clinical Center (MIRECC). All participants were evaluated for the capacity to give informed consent, and they provided written, informed consent after all procedures were fully explained, according to procedures approved by the Institutional Review Boards at UCLA and the VAGLAHS.

Design and procedure

The belief attribution task was modeled after Saxe and Kanwisher (2003) and was composed of three conditions: a false belief condition, a false photograph condition, and a simple reading condition (see examples in Table 1). In the false belief condition, subjects were presented with vignettes in which they needed to infer the beliefs of a character, even when these beliefs were different from the actual state of affairs. The false photograph vignettes had the same story structure and required the same level of complex reasoning as the false belief condition, but lacked the belief attribution component. The simple reading condition controlled for the process of reading the false belief condition required, and consisted of stories describing nonhuman objects.

All subjects were presented with 12 vignettes (average number of words = 32) for each condition, and each vignette was accompanied by a single, two-alternative, forced-choice "fill-in-the-blank" question. The fill-in-the-blank question consisted of a single sentence with a word missing, presented above two alternative completions. For the false belief and false photograph conditions, 50% of the questions asked about the content of false representation; the other 50% asked about the actual outcome of the story. For the reading condition, all of the questions were related to description of nonhuman objects.

At the onset of each trial, a vignette was presented for 12 s; next a fill-in-the-blank question was presented for 10 s while the vignette was still visible. After the vignette and question disappeared, a probe was presented for 3 s, prompting the subject's response. Subjects responded by pressing the corresponding button with their dominant hand. Then, an intertrial interval (ITI) that was jittered between 12 and 18 s ensued. The belief attribution task consisted of six runs, each lasting 4 min and 28 s, with six trials per run (two trials of each condition). All tasks were presented through MR-compatible LCD goggles (Resonance Technology, Northridge, CA, USA). A schematic diagram of the procedure is shown in Figure 1.

fMRI data acquisition

All scanning was conducted on a 3T scanner (Siemens Trio, Erlangen, Germany) located in the UCLA Ahmanson Lovelace Brain Mapping Center. For anatomical reference, a high-resolution, echo planar axial T2-weighted series was obtained for each subject prior to functional scanning (TR = 5000 ms, TE = 30 ms, flip angle = 90°, 33 slices, FOV = 22 cm). A T2*-weighted gradient-echo sequence was used to detect blood-oxygen-level-dependent (BOLD) signal (TR = 2000 ms, TE = 30 ms, flip angle = 75°, voxel size of 3.4 × 3.4 × 4.00 mm), acquiring 33 slices parallel to the AC–PC plane.

fMRI data analysis

Imaging data were analyzed by the FEAT (fMRI Expert Analysis Tool), Version 5.98, part of the FMRIB Software Library (FSL) (Smith et al., 2004). The pre-statistics processing included motion correction by MCFLIRT (Motion Correction using FMRIB's Linear Imaging Registration Tool) (M. Jenkinson, Bannister, Brady, & Smith, 2002), non-brain removal

TABLE 1
Examples of vignettes

Condition	Vignette	Question
False belief	David knows that Ethan is very scared of spiders. Ethan, alone in the attic, sees a shadow move and thinks it is a burglar. David hears Ethan cry for help.	David assumes that Ethan thinks he has seen ___. a spider a burglar
False photograph	Amy made a drawing of a treehouse three years ago. That was before the storm. We built a new treehouse last summer, but we painted it red instead of blue.	The treehouse in Amy's drawing is ___. red blue
Simple reading	A lemon tree can grow up to 20 feet high, but they are usually smaller. The branches are thorny, and the leaves are green and shiny. Flowers are white on the outside with a violet-streaked interior.	The flower of a lemon tree is ___ on the outside. white violet

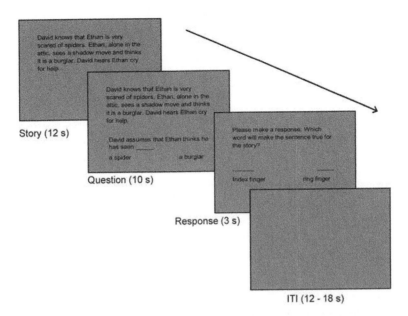

Figure 1. A schematic diagram of the belief attribution task showing the temporal sequence of a single event. At the beginning of each trial, a vignette was presented for 12 s. A fill-in-the-blank question was presented for 10 s while the vignette was still visible. After the vignette and question disappeared, a response probe was presented for 3 s. The intertrial interval (ITI) was jittered between 12 and 18 s.

by BET (Brain Extraction Tool) (Smith, 2002), spatial smoothing with a Gaussian kernel of the full width at half-maximum (FWMH) 5 mm, grand-mean intensity normalization by a single multiplicative factor, and high-pass temporal filtering (Gaussian weighted LSF straight line fitting with sigma = 50.0 s). To facilitate multisubject analyses, statistical images created for each subject were normalized into a standard space from the Montreal Neurological Institute (MNI), using affine transformation with FLIRT (FMRIB's Linear Image Registration Tool) (M. Jenkinson & Smith, 2001)(Jenkinson & Smith, 2001).

Functional images were analyzed with FLIM (FMRIB's Improved Linear Model) with local autocorrelation correction (Woolrich, Ripley, Brady, & Smith, 2001). First, for each run and each subject, data from all three conditions were modeled by convolving them with a canonical hemodynamic response function, and temporal derivatives were included as covariates of no interest to increase statistical sensitivity. We computed three main contrasts of interest for each run and each subject: false belief versus simple reading, false photograph versus simple reading, and false belief versus false photograph. Second, to average across the six runs, we completed a second-level analysis, using a fixed-effects model, by forcing the random effect variance to zero in FLAME (FMRIB's Local Analysis of Mixed Effects) (Beckmann, Jenkinson, & Smith, 2003; Woolrich, Behrens, Beckmann, Jenkinson, & Smith, 2004). Third, to characterize functional activations in each

group separately and directly compare activations of patients to those of controls for each contrast of interest, a mixed-effects model (FLAME stages 1 + 2) (Beckmann et al., 2003; Woolrich, et al., 2004) was performed, and the resulting statistical images were thresholded with a z value of >2.3 and a cluster probability of $p = .05$, corrected for whole-brain multiple comparisons by Gaussian random field theory (Worsley, 2001) unless otherwise noted.

RESULTS

Two patients were excluded from analyses for below-chance-level performance during the belief attribution task (below 50% accuracy across the three conditions), and one control was excluded due to technical problems during scanning. Therefore, 12 schizophrenia patients (2 female) and 13 healthy controls (3 female) were included in the following analyses.

Demographic information and behavioral performance

Table 2 presents demographic information and behavioral performance during the belief attribution task. Schizophrenia patients and healthy controls were comparable in terms of age, parental education, and gender—age, $t(23) = -1.1, p = .27$; parental education,

TABLE 2
Demographic information and behavioral performance

	Schizophrenia patients (n = 12)	Healthy controls (n = 13)
Age	38.3 (10.7)	42.5 (7.7)
Gender (female / male)	2 / 10	3 / 10
Personal education (years)	12.5 (2.3)	14.2 (1.3)
Parental education	13.1 (2.9)	14.3 (2.9)
Belief attribution task*		
False belief	8.5 (1.3)	10.5 (1.3)
False photograph	7.6 (1.8)	9.4 (1.1)
Simple reading	9.9 (1.5)	11.2 (0.7)

Notes: Values are given as mean (SD).*Accuracy = number of correct responses out of 12.

$t(23) = -1.09$, $p = .28$; gender, $\chi^2 = 1.60$, $p = .68$—but not in their own education levels, $t(23) = -2.19$, $p < .05$.

Behavioral performance of both groups during the belief attribution task was examined using a 3 (condition) by 2 (group), repeated-measures ANOVA with condition as a within-subject factor and group as a between-subject factor. We found a significant main effect of condition, $F(2, 46) = 24.32$, $p < .001$, and a significant main effect of group, $F(1, 23) = 16.31$, $p < .01$. The condition by group interaction was not significant. Schizophrenia patients performed worse than controls across all conditions. Across groups, performance was best on the simple reading condition, intermediate on the false belief condition, and worst on the false photograph condition. Each condition was significantly different ($p < .05$) from the other after correction for multiple comparisons.

fMRI activations during the belief attribution task

Table 3 lists brain areas (local maxima of the clusters) that exhibited above-threshold activations for each contrast of interest in the healthy control and schizophrenia patient groups. For the false belief versus simple reading contrast (see Figure 2), healthy controls showed increased activations in several areas, including the bilateral TPJ, bilateral middle temporal gyri, bilateral angular gyri, precuneus, right middle frontal gyrus, and medial prefrontal cortex. Schizophrenia patients showed increased activations in the right TPJ, precuneus, and right supramarginal gyrus. When the groups were compared directly to each other, controls showed significantly more activation relative to patients on the bilateral TPJ, left middle temporal gyri, right medial prefrontal cortex, putamen, globus pallidus, and right amygdala.

For the contrast of the false photograph versus simple reading conditions, controls showed activations in the bilateral TPJ, right supramarginal gyrus, and left middle temporal gyrus, but patients activated only the anterior portion of superior frontal gyrus and middle frontal gyrus. However, a direct group comparison did not reveal any brain regions significantly different between groups above the threshold.

Finally, we examined the contrast between the false belief and false photograph conditions (Figure 3). Healthy controls activated the bilateral TPJ, bilateral angular gyri, middle temporal gyrus, medial prefrontal cortex, posterior cingulate gyrus, and precuneus. Schizophrenia patients, on the other hand, only showed increased activations in the precuneus and posterior cingulate cortex. Direct group comparison revealed that relative to patients, controls showed significantly increased activations in several brain regions, including the medial prefrontal cortex, posterior cingulate cortex, precuneus, and parahippocampus.

DISCUSSION

This study examined neural correlates of ToM in schizophrenia. Using three conditions from the belief attribution task (false belief, false photograph, and simple reading), we evaluated neural activation that was specific to belief attribution while controlling for neural processes associated with general reasoning or reading ability. For the contrast between the false belief and simple reading conditions, schizophrenia patients exhibited significantly less activation than controls in several brain regions including the bilateral TPJ and right medial prefrontal cortex. However, both schizophrenia patients and healthy controls showed comparable patterns of neural activations for the contrast between the false photograph, which required reasoning similar to the false belief condition but without mental state attribution, and the simple reading conditions. Finally, when comparing the false belief and false photograph conditions, we observed significantly less activation of schizophrenia patients

TABLE 3
Brain areas that showed significant activation above threshold for the contrasts of interest

	Areas	Hemisphere	x	y	z	z statistics
False belief vs. simple reading						
Controls	Angular gyrus	R	42	−54	24	7.08
	Angular gyrus	L	−40	−54	24	5.34
	Medial prefrontal cortex	R	14	26	42	3.58
	Middle frontal gyrus	R	40	10	42	4.37
	Middle temporal gyrus	R	60	−26	−8	5.14
	Middle temporal gyrus	L	−52	−60	14	5.33
	Middle temporal gyrus	L	−64	−18	−8	4.26
	Precentral gyrus	R	38	22	34	3.89
	Precuneus	L	−8	−54	40	6.72
	Precuneus	R	4	−56	34	5.58
	Superior frontal gyrus	R	10	40	42	3.37
	Superior temporal gyrus	L	−44	−54	24	5.22
	Temporoparietal junction	R	50	−56	26	7.75
	Temporoparietal junction	L	−52	−58	34	5.17
Patients	Precuneus	R	12	−56	44	3.72
	Precuneus	L	−6	−50	42	3.55
	Superior temporal gyrus	R	66	−42	6	3.74
	Supramarginal gyrus	R	62	−54	24	3.53
	Temporoparietal junction	R	52	−54	24	4.3
	Temporoparietal junction	R	54	−54	18	3.71
Controls > Patients	Amygdala	R	24	−2	−12	3.27
	Globus pallidus	R	28	−14	−2	3.63
	Medial prefrontal cortex	R	14	36	40	3.8
	Middle temporal gyrus	L	−48	−66	18	4.45
	Middle temporal gyrus	L	−62	−32	−12	4.23
	Putamen	R	22	12	−10	3.28
	Superior frontal gyrus	R	20	24	42	3.28
	Superior lateral occipital cortex	L	−54	−68	32	4.16
	Superior lateral occipital cortex	R	52	−64	26	3.99
	Superior temporal gyrus	L	−52	−58	34	4.37
	Supramarginal gyrus	R	52	−54	36	3.92
	Temporoparietal junction	L	−50	−58	30	4.52
	Temporoparietal junction	R	48	−54	30	3.58
False photograph vs. simple reading						
Controls	Middle temporal gyrus	L	−58	−58	16	3.52
	Superior temporal gyrus	L	−56	−48	18	3.99
	Supramarginal gyrus	R	52	−50	30	4.12
	Temporoparietal junction	R	56	−54	26	4.17
	Temporoparietal junction	L	−56	−56	26	4.6
Patients	Middle frontal gyrus	R	36	42	16	3.31
	Superior frontal gyrus	R	30	56	18	3.85
False belief vs. false photograph						
Controls	Angular gyrus	R	44	−66	38	4.94
	Angular gyrus	L	−50	−70	40	4.83
	Angular gyrus	L	−48	−70	44	3.81
	Cuneus	R	4	−64	38	5.12
	Medial prefrontal cortex	R	22	28	36	4.01
	Middle temporal gyrus	L	−44	−68	26	3.9
	Middle temporal gyrus	R	62	−8	−14	4.39
	Posterior cingulate gyrus	R	18	−42	32	4.75
	Precuneus	R	12	−52	34	5.35
	Superior frontal gyrus	R	14	38	40	4.09
	Superior lateral occipital gyrus	R	54	−62	28	4.65
	Superior temporal gyrus	R	42	−52	22	4.9
	Superior temporal gyrus	L	−38	−54	26	4.51
	Superior temporal gyrus	R	46	12	−30	3.54

(Continued)

TABLE 3
(Continued)

	Areas	Hemisphere	x	y	z	z statistics
Patients	Temporoparietal junction	R	56	−54	24	4.87
	Temporoparietal junction	L	−60	−58	26	4.28
	Cingulate gyrus	L	−4	−56	30	3.37
	Posterior cingulate cortex	R	6	−56	26	3.6
	Precuneus	L	−2	−62	34	3.83
	Precuneus	R	6	−50	40	3.75
Controls > patients	Angular gyrus	L	−48	−72	38	3.48
	Cingulate gyrus	L	−8	−22	32	3.74
	Cingulate gyrus	R	4	8	32	3.32
	Culmen	R	26	−46	−16	3.42
	Fusiform gyrus	R	32	−46	−16	3.17
	Globus pallidus	R	28	−12	−2	3.84
	Inferior parietal lobule	L	−52	−58	46	3.18
	Medial prefrontal cortex	L	−2	0	50	3.05
	Middle temporal gyrus	L	−48	−66	16	3.17
	Paracentral lobule		0	−12	48	3.04
	Parahippocampus	R	32	−10	−20	3.98
	Postcentral gyrus	L	−38	−16	32	3.34
	Postcentral gyrus	R	50	−2	30	4.26
	Posterior cingulate cortex	R	8	−58	8	4.3
	Precentral gyrus	L	−50	−12	36	3.12
	Precuneus		0	−70	46	3.38
	Precuneus	L	−28	−78	48	3.24
	Superior lateral occipital gyrus	L	−52	−74	26	3.36
	Thalamus	R	4	−12	12	3.81
	Thalamus	L	−8	−12	12	3.45

relative to controls in several regions, including the medial prefrontal cortex, anterior cingulate cortex, and precuneus. Our findings show reduced task-related neural activations in schizophrenia in several brain regions that have been associated with ToM tasks in healthy controls.

The patients showed lower accuracy than controls on the false belief condition. This performance difference, however, is unlikely to explain the differential patterns of neural activation schizophrenia patients exhibited during the false belief condition. Performance differences were seen across all conditions, and the group by condition interaction was not significant. In addition, both groups performed less accurately during the false photograph condition than during the false belief condition (i.e., both patients and controls found the former more difficult than the latter.) In spite of such behavioral findings, we observed greater fMRI activation differences between groups in the false belief condition than in the false photograph condition. Hence, reduced task-related neural activation in schizophrenia appears to be a specific abnormality in belief attribution as opposed to a result of other nonspecific task-related factors. The current findings suggest that reduced neural activations in key

regions of ToM network are associated with impaired ToM in schizophrenia.

We observed that healthy controls showed increased activations in the TPJ bilaterally, medial prefrontal cortex, and precuneus when belief attribution was compared with general reading ability, as well as when it was compared with nonsocial reasoning. Previous studies in healthy individuals showed the critical role of the right TPJ—and to a lesser extent left TPJ medial prefrontal cortex, and precuneus in belief attribution (Saxe & Powell, 2006; Saxe & Wexler, 2005). The current finding further supports the critical role of these areas in ToM processes (Frith & Frith, 2006; Gallagher & Frith, 2003; Saxe & Kanwisher, 2003; Saxe & Powell, 2006; Siegal & Varley, 2002). Among these areas, schizophrenia patients showed increased activations in the right TPJ and precuneus when belief attribution was compared with reading ability. However, when belief attribution was compared with nonsocial reasoning, schizophrenia patients showed increased activation in precuneus only. The current findings of schizophrenia patients suggest that patients may not have a fully functionally specialized neural network for inferring mental state as controls do.

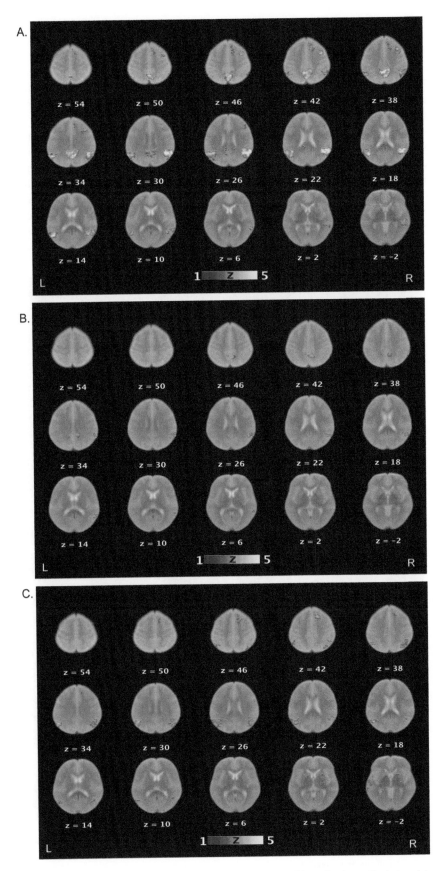

Figure 2. Brain activations for the contrast of false belief versus simple reading conditions. Sections of brain templates with overlaid group analysis results of significant increase in signal intensity during false belief versus simple reading conditions in (A) controls, (B) schizophrenia patients, and (C) controls versus schizophrenia patients.

163

In the few previous reports of TPJ activity during ToM tasks between schizophrenia patients and controls, findings have been inconsistent, including reports of hypoactivations of patients in bilateral TPJ (Walter et al., 2009), hyperactivations of patients in left TPJ (Brune et al., 2008), and no group differences (Brunet, Sarfati, Hardy-Bayle, & Decety, 2003). It is possible that these inconsistent findings of TPJ activation in schizophrenia may arise because the ToM tasks that were used varied in the extent to which they were focused on belief as opposed to other types of mental states (e.g., intention, affective state, preference). Among the ToM network, the TPJ and precuneus have been associated more narrowly with

belief attribution (Saxe, Moran, Scholz, & Gabrieli, 2006; Saxe & Wexler, 2005). Hence, studies using ToM tasks focusing on belief attribution, such as ours, are likely to probe the ToM-related neural activations in the TPJ. Our findings suggest that the hypoactivation in the TPJ may underlie impaired belief attribution in schizophrenia.

It is also noteworthy that schizophrenia patients failed to show increased activations in the medial prefrontal cortex during the belief attribution, in contrast to controls. The medial prefrontal cortex has been associated with ToM beyond belief attribution (e.g., intent, affective states, preference, ambiguous mental state) (A. C. Jenkinson & Mitchell, 2010; van

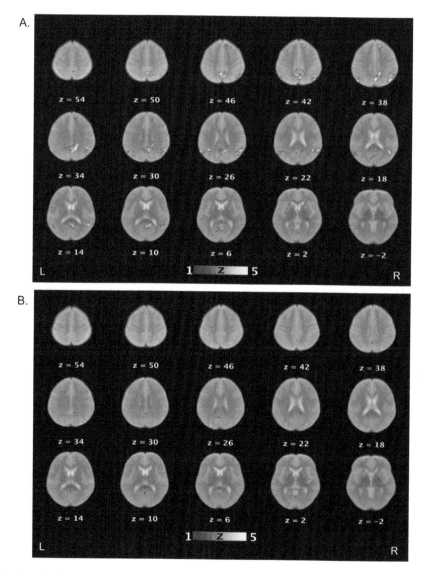

Figure 3. Brain activations for the contrast of false belief versus false photograph conditions. Sections of brain templates with overlaid group analysis results of significant increase in signal intensity during false belief versus false photograph conditions in (A) controls, (B) schizophrenia patients, and (C) controls versus schizophrenia patients.

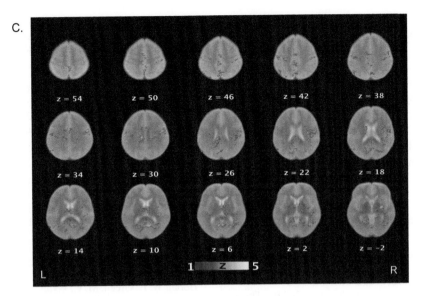

Figure 3. (*Continued*).

Overwalle, 2009). In addition, the medial prefrontal cortex has been also involved in other areas of social cognition in which schizophrenia patients showed impairments (Harvey, Lee, Horan, Ochsner, & Green, 2011; Sergi & Green, 2003), such as self-related information processing and integration of diverse social cues (Kelley et al., 2002; Mitchell, Macrae, & Banaji, 2006). It might be possible that the medial prefrontal cortex is also dysfunctional during these social cognitive processes in schizophrenia. It will be of great importance to carefully examine the extent to which the dysfunctional medial prefrontal cortex underlies social cognitive impairments in schizophrenia.

Recently, there has been growing debate on whether some areas of ToM network, especially the medial prefrontal cortex and TPJ, are exclusively involved with ToM or also involved with other cognitive processing, such as the attentional process or inhibitory control, that is often necessary for ToM. For example, Mitchell (2008) showed overlapping activation in the right TPJ between the ToM task and a selective attention task. A recent study by Rothmayr and colleagues (2011) also found that the medial prefrontal cortex and the right TPJ showed increased activation for both a ToM task and an inhibitory control task, whereas the left TPJ exclusively activated for a ToM task. In contrast, two recent studies (Decety & Lamm, 2007; Scholz et al. 2009) showed that, despite a small overlap between ToM and attentional activation tasks in the right TPJ, each task also activated neighboring, but separate, regions in the right TPJ. Finally, studies with brain-damaged patients showed that patients with damage to the left TPJ showed impaired performance during the ToM task, but not during a control task

requiring general reasoning ability (Samson, Apperly, Chiavarino, & Humphreys, 2004). It appears that the so-called ToM network is indeed involved with ToM ability, and it may be distinct from neighboring regions that are associated with attentional processing. Further studies with high-resolution, functional neuroimaging techniques may clarify the specific roles of subregions within the ToM network.

In summary, we found that schizophrenia patients showed aberrant activation in the brain network associated with ToM when they inferred belief of another person, but not when they preformed tasks that required similar reasoning demands but did not involve a belief attribution component. Complementing a large number of other studies of ToM in schizophrenia that used a performance task, this study demonstrates abnormal neural activations related to ToM processing in schizophrenia. Finally, the findings of the current study also open several promising avenues for future exploration. First, we did not assess cognitive abilities that may be related to performance on the false belief task in schizophrenia (e.g., executive function, inhibitory control, verbal intelligence). Although we included the false-photograph condition and a simple reading condition to control for nonsocial general reasoning ability and a basic reading ability, it would be helpful to include a wide range of neuropsychological assessments in future studies to elucidate the extent to which these abilities are related to belief attribution in schizophrenia. Second, this study focused on a specific component of ToM in schizophrenia, namely belief attribution. Third, we did not assess negative symptoms or community functioning of schizophrenia patients in this study. Deficits in ToM performance have been associated with poor community functioning of schizophrenia (Couture, Penn, & Roberts, 2006; Fett et al., 2011; Pijnenborg et al., 2009; Roncone et al., 2002). Abnormal patterns of neural activation reported here provide a clearer view of the neural basis for social cognitive processes that are related to community functioning in schizophrenia. With larger subject samples, fMRI indices could be entered into a statistical model to test their role as physiological determinants of impaired community functioning.

REFERENCES

Beckmann, C. F., Jenkinson, M., & Smith, S. M. (2003). General multilevel linear modeling for group analysis in fMRI. *NeuroImage, 20,* 1052–1063.

Bora, E., Yucel, M., & Pantelis, C. (2009). Theory of mind impairment in schizophrenia: Meta-anlaysis. *Schizophrenia Research, 109*(1–3), 1–9.

Brune, M., Lissek, S., Fuchs, N., Witthaus, H., Peters, S., Nicolas, V., et al. (2008). An fMRI study of theory of mind in schizophrenia patients with "passivity" symptoms. *Neuropsychologia, 46*(7), 1992–2001.

Brunet, E., Sarfati, Y., Hardy-Bayle, M. C., & Decety, J. (2003). Abnormalities of brain function during a nonverbal theory of mind task in schizophrenia. *Neuropsychologia, 41*(12), 1574–1582.

Couture, S. M., Penn, D. L., & Roberts, D. L. (2006). The functional significance of social cognition in schizophrenia: A review. *Schizophrenia Bulletin, 32*(Suppl 1), S44–63.

Davidson, M., Reichenberg, A., Rabinowitz, J., Weiser, M., Kaplan, Z., & Mark, M. (1999). Behavioral and intellectual markers for schizophrenia in apparently healthy male adolescents. *American Journal of Psychiatry, 156*(9), 1328–1335.

Decety, J., & Lamm, C. (2007). The role of the right temporoparietal junction in social interaction: how low-level computational processes contribute to metacognition. *Neuroscientist, 13*(6), 580–593.

Dennett, D. (1978). Beliefs about beliefs. *Behavioral and Brain Sciences, 1,* 568–570.

Fett, A. K., Viechtbauer, W., Dominguez, M. D., Penn, D. L., van Os, J., & Krabbendam, L. (2011). The relationship between neurocognition and social cognition with functional outcomes in schizophrenia: a meta-analysis. *Neuroscience and Biobehavioral Reviews, 35*(3), 573–588.

First, M. B., Gibbon, M., Spitzer, R. L., Williams, J. B. W., & Benjamin, L. (1996). *Structured Clinical Interview for DSM-IV Axis II Personality Disorders.* New York, NY: New York State Psychiatric Institute.

First, M. B., Spitzer, R. L., Gibbon, M., & Williams, J. B. W. (1997). *Structured Clinical Interview for DSM-IV Axis I Disorders – Patient Edition.* New York, NY: New York State Psychiatric Institute.

Frith, C. D., & Corcoran, R. (1996). Exploring 'theory of mind' in people with schizophrenia. *Psychological Medicine, 26*(3), 521–530.

Frith, C. D., & Frith, U. (2006). How we predict what other people are going to do. *Brain Research, 1079*(1), 36–46.

Gallagher, H. L., & Frith, C. D. (2003). Functional imaging of 'theory of mind'. *Trends in Cognitive Sciences, 7*(2), 77–83.

Green, M. F. (1996). What are the functional consequences of neurocognitive deficits in schizophrenia? *American Journal of Psychiatry, 153*(3), 321–330.

Green, M. F., Bearden, C. E., Cannon, T. D., Fiske, A. P., Hellemann, G., Horan, W. P., et al. (in press). Social cognition across phases of illness in schizophrenia. *Schizophrenia Bulletin.*

Green, M. F., Kern, R. S., Braff, D. L., & Mintz, J. (2000). Neurocognitive deficits and functional outcome in schizophrenia: are we measuring the "right stuff"? *Schizophrenia Bulletin, 26*(1), 119–136.

Harvey, P. D., Lee, J., Horan, W. P., Ochsner, K., & Green, M. F. (2011). Do patients with schizophrenia benefit from a self-referential memory bias? *Schizophrenia Research, 127*(1–3), 171–177.

Horan, W. P., Green, M. F., Wynn, J. K., Lee, J., Castelli, F., & Nuechterlein, K. H. (2009). Disturbances in the spontaneous attribution of social meaning in schizophrenia. *Psychological Medicine, 39*(4), 679–687.

Jenkinson, A. C., & Mitchell, J. P. (2010). Mentalizing under uncertainty: Dissociated neural responses to ambiguous and unambiguous mental state inferences. *Cerebral Cortex, 20,* 404–410.

Jenkinson, M., Bannister, P., Brady, M., & Smith, S. (2002). Improved optimization for the robust and accurate linear registration and motion correction of brain images. *NeuroImage, 17*(2), 825–841.

Jenkinson, M., & Smith, S. (2001). A global optimisation method for robust affine registration of brain images. *Medical Image Analysis, 5,* 143–156.

Kelley, W. M., Macrae, C. N., Wyland, C. L., Caglar, S., Inati, S., & Heatherton, T. F. (2002). Finding the self? An event-related fMRI study. *Journal of Cognitive Neuroscience, 14*(5), 785–794.

Kern, R. S., Green, M. F., Fiske, A. P., Kee, K. S., Lee, J., Sergi, M. J., et al. (2009). Theory of mind deficits for processing counterfactual information in persons with chronic schizophrenia. *Psychological Medicine, 39*(4), 645–654.

Kerr, S. L., & Neale, J. M. (1993). Emotion perception in schizophrenia: Specific deficit or further evidence of generalized poor performance? *Journal of Abnormal Psychology, 102*(2), 312–318.

Langdon, R., Coltheart, M., Ward, P. B., & Catts, S. V. (2001). Mentalising, executive planning and disengagement in schizophrenia. *Cognitive Neuropsychiatry, 6,* 81–108.

Leslie, A. M. (1987). Pretense and representation – the origins of theory of mind. *Psychological Review, 94*(4), 412–426.

Mazza, M., De Risio, A., Surian, L., Roncone, R., & Casacchia, M. (2001). Selective impairments of theory of mind in people with schizophrenia. *Schizophrenia Research, 47*(2–3), 299–308.

Milev, P., Ho, B. C., Arndt, S., & Andreasen, N. C. (2005). Predictive values of neurocognition and negative symptoms on functional outcome in schizophrenia: A longitudinal first-episode study with 7-year follow-up. *American Journal of Psychiatry, 162*(3), 495–506.

Mitchell, J. P. (2008). Activity in the right temporo-parietal junction is not selective for theory-of-mind. *Cerebral Cortex*, *18*(2), 262–271.

Mitchell, J. P., Macrae, C. N., & Banaji, M. R. (2006). Dissociable medial prefrontal contributions to judgments of similar and dissimilar others. *Neuron*, *50*, 655–663.

Mitchley, N. J., Barber, J., Gray, Y. M., Brooks, N., & Livingston, M. G. (1998). Comprehension of irony in schizophrenia. *Cognitive Neuropsychiatry*, *3*, 127–137.

Murray, C. J., & Lopez, A. D. (1997). Global mortality, disability, and the contribution of risk factors: Global Burden of Disease Study [research support, non-US government]. *Lancet*, *349*(9063), 1436–1442.

Penn, D. L., Corrigan, P. W., Bentall, R. P., Racenstein, J. M., & Newman, L. (1997). Social cognition in schizophrenia. *Psychology Bulletin*, *121*(1), 114–132.

Penn, D. L., Ritchie, M., Francis, J., Combs, D., & Martin, J. (2002). Social perception in schizophrenia: The role of context. *Psychiatry Research*, *109*(2), 149–159.

Pickup, G. J., & Frith, C. D. (2001). Theory of mind impairments in schizophrenia: Symptomatology, severity and specificity. *Psychological Medicine*, *31*(2), 207–220.

Pijnenborg, G. H., Withaar, F. K., Evans, J. J., van den Bosch, R. J., Timmerman, M. E., & Brouwer, W. H. (2009). The predictive value of measures of social cognition for community functioning in schizophrenia: Implications for neuropsychological assessment [research support, non-US government]. *Journal of the International Neuropsychological Society*, *15*(2), 239–247.

Premack, D., & Woodruff, G. (1978). Chimpanzee problem-solving: A test for comprehension. *Science*, *202*(4367), 532–535.

Roncone, R., Falloon, I. R., Mazza, M., De Risio, A., Pollice, R., Necozione, S., et al. (2002). Is theory of mind in schizophrenia more strongly associated with clinical and social functioning than with neurocognitive deficits? *Psychopathology*, *35*(5), 280–288.

Rothmayr, C., Sodian, B., Hajak, G., Dohnel, K., Meinhardt, J., & Sommer, M. (2011). Common and distinct neural networks for false-belief reasoning and inhibitory control. *Neuroimage*, *56*(3), 1805–1713.

Russell, T. A., Rubia, K., Bullmore, E. T., Soni, W., Suckling, J., Brammer, M. J., et al. (2000). Exploring the social brain in schizophrenia: Left prefrontal underactivation during mental state attribution. *American Journal of Psychiatry*, *157*(12), 2040–2042.

Samson, D., Apperly, I. A., Chiavarino, C., & Humphreys, G. W. (2004). Left temporoparietal junction is necessary for representing someone else's belief. *Nature Neuroscience*, *7*(5), 499–500.

Sarfati, Y., Hardy-Bayle, M. C., Besche, C., & Widlocher, D. (1997). Attribution of intentions to others in people with schizophrenia: A non-verbal exploration with comic strips. *Schizophrenia Research*, *25*(3), 199–209.

Saxe, R., & Kanwisher, N. (2003). People thinking about thinking people. The role of the temporo-parietal junction in "theory of mind". *NeuroImage*, *19*(4), 1835–1842.

Saxe, R., Moran, J. M., Scholz, J., & Gabrieli, J. (2006). Overlapping and non-overlapping brain regions of theory of mind and self reflection in individual subjects. *Social Cognitive and Affective Neuroscience*, *1*, 229–234.

Saxe, R., & Powell, L. J. (2006). It's the thought that counts: Specific brain regions for one component of theory of mind. *Psychological Science*, *17*(8), 692–699.

Saxe, R., & Wexler, A. (2005). Making sense of another mind: The role of the right temporo-parietal junction. *Neuropsychologia*, *43*(10), 1391–1399.

Schenkel, L. S., Spaulding, W. D., & Silverstein, S. M. (2005). Poor premorbid social functioning and theory of mind deficit in schizophrenia: Evidence of reduced context processing? *Journal of Psychiatric Research*, *39*(5), 499–508.

Scholz, J., Triantafyllou, C., Whitfield-Gabrieli, S., Brown, E. N., & Saxe, R. (2009). Distinct regions of right temporo-parietal junction are selective for theory of mind and exogenous attention. *PLoS One*, *4*(3), e4869.

Sergi, M. J., & Green, M. F. (2003). Social perception and early visual processing in schizophrenia. *Schizophrenia Research*, *59*(2–3), 233–241.

Siegal, M., & Varley, R. (2002). Neural systems involved in "theory of mind". *Nature Reviews Neuroscience*, *3*(6), 463–471.

Smith, S. M. (2002). Fast robust automated brain extraction. *Human Brain Mapping*, *17*(3), 143–155.

Smith, S. M., Jenkinson, M., Woolrich, M. W., Beckmann, C. F., Behrens, T. E., Johansen-Berg, H., et al. (2004). Advances in functional and structural MR image analysis and implementation as FSL. *NeuroImage*, *23*(Suppl 1), S208–219.

van Overwalle, F. (2009). Social cognition and the brain: A meta-analysis. *Human Brain Mapping*, *30*, 829–858.

Walter, H., Ciaramidaro, A., Adenzato, M., Vasic, N., Ardito, R. B., Erk, S., et al. (2009). Dysfunction of the social brain in schizophrenia is modulated by intention type: An fMRI study. *Social Cognitive and Affective Neuroscience*, *4*(2), 166–176.

WHO (2008). *The global burden of disease: 2004 update*. Geneva, Switzerland: WHO.

Woolrich, M. W., Behrens, T. E., Beckmann, C. F., Jenkinson, M., & Smith, S. M. (2004). Multi-level linear modeling for fMRI group analysis using Bayesian inference. *NeuroImage*, *21*, 1732–1747.

Woolrich, M. W., Ripley, B. D., Brady, M., & Smith, S. M. (2001). Temporal autocorrelation in univariate linear modeling of fMRI data. *NeuroImage*, *14*, 1370–1386.

Worsley, K. J. (2001). Statistical analysis of activation images. In P. Jezzard, P. M. Matthews, & S. M. Smith (Eds.), *Functional MRI: An introduction to methods*. Oxford, UK: Oxford University Press.

Zaitchik, D. (1990). When representations conflict with reality: The preschooler's problem with false beliefs and "false" photographs. *Cognition*, *35*(1), 41–68.

Brain activation during a social attribution task in adolescents with moderate to severe traumatic brain injury

Randall S. Scheibel[1,2], Mary R. Newsome[1,2], Elisabeth A. Wilde[1–4], Michelle M. McClelland[5,6], Gerri Hanten[2], Daniel C. Krawczyk[5–7], Lori G. Cook[5], Zili D. Chu[3,8], Ana C. Vásquez[2], Ragini Yallampalli[1,2], Xiaodi Lin[1,2], Jill V. Hunter[3,8], and Harvey S. Levin[1,2,4]

[1]Michael E. DeBakey, Veterans Affairs Medical Center, Houston, TX, USA
[2]Department of Physical Medicine and Rehabilitation, Baylor College of Medicine, Houston, TX, USA
[3]Department of Radiology, Baylor College of Medicine, Houston, TX, USA
[4]Department of Neurology, Baylor College of Medicine, Houston, TX, USA
[5]Center for Brain Health, University of Texas at Dallas, Dallas, TX, USA
[6]School of Behavioral and Brain Sciences, University of Texas at Dallas, Richardson, TX, USA
[7]Department of Psychiatry, University of Texas Southwestern Medical Center at Dallas, Dallas, TX, USA
[8]Department of Pediatric Radiology, Texas Children's Hospital, Houston, TX, USA

The ability to make accurate judgments about the mental states of others, sometimes referred to as theory of mind (ToM), is often impaired following traumatic brain injury (TBI), and this deficit may contribute to problems with interpersonal relationships. The present study used an animated social attribution task (SAT) with functional magnetic resonance imaging (fMRI) to examine structures mediating ToM in adolescents with moderate to severe TBI. The study design also included a comparison group of matched, typically developing (TD) adolescents. The TD group exhibited activation within a number of areas that are thought to be relevant to ToM, including the medial prefrontal and anterior cingulate cortex, fusiform gyrus, and posterior temporal and parietal areas. The TBI subjects had significant activation within many of these same areas, but their activation was generally more intense and excluded the medial prefrontal cortex. Exploratory regression analyses indicated a negative relation between ToM-related activation and measures of white matter integrity derived from diffusion tensor imaging, while there was also a positive relation between activation and lesion volume. These findings are consistent with alterations in the level and pattern of brain activation that may be due to the combined influence of diffuse axonal injury and focal lesions.

Keywords: Traumatic brain injury; fMRI; Social cognition; Adolescents; Diffusion tensor imaging.

Correspondence should be addressed to: Randall S. Scheibel, Cognitive Neuroscience Laboratory, Baylor College of Medicine, 1709 Dryden Road, Suite 725, Houston, TX 77030, USA. E-mail: scheibel@bcm.tmc.edu

This research was supported by National Institutes of Health grant NS021889. We thank the adolescents and their families for their participation, Xiaoqi Li for statistical assistance, and Dr Sandra Chapman for advice regarding the design of this study. Recruitment and imaging of the subjects were facilitated by the General Clinical Research Center at Texas Children's Hospital and Ben Taub General Hospital in Houston, as well as the Children's Medical Center and Our Children's House at the Baylor Medical Center in Dallas and the Advanced Imaging Center at the University of Texas Southwestern Medical Center. The South Central Mental Illness Research, Education, and Clinical Center (MIRECC) and the Michael E. DeBakey Veteran's Affairs Medical Center provided access to laboratory facilities used for the analysis of the image data.

Traumatic brain injury (TBI) is a common cause of disability among children and adolescents (Babikian & Asarnow, 2009) which can have long-term effects on academic achievement (Ewing-Cobbs et al., 2004), cognition (Levin et al., 1997), and quality of life (Sesma, Slomine, Ding, & McCarthy, 2008). Those with moderate to severe TBI often exhibit persistent impairments in executive functions (e.g., Levin et al., 1997), and studies have also revealed problems with social functioning (e.g., Janusz, Kirkwood, Yeates, & Taylor, 2002; Yeates et al., 2007). Adolescence, in particular, is an especially challenging period with multiple social transitions, and individuals who are injured before or during this stage may fail to develop social competence (Max et al., 2006; Turkstra, Dixon, & Baker, 2004).

Social cognition is the ability to recognize, manipulate, and respond appropriately to socially relevant information, and it includes components of perception, motivation, and emotion, as well as intentionality and perspective-taking (Adolphs, 2001). Deficits in processing intentions and emotions or in taking the perspectives of others may contribute to difficulties with social adjustment (Schmidt, Hanten, Li, Orsten, & Levin, 2010). Within the broader domain of social cognition, the ability to make accurate judgments about the mental state of people, including their intentions, desires, and beliefs, is important for predicting the behavior of others and for facilitating positive social interactions (Blakemore, 2008). These particular skills and the associated mental processes have often been referred to collectively as mentalizing or theory of mind (ToM) (Blakemore, 2008). Mentalizing encompasses the states in which an individual has a mental representation about another person's mental representation; that is, meta-representation. The vast literature on ToM addresses a variety of skills requiring meta-representation that include, but are not limited to, first- and second-order false belief (e.g., Baron-Cohen, Leslie, & Frith, 1985; Bowler, 1992; Hughes, Ensor, & Marks, 2010), comprehension of irony, sarcasm, and deception (e.g., Dennis, Purvis, Barnes, Wilkinson, & Winner, 2001), faux pas recognition (e.g., Geraci, Surian, Ferraro, & Cantagallo, 2010; Stone, Baron-Cohen, & Knight, 1998), and various forms of perspective taking (e.g., Aldrich, Tenenbaum, Brooks, Harrison, & Sines, 2011; Chevallier, Noveck, Happé, & Wilson, 2011; Stocks, Lishner, Waits, & Downum, 2011; van den Bos, van Dijk, Westenberg, Rombouts, & Crone, 2011).

Studies of adult TBI have found impairments on tests of ToM requiring the inferring of others' thoughts or emotional states from stories, pictures, and animations (e.g., Henry, Phillips, Crawford, Ietswaart, & Summers, 2006; Muller et al., 2010), and on identification of social faux pas (Channon & Crawford, 2010; Geraci et al., 2010). Relative to healthy subjects, the findings generally suggest sparing on simple, first-order ToM tasks, but increasing impairment with the complexity of the tasks and the skills needed to complete them (for meta-analysis, see Martín-Rodríguez & León-Carrión, 2010).

Much the same story is found in children. Snodgrass and Knott (2006) found that children with moderate to severe TBI performed as well as control subjects on a simple, first-order ToM task that required the recognition of a character's false belief, but more advanced aspects of ToM were impaired. Maureen Dennis and her colleagues examined understanding of irony and deceptive praise in children with mild and severe TBI, as compared to typically developing (TD) uninjured children, and found that these groups did not differ on first-order tasks (Dennis et al., 2001). However, older children with severe TBI were significantly impaired on second-order intentionality tasks. More recently, Dennis, Agostino, Roncadin, and Levin (2009) reported finding ToM deficits in children with TBI that were related to reductions in cognitive inhibition. Thus, some of the impairments in social cognition following brain injury may be due, at least in part, to impairments in more general executive functions. It is not clear from the literature, though, to what extent problems with social cognition following TBI (e.g., impaired ToM) are domain-specific or whether they reflect more general executive deficits in cognitive control.

Functional neuroimaging research has identified brain structures that appear to be part of a network that mediates ToM in healthy individuals, including the temporoparietal junction, superior temporal sulcus, precuneus, and medial prefrontal cortex (Carrington & Bailey, 2009; Dodrell-Feder, Koster-Hale, Bedny, & Saxe, 2011). A study by Schultz et al. (2003) also identified the fusiform face area as being an additional structure that exhibits activation during mentalizing. Research has started to explore the role of these various network components, and Saxe and colleagues (e.g., Saxe & Powell, 2006) have suggested that the medial prefrontal cortex is generally involved in social cognition, rather than ToM specifically, while the superior temporal sulcus may be involved in the detection of motion cues that are useful for understanding another person's mental state (Blakemore et al., 2003; Gobbini, Koralek, Bryan, Montgomery, & Haxby, 2007). There is some evidence, however, that the temporoparietal junction is recruited for thinking about the thoughts of others (i.e., meta-representation) and that it may have a specific and central role in mentalizing (Saxe & Kanwisher, 2003; Saxe & Powell, 2006).

Little is currently known about how activation during social cognition, including ToM, changes in response to acquired neurological disorders. Most functional imaging studies utilizing nonsocial cognitive tasks have found that individuals with moderate to severe TBI have activation that is more intense and diffuse (e.g., Christodoulou et al., 2001; Hillary, 2008; Scheibel et al., 2009). These alterations in the level and pattern of brain activation during cognitive activity may reflect decreases in neural resources or neural inefficiency due to diffuse axonal injury (DAI) (e.g., Huang et al., 2009; Scheibel et al., 2009). However, research examining injury-related alterations in activation during social cognition has been limited. Newsome et al. (2010) reported that adolescents with moderate to severe TBI had posterior brain activation that was greater and more diffuse, relative to control subjects, when they evaluated trait attributions about the self from a third-person perspective. According to these previous findings, ToM-related activation following TBI might be expected to be more intense and to include brain areas that are not typically activated in uninjured individuals.

Animated geometric shapes that move in ways that suggest social interaction and personal agency have also been used to study brain activation associated with ToM (Castelli, Frith, Happé, & Frith, 2002; Heider & Simmel, 1944; Schultz et al., 2003). In normal adults, Schultz et al. (2003) found that such a procedure engaged the right and left dorsal medial prefrontal cortex, the right and left inferior frontal gyri, the orbital frontal cortex, the right temporal pole and amygdala, the right fusiform gyrus, and tissue around the right and left superior temporal sulci. Similarly, Moriguchi, Ohnishi, Mori, Matsuda, and Komaki (2007) used a ToM paradigm depicting interactions among two animated triangles and found that normally developing children and adolescents had activation bilaterally around the superior temporal sulcus, the temporal pole, the amygdala, and the medial prefrontal cortex. Their data also showed that activation shifted from the ventral to the dorsal area of the medial prefrontal cortex during late childhood and adolescence. Moriguchi et al. (2007) interpreted these age-related findings as being consistent with the maturation of prefrontal cortex and the associated development of cognitive functions.

The present study utilized an animated social attribution task (SAT), modified from the version used by Schultz et al. (2003), to examine alterations in ToM-related brain activation associated with TBI. The term 'ToM' is used here in the broader sense to indicate meta-representation, but not necessarily restricted to meta-representation of beliefs and knowledge. The study design included adolescents with moderate to severe TBI and a comparison group of matched, TD adolescents. We hypothesized that the SAT would activate brain structures that previous research has identified as being relevant to mentalizing (Carrington & Bailey, 2009; Moriguchi et al., 2007; Schultz et al., 2003), and that, in subjects with TBI, activation would be greater and would include structures that are not normally activated during ToM. In addition, we hypothesized that there would be significant correlations between ToM-related activation and variables reflecting TBI neuropathology, including lesion volume and measures of white matter integrity derived from the results of diffusion tensor imaging (DTI).

METHODS

Subjects

Nine adolescents with chronic moderate to severe TBI, as defined by a post-resuscitation score of 3–12 on the Glasgow Coma Scale (GCS) (Teasdale & Jennett, 1974), were selected from a larger cohort of TBI patients followed in Houston and Dallas (see Table 1). Their mean GCS score was 5.56 ($SD = 3.13$), and the mean interval between time of injury and the fMRI study was 3.65 years ($SD = 0.56$). Selection criteria included availability for participation, age, severity of injury, and the ability to follow the task instructions and restrain movement during scanning. Nine TD adolescents served as comparison subjects, and they were individually matched to the TBI patients on gender (5 boys) and age. The Wilcoxon two-sample test with normal approximation indicated no significant between-group differences for age at the time of assessment ($p < .86$, TBI mean = 16.32 years, $SD = 2.50$, range = 12.38–19.70; TD mean = 16.84, $SD = 2.24$, range = 13.19–19.94 years) or the mother's level of education ($p < .62$, TBI mean = 13.11 years, $SD = 3.52$; TD mean = 14.22 years, $SD = 2.05$). Fisher's exact test indicated no significant between-group differences for ethnicity ($p < .47$, TBI = 5 Caucasian, 1 Caucasian-Asian, 3 Hispanic; TD = 1 African-American, 3 Caucasian, 5 Hispanic). All subjects were right-handed (Oldfield, 1971), none were taking psychoactive medications at the time of assessment, and none had a history of previous neurologic or psychiatric disorder. Eight TBI patients had focal frontal lobe lesions, and six had temporal lesions on structural MRI (see Table 1). The lesions were measured by the

TABLE 1

Demographic and injury characteristics of nine adolescents with traumatic brain injury

Age at testing	Age at injury	Post-injury interval (years)	Mother's education	Gender	Mechanism of injury	GCS score	Total lesion volume* (cc)	Lesion sites
19.7	16.7	3.0	16	M	MVA	10	1.7	R MFG, L MedFG, B SFG, B MTG, B Temp pole, L operculum
15.4	11.3	4.2	GED	M	Bicycle accident	3	20.2	B GyrRect, B OrbG, B MFG, B SFG, B CC (body), L MTG, L ITG, L Temp pole
18.0	14.7	3.3	14	F	Fall	9	0	L MTG
17.0	12.9	4.2	16	M	Fall	3	1.2	B SFG, L IPC, L hippocampus
19.7	16.0	3.8	14	M	MVA	3	0.6	L IFG, B MFG, R SFG, B SPC, L putamen, R Thal
13.8	9.4	4.4	6	F	Fall	9	0.6	B OrbG, R IFG, R MFG, L Cblm
12.4	9.2	3.2	15	F	MVA	3	6.2	B IFG, R MFG, R SFG, R STG, R MTG, R ITG, R Temp stem, R IPC, L caudate, L CC (body), L Cblm
15.6	12.7	2.9	16	F	MVA	7	34.7	B OrbG, L MFG, B SFG, L STG, R MTG, R ITG, R Temp Pole, B Ant Temp Pole, midline Ant CC
15.4	11.5	3.9	14	M	MVA	3	4.2	L IFG, B MFG, R SFG, B SPC, R Thal, R putamen

Notes: Ant = anterior, B = both sides, L = left, Mid = middle, R = right; Cblm = cerebellum, CC = corpus callosum, GyrRect = gyrus rectus, IFG = inferior frontal gyrus, IPC = inferior parietal cortex, ITG = inferior temporal gyrus, MedFG = medial frontal gyrus, MFG = middle frontal gyrus, MTG = medial temporal gyrus, OccLobe = occipital lobe, OrbG = orbital gyrus, SFG = superior frontal gyrus, SPC = superior parietal cortex, Temp = temporal, Thal = thalamus, GED = graduate equivalency degree, MVA = motor vehicle accident.

*Total lesion volume within brain areas thought to be especially important for social cognition, including the IFG, MFG, SFG, Temp Pole, MedFG, GyrRect, OrbG, and SPC. Note that one subject had a single lesion in the MTG, but this was not included in the lesion volume variable because it was within a portion of the temporal lobe that is not thought to be especially relevant to social cognition.

neuroradiologist on coronal T2-weighted fluid atten- uated inversion recovery (FLAIR) images at the time the anatomical scans were reviewed, and these often consisted of relatively small hemosiderin deposits or areas of gliosis. Total lesion volume within brain areas thought to be especially relevant to social cognition was calculated for each subject, and the mean was 7.7 cc ($SD = 12.0$) (see Table 1). Child assent and parental consent were obtained, and the study was approved by the institutional review boards at Baylor College of Medicine, the University of Texas Southwestern Medical School at Dallas, and the University of Texas at Dallas.

Behavioral measures

All of the subjects were assessed by the Wechsler Abbreviated Scale of Intelligence (WASI) (Wechsler, 1999) to provide an intelligence quotient (IQ) and to assess verbal knowledge and spatial processing. The Gray Oral Reading Test (GORT) (Weiderholt & Bryant, 2001) was also administered.

fMRI task

The SAT was similar to the fMRI paradigm used by Schultz et al. (2003), with films illustrating interac- tions among geometric shapes programmed in E-Prime (www.pstnet.com/eprime) and presented by an MRA fMRI stimulus delivery system (www.mra1.com), but each film had a duration of 17 s instead of 15.1 s. There were eight films for each of two conditions, and each film contained the same three white geometric figures (i.e., triangle, circle, diamond) that moved against a black background. During the social condition, there was a box in the center of this background, with one side that opened as if it were a door, and the shapes moved as if they were able to open or shut the door, enter the box, and chase or drag other shapes inside. Subjects viewed the film and then pressed a button with their right index finger if they judged that the shapes were all friends and pressed with their left fin- ger if the shapes were not friends. In this case, the implied question for the condition of interest (social condition) is, "Do you think the figures are friends?" To answer this question, the subject must engage in meta-representational thought, although not necessar- ily restricted to knowledge or belief. That is, the subject creates a representation of the figures' repre- sentation of each other. However, because the design does not allow for examining the exact process the subject uses to determine whether or not the figures

are friends, we cannot know the nature or complex- ity of the meta-representation. Therefore, in reference to this study and this task, we use the term "ToM" to mean, broadly, engaging in meta-representation. There was also a "bumper car" control condition in which the same shapes moved around the same box in the center of the background, as if they were bumper cars. After viewing the film, the subjects pressed the button on the right if they thought the objects represented in the film were all the same weight. They pressed the left button if they thought the weights differed. Prescan training, including an explanation of the instructions and a sin- gle practice run, was performed outside the scanner environment to familiarize subjects with the task.

Block-design fMRI was performed in the scanner, using the SAT with four runs that each consisted of two films from each condition. Order of presentation of the conditions (social versus "bumper car") was counter- balanced across the four runs and all started with a 3-s screen that identified the condition for the film that was to follow ("PEOPLE, ALL FRIENDS?" or "BUMPER CARS, SAME WEIGHT?"), and then there was a 17- s film, and this was followed by a 13-s fixation cross. Subjects had been instructed during prescan training to withhold their response until the cross appeared. The amount of time to present one combination of the instructions, film, and fixation cross was 33 s, and the total duration of each run was 132 s. Both accuracy and reaction time (RT) were recorded.

Scanning protocol

Whole-brain imaging data were acquired with a multichannel SENSE headcoil on a 3.0 T Philips Achieva scanner. Blood-oxygen-level-dependent (BOLD) T2* weighted, single-shot, gradient-echo echoplanar images (EPI) were acquired in 32 axial slices of 3.75 mm thickness with a 1.0-mm gap, using a 240 mm × 240 mm field of view (FOV), 64 × 64 matrix, a repetition time (TR) of 1500 ms, echo time (TE) of 25 ms, 60° flip angle, and a SENSE factor of 2.0. After the functional scans, a set of high- resolution, T1-weighted, 3D-Turbo Field Echo (TFE) anatomical images was acquired in 132 axial slices of 1.0-mm thickness (no gap) with 240 mm × 240 mm FOV, 256 × 256 matrix, TR of 9.9 ms, TE of 4.6 ms, 8.0° flip angle, and a SENSE factor of 1.5. These parameters produced 1-mm isotropic voxels for the anatomical data. Weisskoff stability measurements (Weisskoff, 1996) (minimum 1/signal to noise ratio index, peak-to-peak and root mean square stability) taken on the day of each scan indicated stability of the scanner over time. The DTI data were acquired with transverse, multislice spin echo, single-shot, EPI sequences (TR = 6161 ms; TE = 51; 2.0-mm slices; no gap). A 224-mm field of view (FOV) (rectangular field of view, RFOV = 100%) was used with a measured voxel size of 1.75 × 1.75 × 2.00 mm and a reconstructed voxel size of 2.00 × 2.00 × 2.00 mm. Diffusion was measured along 32 directions (number of b-value = 2, low b-value = 0, and high b-value = 1000 s/mm^2). To improve the signal to noise ratio, high-b images were acquired twice and were averaged. Each DTI acquisition took approximately 5 min, and 70 slices were obtained. For lesion analysis, a coronal T2-weighted FLAIR sequence was used (1100-ms TR, 140-ms TE, 5.0-mm slices). This sequence had a 220-mm FOV with a reconstructed voxel size of 0.86 × 0.86 × 5.0 mm.

DTI procedures

The Philips diffusion affine registration tool was used to remove shear and eddy current distortion and head motion prior to calculating fractional anisotropy (FA) maps with Philips fiber-tracking 4.1V3 Beta 4 software (Netsch & van Musiwinkel, 2004). A quantitative DTI tractography approach was used where mean FA of the fiber system was used as the measure for DTI variables. The algorithm for fiber tracking is based upon the fiber assignment by the continuous tracking (FACT) method (Mori, Crain, Chacko, & van Zijl, 1999). For each region of interest (ROI), standard parameters were used where tracking terminated if the FA in the voxels decreased below 0.2 or if the angle between adjacent voxels along the tract was greater than 6.75°. Tracts or ROIs were selected by their reproducibility, using the specified protocols, coverage of major white matter regions of the brain, and/or their association with cortical regions implicated in social cognition. Tracts or ROIs included: (1) ventromedial and dorsolateral frontal regions (containing white matter underlying the medial prefrontal areas and inferior frontal gyrus); (2) temporal lobe regions, including the arcuate fasciculus, inferior fronto-occipital longitudinal fasciculus (connecting the ventromedial areas to the occipital regions via the temporal stem), inferior longitudinal fasiculus (connecting the temporal pole to the parietal and occipital areas, and underlying the superior temporal sulcus), and uncinate fasiculus (connecting the frontal and temporal poles); (3) corpus callosum (connecting the hemispheres); (4) cingulum bundle (which underlies the cingulate cortex and contains projections to several areas including the frontal, temporal, and parietal regions; and (5) anterior (connecting the thalamus to the frontal areas) and posterior

(connecting the parietal areas to the brain stem) limbs of the internal capsule (Levin et al., 2008; Oni et al., 2010; Wakana et al., 2007; Wilde et al., 2006, 2009, 2010). Estimates of intraoperator and interoperator reproducibility were obtained for each of these ROIs, using intraclass correlation coefficients. All intraclass correlations exceeded .97.

We elected to investigate the impact of overall white matter integrity by creating a composite of all DTI tracts or regions of interest (whole-brain FA composite): This was created by averaging FA z-scores from each of the ROIs. A second composite score, the social brain FA composite, was created from ROIs that were thought to be especially relevant for social functions (i.e., genu, bilateral uncinate, and inferior longitudinal fasciculi), because these connect cortical areas that have been implicated in social cognition (Blakemore, 2008), including ToM (Carrington & Bailey, 2009).

Functional image processing and analysis

The fMRI data were subjected to voxel by voxel analyses using Statistical Parametric Mapping (SPM) 5 software (Wellcome Department of Cognitive Neurology, University College, London, UK) implemented in Matlab (Mathworks, Inc., Sherborn, MA, USA). Anatomical and EPI functional image data were first imported into SPM5. Slice timing correction was applied, and the EPI data sets were then realigned and checked for excessive head movement. There were no runs with head motion greater than 2 mm translation or 2° rotation. Each subject's own high-resolution, anatomical T1-weighted scan was then coregistered (mutual information coregistration) to their EPI images, and then the anatomical and EPI images were spatially normalized to MNI space, using the unified segmentation approach within SPM5. This method implicitly models lesions as part of the segmentation and normalization process, and it provides superior normalization results when compared to explicit lesion-masking procedures, such as cost function masking (Crinion et al., 2007). Spatial smoothing was performed by convolving the EPI data with a 6-mm, full-width at half-maximum (FWHM) Gaussian filter.

The first six images in each fMRI time series were eliminated to allow equilibrium in magnetization to occur. The BOLD hemodynamic response to each behavioral condition was then modeled in a boxcar design convolved with the SPM5 canonical hemodynamic response function. The SPM5 autocorrelation correction of the time series was conducted and a

128-s, high-pass, temporal filter was used to reduce low-frequency noise. After specifying the appropriate design matrix, the effects were estimated for each individual subject according to the general linear model (GLM) at each voxel. The contrast of interest for this first-level (fixed effects) analysis was the SAT social block minus the "bumper car" block. Thus, for these and all subsequent analyses, brain activation was defined as greater differential activation during the social interaction condition, relative to the "bumper car" control condition. Then SPM5 second-level random effects procedures utilized the contrast images from the individual subjects to perform within- and between-group analyses. These second-level analyses consisted of within-group t-tests for the TD and TBI groups, a between-group t-test, and separate SPM5 simple regression analyses relating SAT brain activation to total lesion volume within areas relevant to social cognition (see Table 1) and to DTI measures reflecting white matter integrity within the whole brain (i.e., whole-brain FA composite score), within white matter tracts that may be especially relevant to ToM (i.e., social brain FA composite score), and within the genu of the corpus callosum.

In SPM image analysis, a cluster is a spatially contiguous group of voxels that all exceed a statistical probability threshold (Friston et al., 1995). The cluster-defining (height) threshold for all analyses was initially set at voxel-level $t = 1.79$. All reported clusters were statistically significant (corrected $p < .05$) at the cluster level of inference, using the random field theory family-wise error (FWE) correction for multiple comparisons over the whole-brain volume. When the cluster size exceeded 2000 voxels, a more stringent cluster-level (height) threshold was used to reduce cluster sizes to 2000 voxels or less. All coordinates from

statistically significant clusters were extracted and transformed to Talairach space (Talairach & Tournoux, 1988), using a Matlab script. The Talairach Daemon (http://ric.uthscsa.edu/projects/talairachdaemon.html) with the single-coordinate query option was then used to provide an anatomical label for each of the transformed coordinates.

RESULTS

Behavioral findings

The WASI IQ score and oral reading performance did not differ between the groups (see Table 2). There were also no significant between-group differences for accuracy or RT for either the social interaction or "bumper car" conditions of the SAT. However, both groups had mean accuracy scores that were under 60% for the "bumper car" portion of the fMRI task (TD mean = 58.33, TBI mean = 44.44).

DTI and lesion volume findings

Between-group comparisons were performed for each of the DTI ROIs while correcting for age, but due to the relatively small sample size, these results were not corrected for multiple comparisons. Least squares means (corrected for age), standard errors, F statistics, p values, and effect sizes (i.e., Cohen's f) are reported in Table 3. Briefly, significant group differences were found in the FA of several ROIs, including the genu of the corpus callosum, the total corpus callosum, the left ventromedial frontal, and the right and left dorsolateral frontal areas. Of these regions, the magnitude

TABLE 2
Behavioral data for the traumatic brain injury and typically developing groups

	n	TD group Mean	SD	Median	n	TBI group Mean	SD	Median	Probability*
WASI IQ	9	111.56	13.06	109.00	9	105.11	11.39	108.00	.307
GORT	6	24.00	4.38	23.50	6	16.33	6.19	16.00	.052
SAT bumper condition									
Accuracy	9	58.33	20.73	62.50	9	44.44	18.86	37.50	.161
Reaction Time	9	1326.66	277.92	1292.00	9	1198.77	644.73	1050.67	.331
SAT social interaction condition									
Accuracy	9	74.60	6.30	71.43	9	74.60	15.61	71.43	.635
Reaction time	9	1571.65	579.79	1561.86	9	1115.71	305.94	1178.00	.077

Notes: GORT = Gray Oral Reading Test, SAT = Social Attribution Task, TBI = traumatic brain injury, TD = typically developing, WASI = Wechsler Abbreviated Scale of Intelligence. *Wilcoxon two-sample test (two-sided probability, normal approximation).

TABLE 3
Between-group DTI results for individual regions of interest

Region	TBI FA LSM (SE)	TD FA LSM (SE)	F-statistic	p value	Effect size (f)
Right ventromedial frontal	0.365 (0.007)	0.383 (0.007)	3.34	.088	0.47
Left ventromedial frontal	0.370 (0.006)	0.398 (0.006)	12.91	.003*	0.93
Right dorsolateral frontal	0.359 (0.006)	0.383 (0.006)	7.58	.015*	0.71
Left dorsolateral frontal	0.370 (0.006)	0.394 (0.006)	6.79	.020*	0.67
Right arcuate fasciculus	0.383 (0.013)	0.389 (0.013)	0.09	.773	0.08
Left arcuate fasciculus	0.435 (0.007)	0.455 (0.007)	4.34	.055	0.54
Right inferior fronto-occipital longitudinal fasciculus	0.419 (0.010)	0.444 (0.010)	2.33	.148	0.39
Left inferior fronto-occipital longitudinal fasciculus	0.438 (0.008)	0.460 (0.008)	3.44	.083	0.48
Right inferior longitudinal fasciculus	0.413 (0.010)	0.400 (0.010)	0.72	.410	0.22
Left inferior longitudinal fasciculus	0.431 (0.009)	0.442 (0.009)	0.61	.445	0.20
Right uncinate fasciculus	0.355 (0.007)	0.361 (0.007)	0.35	.561	0.15
Left uncinate fasciculus	0.367 (0.009)	0.381 (0.009)	1.30	.274	0.30
Right cingulum bundle	0.403 (0.007)	0.419 (0.007)	2.76	.117	0.43
Left cingulum bundle	0.427 (0.011)	0.447 (0.011)	1.76	.204	0.34
Right anterior limb of the internal capsule	0.363 (0.006)	0.380 (0.006)	4.01	.064	0.52
Left anterior limb of the internal capsule	0.375 (0.006)	0.386 (0.006)	1.60	.226	0.33
Right posterior limb of the internal capsule	0.443 (0.006)	0.452 (0.006)	1.20	.290	0.28
Left posterior limb of the internal capsule	0.467 (0.007)	0.476 (0.007)	0.71	.414	0.22
Genu corpus callosum	0.428 (0.006)	0.459 (0.006)	14.44	.002*	0.98
Body corpus callosum	0.444 (0.010)	0.472 (0.010)	4.34	.055	0.54
Splenium corpus callosum	0.501 (0.007)	0.518 (0.007)	3.04	.102	0.45
Total corpus callosum	0.453 (0.006)	0.483 (0.006)	13.36	.002*	0.94

Notes: For Cohen's f, 0.1 = small, 0.25 = moderate and 0.40 = large effect size.
*$p < .05$. LSM = least squares mean (corrected for age); SE = standard error.

of the between-group difference was greatest for the genu of the corpus callosum, and the relation between this measure and the fMRI data was examined further in an SPM regression analysis (see 'Brain activation and white matter integrity').

Total lesion volume was not significantly correlated with the whole-brain FA composite score (rho = −.10, $p < .80$), the social brain FA composite score (rho = .02, $p < .97$), or genu FA (rho = −.13, $p < .73$).

fMRI results

Tables 4 and 5 present a summary of the fMRI findings in relation to major anatomical structures, while the text includes selected Brodmann areas (BA). Anatomical information presented within the following text is based upon detailed output representing all voxels within each significant cluster, but to conserve space a general description is provided for each cluster's location.

Within-group analysis for control subjects

The TD control subjects had extensive activation during the SAT (social interaction minus "bumper car" contrast) with 11 clusters that included a number of brain areas that have been implicated in ToM (see Table 4 and Figure 1, panel A). There was a bilateral cluster within medial prefrontal areas that included the medial (right BA 10, bilateral BA 11) and superior frontal (right BA 10, left BA 11) gyri. A second medial frontal cluster was more superior in location and also included the medial and superior frontal gyri. Both medial prefrontal clusters were centered further anterior than the dorsal medial prefrontal ROIs of Schultz et al. (2003), but this activation appeared to overlap some with the location of a medial prefrontal cluster (−4, 60, 32) that was reported by Castelli, Happé, Frith, and Frith (2000) for a similar animated ToM task. Two other clusters were located within a more lateral area of the right frontal lobe, the first of which included part of the inferior (BA 11, BA 47) and middle frontal (BA 11) gyri, and the second was located within

TABLE 4
Summary of fMRI findings for within- and between-group comparisons organized according to major anatomical region[a]

	Within-group t-tests		Between-group t-test
	TD group	*TBI group*	*TBI group > TD group*
Cluster-defining threshold[b]	3.66	3.92	1.93
Number of clusters	11	7	3
Anatomical region			
Frontal lobe			
Inferior frontal gyrus	R	B	R
Medial frontal gyrus	B		
Middle frontal gyrus	R	B	R
Superior frontal gyrus	B	R	
Precentral gyrus		R	R
Paracentral lobule			
Temporal lobe			
Inferior temporal gyrus	B	R	R
Middle temporal gyrus	B	B	B
Superior temporal gyrus	B	B	R
Transverse temporal gyrus			R
Fusiform gyrus	B	L	B
Parietal lobe			
Postcentral gyrus			R
Paracentral lobule			
Angular gyrus	B		
Supramarginal gyrus	R	R	
Inferior parietal lobule	R	R	R
Superior parietal lobule			
Precuneus	B	B	B
Occipital lobe			
Cuneus			B
Precuneus	B		B
Fusiform gyrus	L	L	L
Inferior occipital gyrus		B	B
Middle occipital gyrus	L	B	B
Superior occipital gyrus	L		
Lingual gyrus		L	B
Cingulate gyrus			
Anterior cingulate gyrus	R		
Posterior cingulate gyrus	B		
Other	B	L	
Parahippocampal gyrus	B		
Insula	R	R	R
Thalamus			
Claustrum			R
Basal ganglia			
Caudate (head)	B		
Caudate (tail)	R		
Cerebellum	B	L	B
Brainstem			

Notes: B = both sides, L = left side, R = right side, TBI = traumatic brain injury, TD = typically developing.

[a]Results reported in this table reflect the location of voxels within significant clusters, as labeled by the Talairach Daemon, but with a focus on major anatomical areas to reduce length. For gyral locations, only those coordinates that are within gray matter are reported here (i.e., subgyral while matter coordinates were excluded from this table).

[b]*T* value threshold that was used to define the cluster. This *T* value was conservatively increased to reduce the size of all significant clusters to 2000 voxels or less.

TABLE 5
Summary of fMRI findings for the regression of SAT activation with DTI measures and lesion volume within the TBI group organized according to major anatomical region[a]

	Whole-brain FA Negative regression	Social brain FA Negative regression	Genu FA Negative regression	Lesion volume[b] Positive regression
Cluster-defining threshold[c]	3.07	2.93	3.00	2.29
Number of clusters	12	14	11	4
Anatomical region				
Frontal lobe				
Inferior frontal gyrus	R	R		
Medial frontal gyrus	B	B	R	
Middle frontal gyrus	B	B		
Superior frontal gyrus	L	B	R	
Precentral gyrus	B	B	B	B
Paracentral lobule	B	B	B	
Temporal lobe				
Inferior temporal gyrus	L	L	R	L
Middle temporal gyrus	R	B	R	B
Superior temporal gyrus	R	B		
Transverse temporal gyrus	R	R		
Hippocampus		L	L	L
Fusiform gyrus	L	B	B	L
Parietal lobe				
Postcentral gyrus	B	B	B	B
Paracentral lobule		B	R	
Angular gyrus		L		
Supramarginal gyrus				
Inferior parietal lobule		L	B	B
Superior parietal lobule		L	B	
Precuneus	B	B	B	R
Occipital lobe				
Cuneus	B	L	B	B
Precuneus	B	B	B	B
Fusiform gyrus	L	B	B	
Inferior occipital gyrus	B		R	
Middle occipital gyrus	B	L	R	B
Superior occipital gyrus	R		L	
Lingual gyrus	B	B	R	B
Cingulate gyrus				
Anterior cingulate gyrus		R		B
Posterior cingulate gyrus	B	B	B	B
Other	B	B		
Parahippocampal gyrus	B	B	B	B
Insula	B			
Thalamus		B		B
Basal ganglia				
Caudate (head)				L
Caudate (body)				L
Caudate (tail)				L
Putamen		R		R
Globus pallidus		R		R
Cerebellum	B	B	B	B
Brainstem		B	B	B

Notes: B = both sides, L = left side, R = right side, FA = fractional anisotropy.
[a]Results reported in this table reflect the location of voxels within significant clusters, as labeled by the Talairach Daemon, but with a focus on major anatomical areas to reduce length. For gyral locations, only those coordinates that are within gray matter are reported here (i.e., subgyral while matter coordinates were excluded from this table).
[b]Total lesion volume within brain areas thought to be especially important for social cognition (see Table 1).
[c]T value threshold that was used to define the cluster. This T value was conservatively increased to reduce the size of all significant clusters to 2000 voxels or less.

T-TESTS

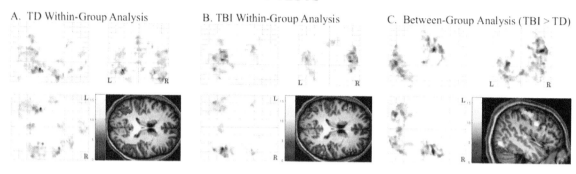

NEGATIVE REGRESSION OF FA WITH BRAIN ACTIVATION

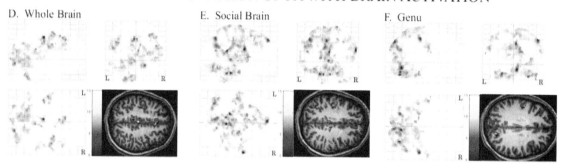

Figure 1. Three-plane maximum intensity projections and image overlays for *t*-test contrasts and simple regression analyses. (A) Activation (social interaction minus "bumper car" contrast) for the TD adolescents. (B) Activation for adolescents with TBI. (C) Areas where SAT-related activation was greater within adolescents with TBI, relative to the TD comparison group. (D) Areas where there was a significant negative relation between activation during the SAT and the whole brain FA composite score. (E) Areas where there was a significant negative relation between SAT-related activation and the social brain FA composite score. (F) Areas where there was a significant negative relation between whole brain activation and FA within the genu of the corpus callosum. L = left side, R = right side.

the middle frontal (BA 8) and superior frontal gyri. There were also seven other clusters that, in combination, included the right anterior cingulate gyrus (BA 25, BA 32) and bilateral parahippocampal and fusiform gyri (e.g., BA 20, BA 36, BA 37), the right temporal pole, and an area around the superior temporal sulcus (e.g., bilateral superior temporal gyri) and temporoparietal junction of both hemispheres. The posterior regions included several areas that Schultz et al. (2003) had also identified as exhibiting activation, such as the right fusiform gyrus and the right and left superior temporal gyri. Both studies also found activation within the right temporal pole, but Schultz et al. (2003) did not report a full set of coordinates for their results, and it is difficult to determine whether our activation includes precisely the same portions of that structure. The current study found no activation within the amygdala.

Within-group analysis for TBI subjects

The TBI subjects had seven significant clusters with SAT activation (see Table 4 and Figure 1, panel B).

A right-sided cluster included part of the middle and superior frontal gyri and was centered more laterally than the right medial prefrontal cluster reported by Schultz et al. (2003). There were also two lateral frontal clusters and one of these included the right precentral gyrus and the inferior (BA 6, BA 9, BA 44, BA 45, BA 46) and middle frontal gyri (BA 6, BA 8, BA 9, BA 46). The other lateral frontal cluster was located within the left inferior (BA 9) and middle frontal gyri (BA 9, BA 46). There was a large left-sided, posterior cluster that included the cerebellum and a portion of the occipital (e.g., inferior and middle occipital gyri) and temporal lobes (e.g., fusiform gyrus and the middle and superior temporal gyri). A right-sided cluster included the insula and parts of the occipital (e.g., inferior and middle occipital gyri), temporal (inferior, middle, and superior temporal gyri), and parietal lobes (e.g., inferior parietal lobule and supramarginal gyri). The other posterior clusters were smaller in size and were located primarily within the right occipital lobe (e.g., middle occipital gyrus) and bilaterally within the precuneus (BA 7).

Between-group comparison

There were no areas where the TD subjects had greater activation than the TBI subjects, but there were three clusters where the TBI group had greater activation than the TD group (see Table 4 and Figure 1, panel C). This included a right-sided cluster located within a posterior and lateral portion of the frontal lobe (i.e., precentral, inferior frontal, and middle frontal gyri), the postcentral gyrus and inferior parietal lobule (BA 40), the insula, the claustrum, and the transverse temporal and superior temporal gyri (BA 13, BA 22, BA 41). Another cluster included the cerebellum and portions of the right occipital (e.g., cuneus, lingual gyrus, and the inferior and middle occipital gyri), parietal (e.g., precuneus), and temporal lobes (e.g., fusiform gyrus and the inferior and middle temporal gyri). There was also a bilateral cluster that was primarily located within the left occipital lobe (e.g., cuneus, lingual gyrus, and the inferior and middle occipital gyri) and that extended into the left temporal lobe (e.g., fusiform and middle temporal gyri), the lingual gyrus of the right occipital lobe, and the cerebellum.

Brain activation and white matter integrity

Separate exploratory SPM5 simple regression analyses were conducted to relate activation during the SAT to (1) the whole-brain FA composite score, which was considered an index for overall white matter integrity; (2) the social brain FA composite score, which was included to contain a limited set of specific tracts considered to be most important for ToM; and (3) the genu of the corpus callosum FA. The genu was selected because this region exhibited the greatest between-group difference in FA, and also because of its potential role in connection of medial prefrontal regions important in social cognition.

There was no significant positive correlation between SAT activation and the whole-brain FA composite score in TBI subjects. However, there were 12 clusters where there was a significant negative relation, including a bilateral medial frontal cluster that included the cingulate (right BA 24, bilateral BA 32), medial (e.g., BA 6, BA 8, right BA32), and superior frontal gyri (left BA 8) (see Table 5 and Figure 1, panel D). Three other clusters were located primarily within more lateral areas of the right frontal lobe and, together, these included the precentral gyrus and the inferior (BA 9, BA 44) and middle frontal gyri (BA 6, BA 8, BA 9, BA 46). There was a large bilateral cluster within both occipital lobes that extended into the cerebellum and part of the left temporal lobe (e.g., fusiform and inferior temporal gyri), the posterior cingulate gyrus (e.g., BA 30), and the parahippocampal gyrus. Together, the other seven clusters were scattered throughout posterior brain areas and included structures within the cerebellum, the right temporal lobe, and both occipital and parietal lobes.

In the TBI subjects, there were 14 significant clusters where there was a negative relation between SAT activation and the social brain FA composite score. In general, these clusters included many of the same structures that were identified in the regression analysis for the whole-brain FA composite score (see Table 5 and Figure 1, panel E). However, a significant negative association between activation and genu FA had a distribution that was more restricted and excluded a number of anterior brain regions identified in the other analyses (see Figure 1, panel F).

Brain activation and lesions

An SPM5 simple regression analysis indicated no significant negative relation between SAT activation and the lesion volume measure, but there were four clusters where there was a significant positive correlation, and one of these was located primarily within the right parietal lobe. A second cluster included parts of the cerebellum, brainstem, diencephalon, and structures within the inferior portion of the left cerebrum (e.g., parahippocampal and fusiform gyri), while another was located primarily within the left parietal lobe (e.g., postcentral gyrus, inferior parietal lobule). There was also a large bilateral cluster that included deep brain structures, such as the midbrain, and that extended into the posterior cingulate gyrus, both occipital lobes (e.g., cuneus, lingual gyrus), and the right middle temporal gyrus (see Table 5).

DISCUSSION

During the SAT, the TD adolescents exhibited an activation pattern that was generally similar to that reported by previous investigations of ToM (Carrington & Bailey, 2009), including structures such as the medial prefrontal cortex, cingulate cortex, and posterior temporal and parietal areas. The TBI subjects had significant activation within many of these same areas, but their activation was generally more intense and excluded medial prefrontal areas that were activated in previous studies of ToM with healthy subjects (e.g., Castelli et al., 2000; Schultz et al., 2003). Brain activation during the SAT was also related to variables reflecting TBI neuropathology, including lesion volume and DTI measures of white matter integrity. As expected, greater neuropathology (i.e., greater lesion volume, decreased white matter integrity as indicated by lower FA) was associated with higher activation levels, and this is generally consistent with previous reports of a relationship between greater TBI severity and increased activation during cognitive tasks (e.g., Scheibel et al., 2009). Overall, these findings provide initial information about differences in activation during ToM in adolescents who have sustained TBI, relative to matched TD adolescents, as well as some preliminary results indicating that both DAI and focal lesions contribute to the activation changes.

Typically developing adolescents had SAT-related activation in many of the same brain areas that activated in normal adults studied by Schultz et al. (2003), but examination of our cluster coordinates in relation to that study's ROIs indicated only partial convergence. For example, our TD adolescents had medial prefrontal clusters that were further anterior than the activation reported by Schultz et al. (2003), but these appear to overlap more with the location of a medial prefrontal cluster from another ToM study (Castelli et al., 2000). The location and role of some components of the ToM network may vary among individuals, perhaps reflecting differences in experience or development (Bird, Castelli, Malik, Frith, & Husain, 2004; Moriguichi et al., 2007), and these factors may account for some minor differences in the precise location of brain activation within various studies. In general, however, the pattern observed in the current study is consistent with prior reports of ToM-related brain activation, and replicated findings of medial prefrontal and superior temporal activation (Carrington & Bailey, 2009; Castelli et al., 2000; Schultz et al., 2003).

Carrington and Bailey (2009) reviewed the findings from functional neuroimaging research and concluded that the medial prefrontal area is the region that is most consistently activated in studies using ToM tasks. However, there is some evidence that the temporoparietal junction has a more specific and central role in ToM (Saxe & Kanwisher, 2003; Saxe & Powell, 2006). Many studies have also found task-related activation in structures around the superior temporal sulcus and, in combination with medial prefrontal cortex, these structures may function as part of a neural network that mediates ToM reasoning (Carrington & Bailey, 2009). Other brain areas (e.g., insula) may also activate during various types of ToM tasks, but activation in these areas has been observed with less consistency and their recruitment may reflect ancillary aspects of the particular task (Carrington & Bailey, 2009). In the within-group analysis, our TBI

subjects had significant activation within a number of the same areas that activated in our TD adolescents, such as the posterior portion of the superior temporal gyrus, but their prefrontal activation had a more lateral distribution.

When directly compared to TD adolescents, the TBI subjects had greater activation within right lateral frontal and parietal areas, and bilateral increases within posterior brain regions. Some of these posterior areas have been implicated in social reasoning, such as the fusiform and superior temporal gyri (Blakemore, 2008; Schultz et al., 2003), while others are not typically considered to be involved in social cognition. A similar finding was reported by Newsome et al. (2010), who found increased posterior activation when children with TBI were asked to think about their own traits from another person's perspective. Based upon the current findings and those of Newsome et al. (2010), it appears that the neural resources utilized for social cognition may be greater following brain injury and that these alterations are not restricted to areas that typically mediate such functions in uninjured individuals. Activation increases during cognitive fMRI paradigms are frequently observed in association with neuropathology, and proposed interpretations of this finding have included the disinhibition of duplicate neural systems, learning-related neuroplasticity, and cognitive reorganization (Price & Friston, 2002). When task performance is equated in comparisons with a control group, the overactivation may reflect a higher level of effort, perhaps as a consequence of inefficient processing or as a form of compensation involving the allocation of additional cognitive and neural resources (Price & Friston, 2002; Ricker, Hillary, & DeLuca, 2001). There has also been some speculation that such alterations in the level and pattern of brain activation following TBI may reflect decreases in neural resources or neural inefficiency due to DAI (e.g., Huang et al., 2009; Scheibel et al., 2009).

A large number of studies have documented changes in white matter following TBI, including postmortem findings (e.g., Adams, Mitchell, Graham, & Doyle, 1977) and the *in vivo* examination of white matter by imaging techniques such as DTI (e.g., Wilde et al., 2006). The present study also used DTI and found evidence for reduced white matter integrity in the current sample of adolescents with moderate to severe TBI, including reduced FA within the genu of the corpus callosum and the total corpus callosum, and left ventromedial and left and right dorsolateral frontal areas. White matter traversing the genu of the corpus callosum connects the right and left medial prefrontal areas, regions previously reported to be important in social cognition (Blakemore, 2008; Carrington &

Bailey, 2009). Additionally, the dorsolateral frontal area, which includes the inferior frontal gyrus, has also been implicated in social cognition. These regions are also known to be vulnerable to TBI-related injury (Povlishock & Katz, 2005). Because FA was examined within a large number of regions, further analysis of these findings was conducted using only the single region where the changes were greatest (i.e., genu) and two composite scores, one of which reflected damage to white matter within the entire brain and another that was specific for regions that are related to ToM.

Regression analyses using the FA composite scores indicated a relation between diffuse decreases in white matter integrity and activation increases within a large posterior portion of the brain, as well as the anterior cingulate cortex and medial prefrontal areas. In contrast, reductions in genu FA were related to increased activation within posterior areas, and there was no significant correlation between genu FA and ToM-related activation within many of the more anterior brain structures. This latter finding was unexpected since, anatomically, the medial prefrontal cortex is in relatively close proximity to the genu and because interhemispheric prefrontal connections pass through the genu and anterior body of the corpus callosum (Zarei et al., 2006). Possibly, the same traumatic forces that injured the genu white matter also produced local injury to the prefrontal cortex, thus forcing greater reliance upon posterior brain areas to perform the SAT. However, another possibility is that damage to white matter in and around the genu may have caused some disconnection among components of the social brain network that are relevant to ToM, including tracts between the medial prefrontal cortex and posterior areas such as the superior temporal gyrus. Disrupted functional connectivity has been proposed as a mechanism for decreased ToM performance in individuals with autism or Asperger's syndrome (Castelli et al., 2002), including the loss of top-down modulation by more anterior brain areas upon the extrastriate cortex. Similarly, decreased top-down modulation may also contribute to the activation increases noted among posterior areas following TBI, and, based upon the current DTI findings, physical injury to the white matter appears to contribute to the changes in brain activation that were observed during the SAT.

The neuropathology associated with TBI is heterogeneous and, in addition to DAI, focal lesions in gray and white matter are also common in individuals with moderate to severe TBI (Povlishock & Katz, 2005). Many lesions in the current TBI sample were relatively small, however, and most subjects had multiple areas of focal pathology, and all but one had at least one frontal lobe lesion. These lesion characteristics and the small sample size made the analysis of lesion location in relation to brain activation or SAT performance variables impractical, but total lesion volume within task-relevant brain areas was used as a variable reflecting the overall severity of focal injury. The regression analysis with lesion volume indicated increased activation, possibly reflecting increased recruitment, of nonfrontal brain structures in association with greater focal neuropathology. However, lesion volume was not correlated with the FA measures, and, since greater lesion volume and reduced white matter integrity were both associated with activation increases, it appears that these different types of neuropathology make separate contributions to the activation changes that occur following TBI.

Although the use of DTI measures and lesion volume to examine the relationship between the severity of TBI neuropathology and brain activation is relatively new, previous studies have reported that activation during cognitive fMRI tasks is associated with other severity measures in individuals with neurological disorders. For example, Mainero et al. (2004) reported that patients with multiple sclerosis had increased activation on attention and memory tasks that was positively correlated with lesion load. Similarly, Scheibel et al. (2009) found that the level of activation during a visual stimulus-response compatibility task was greater when adults had more severe TBI, as indicated by lower initial scores on the Glasgow Coma Scale. In that study the activation pattern changed with increasing severity so that additional brain areas, such as the left lateral frontal cortex, exhibited significant increases only when the TBI was more severe. The current results are generally consistent with those previous findings, since changes in brain activation were noted in association with the degree of injury, but in the present investigation the design focused on ToM. Alterations in the level and pattern of activation that were observed in association with neuropathology may reflect higher resource utilization to perform the ToM task, perhaps as a form of compensation for reduced neural efficiency (Scheibel et al., 2009). However, it is also possible that some activation increases and diffusivity associated with TBI reflect pathological processes, such as the loss of functional connectivity and decreased modulatory control (Castelli et al., 2002).

Performance on ToM tasks is related to deficits in social behavior following TBI (Hynes, Stone, & Kelso, 2011 this issue). However, in the present study, the prescan training allowed both groups to complete the social interaction condition of the

SAT with \hbox{accuracy} that exceeded 70%. Accuracy during the "bumper car" control condition was under 60% for both groups despite the fact that the overall level of intellectual functioning, as reflected by the WASI IQ, was within the average to high-average range and did not differ between the groups. This accuracy level for the control condition is lower than that found by Schultz et al. (2003), who reported 74% correct for the "bumper car" condition with a nearly identical task. The sample used by Schultz et al. (2003) consisted entirely of normal adults, however, and the bumper car condition can be relatively challenging. It is possible that adolescents in the current study had not yet acquired the skills required to accurately comprehend the relationships among velocity, direction, and mass that are needed to perform well during this control condition (Jacobs, Michaels, & Runeson, 2000).

The current study's design has several limitations and produced some unexpected findings. The first of these, as noted above, was lower than expected accuracy during the control (i.e., "bumper car") condition of the fMRI task. However, the activation pattern within the group of TD adolescents was generally consistent with the results from previous studies of ToM function (Carrington & Bailey, 2009), including research using similar animated stimuli (e.g., Castelli et al., 2000). Amygdala activation was observed in some previous studies (e.g., Schultz et al., 2003), and this was not found in the current investigation, but a limitation of BOLD imaging methods is the weak signal within the amygdala (LaBar, Gitelman, Mesulam, & Parrish, 2001), and activation during ToM has not been reported consistently within this brain area (Carrington & Bailey, 2009). In addition, one study examining ToM in normal individuals found greater amygdala activation in adults than in a group of children (Kobayashi, Glover, & Temple, 2007), suggesting that lack of amygdala activation within the current sample may reflect maturational differences.

Another limitation of the current study is that the sample size did not allow for corrections for multiple comparisons, and the design was not optimized for exploring relationships among brain activation, neuropathology variables (e.g., lesion location), SAT performance, and additional measures of cognitive function and social competence. Thus, it is not clear from the present results how focal injury to different brain structures relates to the SAT or to what degree the fMRI findings reflect alterations in executive functions and other general cognitive skills, as opposed to abilities more specially associated with ToM. A larger study is currently underway to explore these issues further.

REFERENCES

Adams, H., Mitchell, D. E., Graham, D. I., & Doyle, D. (1977). Diffuse brain damage of the immediate impact type. Its relationship to 'primary brain-stem damage' in head injury. *Brain, 100*(3), 489–502.

Adolphs, R. (2001). The neurobiology of social cognition. *Current Opinion in Neurobiology, 11*, 231–239.

Aldrich, N. J., Tenenbaum, H. R., Brooks, P. J., Harrison, K. S., & Sines, J. (2011). Perspective taking in children's narratives about jealousy. *British Journal of Developmental Psychology, 29*(1), 86–109.

Babikian, T., & Asarnow, R. (2009). Neurocognitive outcomes and recovery after pediatric TBI: Meta-analytic review of the literature. *Neuropsychology, 23*(3), 283–296.

Baron-Cohen, S., Leslie, A. M., & Frith, U. (1985). Does the autistic child have a "theory of mind"? *Cognition, 21*(1), 37–46.

Bird, C. M., Castelli, F., Malik, O., Frith, U., & Husain, M. (2004). The impact of extensive medial frontal lobe damage on "theory of mind" and cognition. *Brain, 127*(4), 914–928.

Blakemore, S. J. (2008). The social brain in adolescence. *Nature Reviews Neuroscience, 9*(4), 267–277.

Blakemore, S. J., Boyer, P., Pachot-Clouard, M., Meltzoff, A., Segebarth, C., & Decety, J. (2003). The detection of contingency and animacy from simple animations in the human brain. *Cerebral Cortex, 13*, 837–844.

Bowler, D. M. (1992). "Theory of mind" in Asperger's syndrome. *Journal of Child Psychology and Psychiatry, 33*(5), 877–893.

Carrington, S. J., & Bailey, A. J. (2009). Are there theory of mind regions in the brain? A review of the neuroimaging literature. *Human Brain Mapping, 30*(8), 2313–2335.

Castelli, F., Frith, C., Happé, F., & Frith, U. (2002). Autism, Asperger syndrome and brain mechanisms for the attribution of mental states to animated shapes. *Brain, 125*(8), 1839–1849.

Castelli, F., Happé, F., Frith, U., & Frith, C. (2000). Movement and minds: A functional imaging study of perception and interpretation of complex intentional movement patterns. *NeuroImage, 12*(3), 314–325.

Channon, S., & Crawford, S. (2010). Mentalising and social problem-solving after brain injury. *Neuropsychological Rehabilitation, 20*(5), 739–759.

Chevallier, C., Noveck, I., Happé, F., & Wilson, D. (2011). What's in a voice? Prosody as a test case for the theory of mind account of autism. *Neuropsychologia, 49*(3), 507–517.

Christodoulou, C., DeLuca, J., Ricker J. H., Madigan, N. K., Bly, B. M., Lange, G., et al. (2001). Functional magnetic resonance imaging of working memory impairment after traumatic brain injury. *Journal of Neurology, Neurosurgery, and Psychiatry, 71*(2), 161–168.

Crinion, J., Ashburner, J., Leff, A., Brett, M., Price, C., & Friston, K. (2007). Spatial normalization of lesioned brains: Performance evaluation and impact on fMRI analyses. *NeuroImage, 37*, 866–875.

Dennis, M., Agostino, A., Roncadin, C., & Levin, H. (2009). Theory of mind depends on domain-general executive functions of working memory and cognitive inhibition in children with traumatic brain injury. *Journal of Clinical and Experimental Neuropsychology, 31*(7), 835–847.

Dennis, M., Purvis, K., Barnes, M. A., Wilkinson, M., & Winner, E. (2001). Understanding of literal truth, ironic criticism, and deceptive praise following childhood head injury. *Brain and Language*, *78*(1), 1–16.

Dodrell-Feder, D., Koster-Hale, J., Bedny, M., & Saxe, R. (2011). fMRI item analysis in a theory of mind task. *NeuroImage*, *55*, 705–712.

Ewing-Cobbs, L., Barnes, M., Fletcher, J. M., Levin, H. S., Swank, P. R., & Song, J. (2004). Modeling of longitudinal academic achievement scores after pediatric traumatic brain injury. *Developmental Neuropsychology*, *25*(1–2), 107–133.

Friston, K. J., Holmes, A., Worsley, K. J., Poline, J. P., Frith, C. D., & Frackowiak, R. S. J. (1995). Statistical parametric maps in functional imaging: A general linear approach. *Human Brain Mapping*, *2*, 189–210.

Geraci, A., Surian, L., Ferraro, M., & Cantagallo, A. (2010). Theory of mind in patients with ventromedial or dorsolateral prefrontal lesions following traumatic brain injury. *Brain Injury*, *24*(7–8), 978–987.

Gobbini, M. I., Koralek, A. C., Bryan, R. E., Montgomery, K. J., & Haxby, J. V. (2007). Two takes on the social brain: A comparison of theory of mind tasks. *Journal of Cognitive Neuroscience*, *19*, 1803–1814.

Heider, F., & Simmel, M. (1944). An experimental study of apparent behavior. *American Journal of Psychology*, *57*(2), 243–259.

Henry, J. D., Phillips, L. H., Crawford, J. R., Ietswaart, M., & Summers, F. (2006). Theory of mind following traumatic brain injury: The role of emotion recognition and executive dysfunction. *Neuropsychologia*, *44*(10), 1623–1628.

Hillary, F. G. (2008). Neuroimaging of working memory dysfunction and the dilemma with brain reorganization hypotheses. *Journal of the International Neuropsychological Society*, *14*(4), 526–534.

Huang, M. X., Theilmann, R. J., Robb, A., Angeles, A., Nichols, S., Drake, A., et al. (2009). Integrated imaging approach with MEG and DTI to detect mild traumatic brain injury in military and civilian patients. *Journal of Neurotrauma*, *26*(8), 1213–1226.

Hughes, C., Ensor, R., & Marks, A. (2010). Individual differences in false belief understanding are stable from 3 to 6 years of age and predict children's mental state talk with school friends. *Journal of Experimental Child Psychology*, *108*(1), 96–112.

Hynes, C. A., Stone, V. E., & Kelso, L. A. (2011). Social and emotional competence in traumatic brain injury: New and established assessment tools. *Social Neuroscience*, *XX*, XXX–XXX.

Jacobs, D. M., Michaels, C. F., & Runeson, S. (2000). Learning to perceive the relative mass of colliding balls: The effects of ratio scaling and feedback. *Perception and Psychophysics*, *62*(7), 1332–1340.

Janusz, J. A., Kirkwood, M. W., Yeates, K. O., & Taylor, H. G. (2002). Social problem-solving skills in children with traumatic brain injury: Long-term outcomes and prediction of social competence. *Child Neuropsychology*, *8*(3), 179–194.

Kobayashi, C., Glover, G. H., & Temple, E. (2007). Children's and adults' neural bases of verbal and nonverbal 'theory of mind'. *Neuropsychologia*, *45*(7), 1522–1532.

LaBar, K. S., Gitelman, D. R., Mesulam, M. M., & Parrish, T. B. (2001). Impact of signal-to-noise on functional MRI of the human amygdala. *Neuroreport*, *12*(16), 3461–3464.

Levin, H. S., Song, J., Scheibel, R. S., Fletcher, J. M., Harward, H., Lilly, M., et al. (1997). Concept formation and problem-solving following closed head injury in children. *Journal of the International Neuropsychological Society*, *3*(6), 598–607.

Levin, H. S., Wilde, E. A., Chu, Z., Yallampalli, R., Hanten, G. R., Li, X., et al. (2008). Diffusion tensor imaging in relation to cognitive and functional outcome of traumatic brain injury in children. *Journal of Head Trauma Rehabilitation*, *23*(4), 197–208.

Mainero, C., Caramia, F., Pozzilli, C., Pisani, A., Pestalozza, I., Borriello, G., et al. (2004). fMRI evidence of brain reorganization during attention and memory tasks in multiple sclerosis. *NeuroImage*, *21*(3), 858–867.

Martín-Rodríguez, J. F., & León-Carrión, J. (2010). Theory of mind deficits in patients with acquired brain injury: A quantitative review. *Neuropsychologia*, *48*(5), 1181–1191.

Max, J. E., Levin, H. S., Schachar, R. J., Landis, J., Saunders, A. E., Ewing-Cobbs, L., et al. (2006). Predictors of personality change due to traumatic brain injury in children and adolescents six to twenty-four months after injury. *Journal of Neuropsychiatry and Clinical Neurosciences*, *18*(1), 21–32.

Mori, S., Crain, B. J., Chacko, V. P., & van Zijl, P. C. (1999). Three-dimensional tracking of axonal projections in the brain by magnetic resonance imaging. *Annals of Neurology*, *45*(2), 265–269.

Moriguchi, Y., Ohnishi, T., Mori, T., Matsuda, H., & Komaki, G. (2007). Changes of brain activity in the neural substrates for theory of mind during childhood and adolescence. *Psychiatry and Clinical Neurosciences*, *61*(4), 355–363.

Muller, F., Simion, A., Reviriego, E., Galera, C., Mazaux, J. M., Barat, M., et al. (2010). Exploring theory of mind after severe traumatic brain injury. *Cortex*, *46*(9), 1088–1099.

Netsch, T., & van Muiswinkel, A. (2004). Quantitative evaluation of image-based distortion correction in diffusion tensor imaging. *IEEE Transactions on Medical Imaging*, *23*(7), 789–798.

Newsome, M. R., Scheibel, R. S., Hanten, G., Chu, Z., Steinberg, J. L., Hunter, J. V., et al. (2010). Brain activation while thinking about the self from another person's perspective after traumatic brain injury. *Neuropsychology*, *24*(2), 139–147.

Oldfield, R. C. (1971).The assessment and analysis of handedness: The Edinburgh inventory. *Neuropsychologia*, *9*(1), 97–113.

Oni, M. B., Wilde, E. A., Bigler, E. D., McCauley, S. R., Wu, T. C., Yallampalli, R., et al. (2010). Diffusion tensor imaging analysis of frontal lobes in pediatric traumatic brain injury. *Journal of Child Neurology*, *25*(8), 976–984.

Povlishock, J. T., & Katz, D. I. (2005). Update of neuropathology and neurological recovery after traumatic brain injury. *Journal of Head Trauma Rehabilitation*, *20*, 76–94.

Price, C. J., & Friston, K. J. (2002). Functional imaging studies of neuropsychological patients: Applications and limitations. *Neurocase*, *8*, 345–354.

Ricker, J. H., Hillary, F. G., & DeLuca, J. (2001). Functionally activated brain imaging (O-15 PET and fMRI) in the study of learning and memory after traumatic brain injury. *Journal of Head Trauma Rehabilitation*, *16*, 191–205.

Saxe, R., & Kanwisher, N. (2003). People thinking about thinking people. The role of the temporo-parietal junction in "theory of mind." *NeuroImage, 19*, 1835–1842.

Saxe, R., & Powell, L. J. (2006). It's the thought that counts: Specific brain regions for one component of theory of mind. *Psychological Science, 17*, 692–699.

Scheibel, R. S., Newsome, M. R., Troyanskaya, M., Steinberg, J. L., Goldstein, F. C., Mao, H., et al. (2009). Effects of severity of traumatic brain injury and brain reserve on cognitive-control related brain activation. *Journal of Neurotrauma, 26*(9), 1447–1461.

Schmidt, A. T., Hanten, G. R., Li, X., Orsten, K. D., & Levin, H. S. (2010). Emotion recognition following pediatric traumatic brain injury: Longitudinal analysis of emotional prosody and facial expression. *Neuropsychologia, 48*, 2869–2877.

Schultz, R. T., Grelotti, D. J., Klin, A., Kleinman, J., Van der Gaag, C., Marois, R., et al. (2003). The role of the fusiform face area in social cognition: Implications for the pathobiology of autism. *Philosophical Transactions of the Royal Society of London. Series B, Biological Sciences, 358*(1430), 415–427.

Sesma, H. W., Slomine, B. S., Ding, R., & McCarthy, M. L. (2008). Executive functioning in the first year after pediatric traumatic brain injury. *Pediatrics, 121*(6), e1686–e1695.

Snodgrass, C., & Knott, F. (2006). Theory of mind in children with traumatic brain injury. *Brain Injury, 20*(8), 825–833.

Stocks, E. L., Lishner, D. A., Waits, B. L., & Downum, E. M. (2011). I'm embarrassed for you: The effect of valuing and perspective taking on empathic embarrassment and empathic concern. *Journal of Applied Social Psychology, 41*(1), 1–26.

Stone, V. E., Baron-Cohen, S., & Knight, R. T. (1998). Frontal lobe contributions to theory of mind. *Journal of Cognitive Neuroscience, 10*(5), 640–656.

Talairach, J., & Tournoux, P. (1988). *Co-planar stereotaxic atlas of the human brain.* New York, NY: Thieme Medical Publishers.

Teasdale, G., & Jennett, B. (1974). Assessment of coma and impaired consciousness. A practical scale. *Lancet, 2*(7872), 81–84.

Turkstra, L. S., Dixon, T. M., & Baker, K. K. (2004). Theory of mind and social beliefs in adolescents with traumatic brain injury. *NeuroRehabilitation, 19*(3), 245–256.

van den Bos, W., van Dijk, E., Westenberg, M., Rombouts, S. A., & Crone, E. A. (2011). Changing brains, changing perspectives: The neurocognitive development of reciprocity. *Psychological Science, 22*(1), 60–70.

Wakana, S., Caprihan, A., Panzenboeck, M. M., Fallon, J. H., Perry, M., Gollub, R. L., et al. (2007). Reproducibility of quantitative tractography methods applied to cerebral white matter. *NeuroImage, 36*(3), 630–644.

Wechsler, D. (1999). *Wechsler Abbreviated Scale of Intelligence.* San Antonio, TX: Psychological Corporation.

Weiderholt, J. L., & Bryant, B. R. (2001). *Gray Oral Reading Test* (4th ed.). Austin, TX: ProEd.

Weisskoff, R. M. (1996). Simple measurement of scanner stability for functional NMR imaging of activation in the brain. *Magnetic Resonance Medicine, 36*(4), 643–645.

Wilde, E. A., Chu, Z., Bigler, E. D., Hunter, J. V., Fearing, M. A., Hanten, G., et al. (2006). Diffusion tensor imaging in the corpus callosum in children after moderate to severe traumatic brain injury. *Journal of Neurotrauma, 23*(10), 1412–1426.

Wilde, E. A., McCauley, S. R., Chu, Z., Hunter, J. V., Bigler, E. D., Yallampalli, R., et al. (2009). Diffusion tensor imaging of hemispheric asymmetries in the developing brain. *Journal of Clinical and Experimental Neuropsychology, 31*(2), 205–218.

Wilde, E. A., Ramos, M. A., Yallampalli, R., Bigler, E. D., McCauley, S. R., Chu, Z., et al. (2010). Diffusion tensor imaging of the cingulum bundle in children after traumatic brain injury. *Developmental Neuropsychology, 35*(3), 333–351.

Yeates, K. O., Bigler, E. D., Dennis, M., Gerhardt, C. A., Rubin, K. H., Stancin, T., et al. (2007). Social outcomes in childhood brain disorder: A heuristic integration of social neuroscience and developmental psychology. *Psychological Bulletin, 133*(3), 535–556.

Zarei, M., Johansen-Berg, H., Smith, S., Ciccarelli, O., Thompson, A. J., & Mathews, P. M. (2006). Functional anatomy of interhemispheric cortical connections in the human brain. *Journal of Anatomy, 209*(3), 311–320.

Social and emotional competence in traumatic brain injury: New and established assessment tools

Catherine A. Hynes[1,2], Valerie E. Stone[1,3], and Louise A. Kelso[1,4]

[1]School of Psychology, University of Queensland, Brisbane, Australia
[2]Rehabilitation Service, Kent and Medway NHS and Social Care Partnership Trust, Maidstone, UK
[3]Answers About Brain Injury LLC, Golden, CO, USA
[4]School of Psychology, Griffith University, Mt Gravatt, Australia

Chronic social/emotional deficits are common in moderate to severe traumatic brain injury (TBI), leading to significant functional difficulties. Objective, quantitative tools for assessing social/emotional competence are an important adjunct to cognitive assessments. We review existing social/emotional measures, conclude that theory of mind tests are not adequate for clinical assessments of social competence, and explain the development and piloting of novel measures in a small group of moderate to severe TBI patients ($N = 16$) and non-brain-damaged controls ($N = 16$). The novel measures are the Global Interpersonal Skills Test (GIST), a questionnaire measuring informant-rated social skills; the Assessment of Social Context (ASC), a video-based task examining understanding of others' emotions, attitudes, and intentions; the Social Interpretations Test, a social framing task based on Heider and Simmel (1944); and Awareness of Interoception, a heartbeat-detection paradigm related to physiological self--awareness. In a MANOVA, other-rated social skills (GIST), ASC, and Awareness of Interoception scores were significantly lower for TBI patients than controls. ASC, $r(31) = .655$, and Social Interpretations, $r(31) = .460$, scores were significantly correlated with informant-rated social skills (GIST). We encourage clinicians to add social/emotional measures to assessments of TBI patients.

Keywords: Traumatic brain injury; Social skills; Emotions; Head injury.

Probably one of the saddest secondary effects of a severe head-injury is loneliness.
(Gronwall, Wrightson, & Waddell, 1998, p. 146)

Sadly, personality change and deficits in social functioning can be common long-term effects of moderate to severe traumatic brain injury (TBI), though they do not happen in all cases (e.g., Engberg & Teasdale, 2004; Stone & Hynes, 2011; Temkin,

Corrigan, Dikmen, & Machamer, 2009). Social deficits following more severe TBI include disinhibited behavior, inappropriate jokes, vulnerability to deception/exploitation, sexually inappropriate behavior, and difficulty with conversational turn-taking and relevance, reading others' expressions, and tracking others' intentions (e.g., Berlin, Rolls, & Kischka, 2004; Channon, Pellijeff, & Rule, 2005; Kim, 2002). TBI survivors with such problems face a difficult

Correspondence should be addressed to: Valerie E. Stone, Answers About Brain Injury LLC, 601-C 16th Street, No. 191, Golden, CO 80401, USA. E-mail: vestone@gmail.com

Original studies reported here were the doctoral dissertation research of Catherine Hynes, PhD (awarded 2009, University of Queensland), and the Honours Thesis research (GIST validation) of Louise Kelso, BPSc Hons (awarded 2006, University of Queensland). This research was supported by a Uniquest Trailblazer Award, an Endeavour International Postgraduate Research Scholarship from the Commonwealth Government of Australia, a University of Queensland (UQ) Research Development Grant, a UQ Social and Behavioral Sciences Tuition Fee Waiver Scholarship, and a UQ International Living Allowance Scholarship. The authors would like to thank the following people for their contributions to this research: Dana Schneider, Katrina Van Der List, Tirta Susilo, and Trudy Sinnamon (data collection); Anna Yuen and James White (acting in video clips); Paul Jackson (programming help); Dr Ron Hazelton (help with recruitment); and Beth O'Brien (statistical assistance).

paradox. If they maintain social connections, they adjust better to life post-injury, but they may suffer rejection by others because of their inappropriate social behavior, and many (particularly men with TBI) experience marital problems, social exclusion, and isolation (Arango-Lasprilla et al., 2008; Blais & Boisvert, 2007; Bond & Godfrey, 1997; Hawthorne, Gruen, & Kaye, 2009; Temkin et al., 2009).

Problems with emotion regulation and responsiveness can also follow moderate to severe TBI (de Sousa et al., 2010, 2011; Milders, Fuchs, & Crawford, 2003; Saunders et al., 2006). Many people with TBI exhibit emotional lability and aggression, and meet criteria for major depression and anxiety (Jorge, 2005). Long-term functional outcomes are poorer for anxious and depressed TBI survivors (Ownsworth & McKenna, 2004; Pagulayan, Hoffman, Temkin, Machamer, & Dikmen, 2008; Ponsford, Draper, & Schonberger, 2008). People with more severe TBI also have difficulty inferring others' emotions from facial or vocal expressions, contributing to poorer social outcomes (e.g., Knox & Douglas, 2009; McDonald, Flanagan, Rollins, & Kinch, 2003).

Brain injuries are classified as mild, moderate, or severe by level of orientation/awareness, and period of loss of consciousness and amnesia post-injury. Personality change and social deficits are seldom described as symptoms of concussions/mild traumatic brain injuries; more severe brain injuries are more likely to produce social deficits (Temkin et al., 2009). These social deficits can be improved with rehabilitation focused on social problems, but effective rehabilitation requires good measures of social/emotional functioning to serve as benchmarks for improvement (Hawley & Newman, 2010; Stone & Hynes, 2011). Neuropsychologists have a diverse toolkit for assessing cognitive dysfunction, including hundreds of tests to measure specific aspects of cognition. They also need a large toolkit for assessing specific aspects of social/emotional deficits.

Non-penetrating TBI, or closed head injury, causes focal damage predominantly in orbitofrontal (also called ventral frontal) and anterior temporal regions (e.g., Devinsky & D'Esposito, 2004; Levine et al., 2008). Anterior regions exhibit the most deformation by being compressed forward against the skull, whereas more posterior regions exhibit stretching, which shears fiber pathways (Bayly et al., 2005). Social neuroscientists have documented the contribution of these regions to social information-processing; TBI survivors with orbitofrontal lesions are at particular risk of social deficits (Stone & Hynes, 2011). TBI also causes diffuse axonal injury, which affects

connections to and from orbitofrontal cortex, as well as corpus callosum (e.g., Wang et al., 2008). Thus, people with TBI may have several sources of deficits in social/emotional competencies and cognition.

Cognitive deficits can contribute to social difficulties (Struchen et al., 2008), because social interaction is cognitively complex. It requires applying memories to a changing behavior stream, tracking changing social context and rewards, selecting behaviors from many options, and flexibly refocusing attention. Such situations make demands on executive functions, a group of cognitive abilities including working memory, distraction suppression, planning, problem solving, and organizing behavioral output. Executive deficits can cause problems in social interaction, just as they do in other complex tasks; for example, a deficit in shifting attention flexibly, though not specifically social, can cause someone to miss social cues.

However, although it is *necessary* to assess executive functions in people with social deficits following TBI, it is not *sufficient*. Performance on many executive tests does *not* predict social performance in people with TBI (e.g., Milders, Ietswaart, Crawford, & Currie, 2003; Rochat, Ammann, Mayer, Annoni, & Van der Linden, 2009). Functional neuroimaging shows that executive functions are associated primarily with dorsolateral frontal activation, whereas reward and social processing activate orbitofrontal regions (Baldo, Schwartz, Wilkins, & Dronkers, 2006; Kringelbach & Rolls, 2003; Nyhus & Barceló, 2009). With evidence that social/emotional deficits after TBI are not caused solely by cognitive and executive functioning deficits, specific assessments of social competence would benefit neuropsychology (Stone & Hynes, 2011).

Social deficits can result from impairments in multiple systems. Defining social cognition rigorously is difficult, because so many processes are involved, just as defining the components of cognition has been a lengthy scientific undertaking. For example, people with TBI may have difficulty both reading other people's emotional expressions and inhibiting inappropriate remarks (Berlin et al., 2004; Kim, 2002; Knox & Douglas, 2009). Although both difficulties can be categorized as deficits in the *social* realm, they may result from impairments in different underlying systems, and would entail different treatments. In order to get a full picture of the many factors influencing inappropriate social behavior, clinicians need to assess cognitive *and* social functions in detail.

Psychological assessments can use any of four methods to evaluate a patient's competence in a particular domain. First, performance measures are objective tests requiring patients to demonstrate competence by solving a problem in a particular domain, as in IQ tests. Second and third, informant reports and self-reports are questionnaires given to caregivers or patients to report on patients' behavior, or qualitative reports of their experiences. Fourth, clinicians use their own qualitative judgments to assess certain behaviors.

All of these methods have value. Some phenomena, such as emotional distress, are inherently subjective, internal experiences, making a self-report necessary to gather this information. Other behaviors, such as easily losing one's temper, may not occur in front of strangers or in formal settings, making them difficult to observe in the clinic or laboratory. Reports about patients' behaviors in multiple settings are a rich source of information. Nevertheless, objective performance measures can more precisely describe particular abilities, and differentiate problems in different domains of social and emotional competence. Social neuroscience offers some measures, but more are required for a complete description of social competence.

EMOTION TASKS AND EMPATHY IN PEOPLE WITH TBI

Emotion recognition may be an important early step in accurately interpreting others' behavior, and a prerequisite for empathy, the ability to emotionally resonate with others' feelings. Several measures of visual emotion recognition have been developed, using either static or video stimuli, including the first subtest of the Awareness of Social Inference Test (TASIT) (McDonald et al., 2003). Recognition of vocal emotion obviously requires dynamic audio stimuli. People with TBI have consistently been found to have difficulty recognizing emotion from both faces and voices, though not all have difficulty with both modalities (e.g., Knox & Douglas, 2009; McDonald et al., 2003; Milders et al., 2008). Emotion recognition impairment seems to be difficult to improve through rehabilitation (McDonald, Bornhofen, & Hunt, 2009), and thus may be a lasting consequence of TBI.

Deficits in emotion recognition are associated with other social functioning difficulties (e.g., McDonald, Flanagan, Martin, & Saunders, 2004; Knox & Douglas, 2009). People with TBI have lower self-rated empathy than controls, and questionnaire measures of empathy correlated with emotion recognition and emotional responsivity (de Sousa et al., 2010, 2011; Muller et al., 2010; Shamay-Tsoory, Tomer, Berger, & Aharon-Peretz, 2003; Williams & Wood, 2010).

Some studies indicate that TBI can alter visceral arousal and subjective experiences of emotion. TBI patients rate their subjective emotional arousal for negative pictures significantly lower than controls, and exhibit significantly less physiological reactivity to negative images, usually tested by showing images from the International Affective Picture Set (de Sousa et al., 2010, 2011; Lang, Bradley, & Cuthbert, 2005; Saunders, McDonald, & Richardson, 2006). Although assessments of empathy, emotion recognition, and subjective emotional arousal are established, emotional functioning is not limited to these three abilities. Furthermore, methods of measuring arousal are cumbersome and not easily adapted to the clinic.

Sarcasm detection in people with TBI

One way of looking at the ability to understand others' intentions is to investigate responses to sarcasm (also called irony); that is, saying the opposite of what one means. For example, it is sarcastic to say, "You're so graceful," when someone drops something. Sarcasm/irony tasks have the participant read stories or watch videos, and answer questions about why someone said what they said, whether the utterance was sarcastic or sincere, or what someone meant by such an utterance. Patients with TBI are consistently impaired in understanding sarcasm as opposed to sincere statements, relative to healthy controls (e.g., Channon et al., 2005; McDonald & Pearce, 1996; McDonald et al., 2003), and performance on the two sarcasm-detection subtests of TASIT correlates highly with other social behavior problems in people with TBI (McDonald et al., 2004). Although we have well-validated assessments of sarcasm detection, social competence includes more than just this ability.

Theory of mind tasks in people with TBI

Several researchers have looked at theory of mind (ToM) in patients with TBI and orbitofrontal damage, with the idea that deficits in ToM might underlie the patients' social difficulties (for a detailed review, see Stone & Hynes, 2011). The term "ToM" is problematic, because different groups of researchers use it to mean different things. In the social neuroscience literature, ToM has often been construed broadly as "the ability to infer others' mental states," including intentions, thoughts, beliefs, desires, emotions, focus of attention, or attitudes. Within developmental and cognitive psychology, however, ToM refers more narrowly to meta-representation only; that is, the ability to understand that mental states of knowledge and belief represent the world, and thus that such mental states can be mistaken.

The different tasks used to assess ToM reflect these differences in terminology. With a narrow developmental/cognitive psychology definition of ToM, the only valid task is a false-belief task. In contrast, with the broader social neuroscience definition of ToM, tasks examining inferences about intention, attention, sarcasm, and empathy, as well as false beliefs, have all been used. Patients with TBI do not seem to be impaired in belief understanding and meta-representation, but they do appear to have difficulty understanding others' intentions and feelings.

Results on false-belief task performance in TBI are mixed, because of inconsistency in controlling for non-ToM factors that affect task performance. Where working memory demands of false-belief tasks are controlled for, as by placing pictures depicting the sequence of events in front of the participant throughout the session, deficits on false-belief tasks are not evident (Bibby & McDonald, 2005; Muller et al., 2010; Stone, Baron-Cohen, & Knight, 1998). Forensic psychologists use below-chance performance on certain tasks to test whether a client is giving full effort during testing (Boone, 2007). With such a consistent lack of deficits on properly controlled false-belief tasks in people with TBI, these tasks could possibly be used to measure suspect effort in testing for social deficits in TBI.

Other mental-state inference tasks

Broadly defined ToM tasks involve listening to or seeing brief stories, cartoons, or pictures, and then inferring characters' intentions, feelings, focus of attention, and beliefs, or recognizing when something awkward has been said. Using this broader definition of ToM and these tasks, patients with TBI do have deficits in inferring mental states.

The "Strange Stories Task" tests participants' ability to infer story characters' thoughts, feelings, and intentions from verbal stories (Happé, 1994). Control stories require non-mentalistic inferences, such as inferences about physical processes. Some researchers have found deficits on questions about mentalistic but not control stories, while others controlling for non-ToM task demands have not (Bibby & McDonald, 2005; Channon et al., 2005; Shamay-Tsoory & Aharon-Peretz, 2007).

Cartoon tasks have the advantage of testing mental-state inferences without requiring verbal comprehension. They use visual cartoons that require inferences about characters' feelings, intentions, focus of attention, or beliefs to understand the joke (Happé, Brownell, & Winner, 1999). Control cartoons are more "slapstick," requiring physical inferences to get the joke (examples in Gallagher et al., 2000, p. 14). As with the stories task, some studies have found deficits in people with moderate to severe TBI relative to neurologically healthy controls, and others have not (Bibby & McDonald, 2005; Milders, Ietswaart, Crawford, & Currie, 2006, 2008).

The Faux Pas Recognition Task involves brief stories in which someone unintentionally says something awkward or insulting; in control stories, a minor conflict occurs, but no faux pas (Gregory et al., 2002; Stone, Baron-Cohen, & Knight, 1998). Several questions assess whether participants understand that something awkward was said, why it was inappropriate, that it was unintentional, and that one character might have felt bad. Control comprehension questions test for general comprehension of story facts apart from mental state understanding. Although introduced as a measure of ToM, this task measures multiple inferential abilities: false-belief understanding, empathy, inferences about intentions, and knowledge of appropriateness. Studies consistently find that people with moderate to severe TBI score lower than either patients with dorsolateral prefrontal lesions or non-brain-injured controls. While answering control comprehension questions correctly, TBI patients fail to identify faux pas or erroneously label non-faux pas as faux pas (Milders, Fuchs, & Crawford, 2003; Milders et al., 2006; Muller et al., 2010; Shamay-Tsoory, Tomer, Berger, Goldsher, & Aharon-Peretz et al., 2005; Stone et al., 1998; Stone & Hynes, 2011). Longitudinal research suggests that the questions about intentions primarily differentiate people with TBI from controls (Milders et al., 2008).

ToM tests, because of their mixed results in TBI patients, are inadequate for assessing social competence. The Faux Pas test, while useful in TBI, might better be replaced with a measure of intention specifically.

Questionnaire measures of behavioral problems in TBI

Existing questionnaires are not tailored specifically to the social deficits of people with TBI. Several rating scales for frontal dysfunction focus on cognitive/executive difficulties, but cognitive abilities do not necessarily predict social competence (Grace & Malloy, 2001; Milders et al., 2003, 2008; Rochat et al., 2009). Other scales for measuring behavioral/neuropsychiatric problems include some interpersonal items or subscales, but do not *primarily* focus on social competence (Barrash, Tranel,

& Anderson, 2000; Cummings, Mega, Gray, & Rosenberg-Thompson, 1994; Kertesz, Davidson, & Fox, 1994; Kreutzer, Marwitz, Seel, & Devany Serio, 1996; Nelson, Drebing, Satz, & Uchiyama, 1998; Yamasoto et al., 2007). There are also scales that focus on social behavior, but were not developed for TBI. Difficulties arise when using measures designed for a different population with TBI: When the Social Performance Survey Schedule, designed to assess social skills in people with intellectual disabilities (Lowe & Cautela, 1978), was tested in people with TBI, it did not differentiate people with TBI from controls (McDonald et al., 2004).

Finally, there are clinician rating instruments that focus specifically on conversational pragmatics in TBI (Drummond & Boss, 2004; Linscott, Knight, & Godfrey, 1996). As with any informant-report measure, they depend on the accuracy of clinicians' ratings, but these are well-validated measures for assessing inappropriate conversational behavior in TBI, and scores improve with rehabilitation (Dahlberg et al., 2007; McGann, Werven, & Douglas, 1997). Their main limitation as social measures is that they do not sample the full range of social competence.

Need for more social/emotional measures

Many social/emotional abilities are simply not covered by existing tests. Tools for assessing conversational pragmatics, including sarcasm detection, physiological responses to emotional pictures, empathy, and emotion recognition, are well developed and have norms available (Lang, Bradley, & Cuthbert, 2005; Linscott et al., 1996; McDonald, Flanagan, & Rollins, 2002). However, deficits in social/emotional competence in TBI go well beyond these, so a broader set of tests is needed. ToM tasks cannot fill this role: without consistent impairment following TBI, their clinical usefulness is questionable. The Faux Pas Recognition Task is promising, but a briefer task asking about intention might provide as much information in less time.

Therefore, it seems timely to develop and validate new tasks that test specifically for understanding intentions, and sample other areas of social/emotional competence. In the current study, we developed four novel measures for use in people with TBI, meant to test different social/emotional competencies than those tapped by existing measures. Our performance-based measures were designed to assess understanding of intentions, attitudes, framing of social situations, and physiological self-awareness. We also developed a questionnaire measuring global social skills,

to provide one measure that would focus on social behavior changes in TBI rather than behavioral and psychiatric changes more generally. We investigated whether social/emotional deficits following TBI could be documented quantitatively using our tasks.

METHODS

Development of the Global Interpersonal Skills Test (GIST)

The GIST was designed to assess social competence specifically, rather than the broader range of psychiatric and behavioral disturbances measured by other questionnaires used in TBI. Measuring social skills in people with TBI must be integrated with knowledge of this population's other cognitive and behavioral characteristics. People with TBI may have impaired self-awareness (Garmoe, Newman, & O'Connell, 2005; Hart, Sherer, Whyte, Polansky, & Novack, 2004; Prigatano, 2005); thus, social skills on the GIST are informant-rated.

Based on the methods Lowe and Cautela (1978) used in the development of the Social Performance Survey Schedule, 57 undergraduates at the University of Queensland from two second-year psychology classes were given 10 min to list traits and behaviors that described either positive or negative social skills, defined as traits or behaviors that make people likable or unlikable, or socially skilled or unskilled. Similar responses were grouped together and tallied across all participants. Responses endorsed by 10 or more participants were included. Items were paired with their positive or negative counterpart on the final questionnaire; for example, "rude" was paired with "polite." Nine additional items that captured norm-violating behavior, which can occur in patients with TBI, were included; for example, "touches others inappropriately" versus "never touches others inappropriately." Examples of items are included in Figure 1.

The GIST contained 69 descriptions of character traits or observable behaviors which were paired with their opposites: The socially undesirable item was listed (e.g., selfish) with its socially desirable counterpart (e.g., generous) on the same line. Informants (e.g., friends, spouses) rated participants ("Please rate the other person on each character trait") on a 7-point Likert scale ranging from −3 ("very much like this"), through 0 ("neither one nor the other"), to +3 ("very much like this"). A total score was summed across all items, with a minimum score of −207 meaning the person was socially unskilled, and +207 meaning the person was socially skilled.

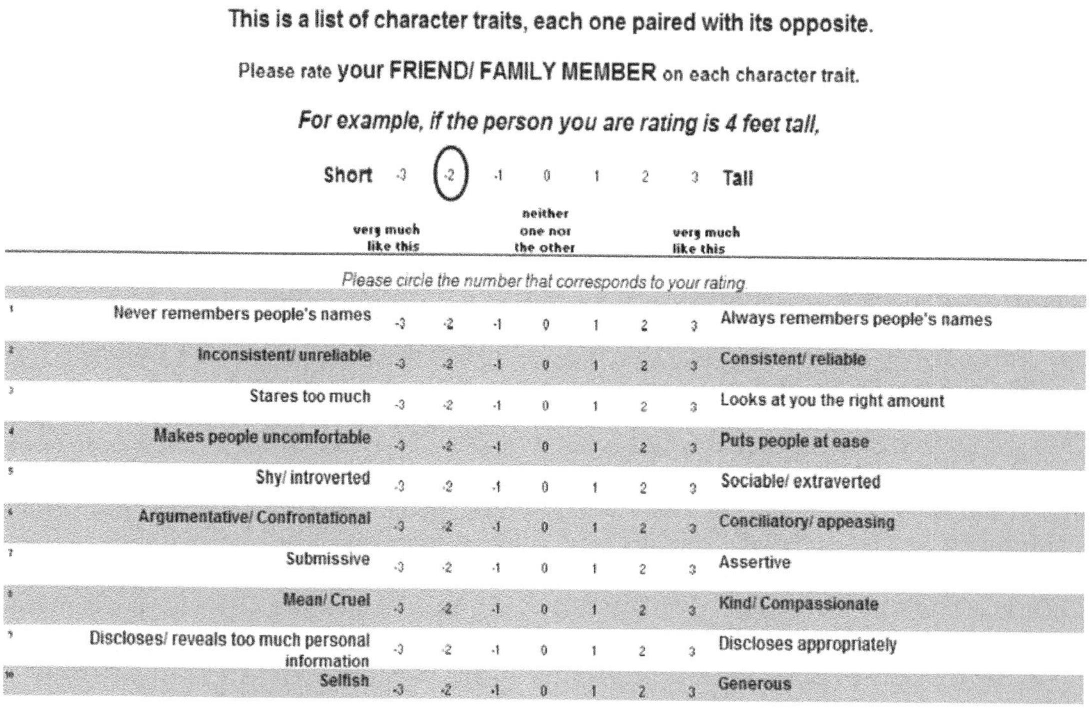

This is a list of character traits, each one paired with its opposite.

Please rate your FRIEND/ FAMILY MEMBER on each character trait.

For example, if the person you are rating is 4 feet tall,

Short -3 ⟨-2⟩ -1 0 1 2 3 Tall

| | very much like this | | neither one nor the other | | very much like this | |

Please circle the number that corresponds to your rating.

1	Never remembers people's names	-3	-2	-1	0	1	2	3	Always remembers people's names
2	Inconsistent/ unreliable	-3	-2	-1	0	1	2	3	Consistent/ reliable
3	Stares too much	-3	-2	-1	0	1	2	3	Looks at you the right amount
4	Makes people uncomfortable	-3	-2	-1	0	1	2	3	Puts people at ease
5	Shy/ introverted	-3	-2	-1	0	1	2	3	Sociable/ extraverted
6	Argumentative/ Confrontational	-3	-2	-1	0	1	2	3	Conciliatory/ appeasing
7	Submissive	-3	-2	-1	0	1	2	3	Assertive
8	Mean/ Cruel	-3	-2	-1	0	1	2	3	Kind/ Compassionate
9	Discloses/ reveals too much personal information	-3	-2	-1	0	1	2	3	Discloses appropriately
10	Selfish	-3	-2	-1	0	1	2	3	Generous

Figure 1. Examples of items on the Global Interpersonal Skills Test (GIST).

Psychometric properties of the questionnaire were investigated in 38 participants (18 female) with a mean age of 22.2 years ($SD = 4.37$) and a mean level of education of 15.6 years ($SD = 1.3$), falling into four groups of friends who knew each other and could rate each other. There was a significant, positive relationship, $r(40) = .66$, $p < .0001$, between GIST scores and how much participants liked the rated person ("How fond are you of this person?", 5-point Likert scale from "hardly at all" to "very much"), such that the more highly people rated a particular participant's social skills, the more people liked that person. Within three of the groups of friends, there was also convergence in how people rated a particular individual [Group 1 intraclass correlation (ICC) = .83, $F(10, 100) = 5.95$, $p < .001$; Group 2, ICC = .72, $F(6, 54) = 3.55$, $p < .005$; Group 3, ICC = .77, $F(8, 64) = 4.39$, $p < .001$; Group 4, ICC = .60, ns].

Development of the Assessment of Social Context (ASC) task

The ASC task was designed to expand upon the ecological validity of tasks available to measure social comprehension and to pinpoint specific types of mental-state inferences more precisely. In general, real-life social interactions are not written down, and involve dynamic faces, bodies, and prosodic cues.

While the TASIT presents dynamic, videotaped stimuli, it assesses only emotion recognition and sarcasm detection, and we wished to sample other types of social inferences. The TASIT provides participants with multiple-choice responses, whereas, in real situations, people do not have multiple-choice options to assist with social interpretations. For this reason, we used open-ended questions where possible on the ASC. The TASIT is a criterion-referenced test, meaning that most people perform close to ceiling (McDonald et al., 2004), and we wished to design a test that would be sensitive to even subtle difficulties in participants' social comprehension.

We designed the ASC to capture these aspects of real-life socializing, by using videotaped stimuli of everyday interactions, and asking open-ended questions where appropriate. Furthermore, this task asks separate questions about the feelings, intentions, and attitudes of people in the videos, thus allowing separate measurement of specific types of mental-state inferences, unlike many ToM tests. These particular types of inferences were chosen because they are essential for both appropriate responses to a social context, and avoiding harmful interactions.

Ten pairs of dyadic interactions were developed into 10-s-long film clips, in order to reduce the working memory load of the task. The verbal content of the interaction was identical in each pair, but nonverbal information differentiated the meaning of the statement. For example, in one pair, a woman said to a man "I haven't seen you in ages," and in one scenario, she was happy to see him, while in the other, she was awkward and unfriendly. Two actors with experience in local theatre were hired to act out the scenarios.

Three questions, pertaining to the target actor's Intention, Emotion, and Attitude toward the other actor were asked after each clip. For example, for the clip described above, the questions were as follows:

Intention: When the woman said, "I haven't seen you in ages," what purpose was she trying to accomplish?
[open-ended response]
Emotion: What emotion or attitude did the woman express when she said, "I haven't seen you in ages"?
[open-ended response; choose one word only]
Attitude: How positive or negative was the woman toward the man?
[–1 (negative), 0 (neutral), 1 (positive)]

Where possible, we used open-ended responses. For the Attitude question only, the options were negative, neutral, or positive, so this was multiple-choice.

An initial 35 participants (23 female; age: $M = 19.4$, $SD = 3.1$) were recruited from the University of Queensland psychology research participation pool to pilot the stimuli, the participants receiving course credit for participation. They viewed 26 video pairs, and responded to questions to determine whether the situations were everyday events, whether the acting was convincing, and whether the paired scenarios were distinguishable on emotions, intentions, and attitudes. A further 20 participants (15 female; age: $M = 21.3$ years, $SD = 3.6$) were recruited from the same participant pool to further refine the remaining 17 scenario pairs. Any plausible response generated by these 20 participants was considered to be correct, because the scenarios, like real-world social interactions, were open to various interpretations. For instance, in the example in which the woman was unfriendly toward the man, responses such as "angry," "awkward," or "hurt" were scored correct, whereas "delighted," "friendly," or "neutral" were scored as incorrect. Pilot testing enabled us to narrow the number of video clips down to 10 pairs, by selecting the most distinguishable clip pairs. Emotion, Intention, and Attitude questions were scored as correct (1) or incorrect (0), and summed, making a maximum score of 60 points, and a minimum of 0 points. Scores were converted to a proportion of correctly answered questions to total questions answered, in order to include data for participants who skipped questions.

Development of the Social Interpretations Task

How one frames and interprets a social interaction can impact the social inferences one draws, and how appropriately one responds. Heider and Simmel (1944) created a task with a brief silent film in which a large triangle, a small triangle, and a small circle move around. Undergraduate participants projected intentions, motives, and emotions onto the geometric figures when describing the film. This social framing occurred spontaneously, despite the absence of social cues: no eyes, faces, limbs, sounds, or language. Such abstract social framing seems important for healthy socializing, as clinical groups with social difficulties have trouble with the task. Children and adults with autism-spectrum disorders attributed fewer mental states (feelings, thoughts, beliefs) to the geometric figures than healthy control participants (Castelli, Frith, Happé, & Frith, 2002; Klin, 2000). Patients with orbitofrontal damage or amygdala damage used fewer affect and social words than healthy controls in describing the geometric figures, relying more on movement-related words (Heberlein & Adolphs, 2004).

We devised the Social Interpretations Task with videos of geometric forms interacting. However, although Heider and Simmel (1944) used a general task, "describe what happens in the film," we focused the task by providing multiple-choice answers. Twenty-four animations were developed. Participants were told that they would watch a social interaction between two geometric figures, and respond to "What happened in the scenario?", with two multiple-choice options. (See Figure 2 for illustration

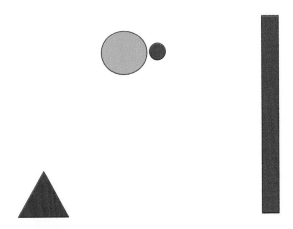

Figure 2. Example diagram of the Social Interpretations Task. Animations consisted of, for example, the red triangle bouncing a ball (purple circle) off a wall, and the green circle stealing the ball and taunting the red triangle with it. Multiple-choice responses were as follows: "1) The other person stole your ball. 2) The other person pitched a ball to you."

and example.) When 38 University of Queensland undergraduate students and members of the community (15 female; age: $M = 21.5$, $SD = 3.7$) were pilot-tested on these stimuli, their mean percent accuracy was 92% ($SD = 6\%$).

Development of the Awareness of Interoception Task

Because self-awareness of one's emotional reactions is a necessary element of emotional regulation, we included a measure of physiological sensitivity, the Awareness of Interoception Task. ("interoception" denotes conscious detection of visceral states.) Different emotions may be distinguished from one another partly by their cardiac signatures; thus, heartbeat detection may be required to detect the onset and quality of an emotion (e.g., Levenson, 1992). Interoception has been measured by heartbeat detection, in which participants distinguish between sets of sounds that are either synchronous or asynchronous with their heartbeats (Critchley, Wiens, Rotshtein, Ohman, & Dolan, 2004). This heartbeat-detection task has been associated with subjective experience of emotion in healthy people, and appears to depend on the insula (Barrett, Quigley, Bliss-Moreau, & Aronson, 2004; Critchley et al., 2004; Wiens, Mezzacappa, & Katkin, 2000). We developed a task requiring participants to distinguish between tones either synchronous or asynchronous with their heartbeats.

The methods for this task were taken from Critchley and colleagues' (2004) examination of interoception. Participants were seated at a computer with a Nonin Xpod 3012 clipped to the index finger of their non-dominant hand to extract heartbeat time course from their pulse. An in-house computer program presented a series of tones that were either synchronous with the participant's heartbeat, or had a 700-ms delay introduced (Figure 3). Participants were asked to decide which trials were synchronous with their heartbeats, and which were not. Trials included two for training, and 14 experimental trials.

Hypotheses

We predicted that people with TBI would have lower scores on all four social/emotional measures than controls. We investigated whether performance on the novel tasks would be related to real-world social functioning, tested with a correlation between task scores and social skill (GIST scores). In addition, to explore emotional changes post-injury qualitatively,

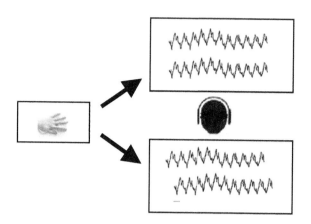

Figure 3. Diagram of the Awareness of Interoception Task. Participants were seated at a computer, and a Nonin Xpod 3012 was clipped to a finger to extract heartbeat time course from their pulse. Over headphones, they were presented with a series of tones that either were synchronous with their own heartbeat (upper diagram) or had a 700-ms delay introduced (lower diagram). Pilot testing determined that a 700-ms delay provided a good balance between ceiling effects and the tasks being too difficult. Participants responded by indicating which trials were synchronous with their heartbeats, and which were not.

participants with TBI were asked, "Have you noticed any changes in your emotions since your brain injury?"

Participants

Ethical approval for this study was granted by the School of Psychology Ethics Review Committee at the University of Queensland, and by the Human Research and Ethics Committees of two hospitals in Brisbane, Australia. People with nonpenetrating moderate to severe TBI were recruited through advertisements on electronic mailing lists and websites, and through two hospitals' rehabilitation wards.

Inclusion criteria were that patients had to be aged 18–59, reside within 150 km of Brisbane, and have no premorbid neurological or psychiatric conditions requiring medical intervention, and no aphasia, because aphasia is likely to cause difficulties with social interaction that were beyond the scope of our investigation. Alcohol and drug misuse is more prevalent among people with TBI than the general population (Graham & Cardon, 2008; Parry-Jones, Vaughan, & Cox, 2006; Ponsford, Draper, & Schonberger, 2007); thus, we adopted a liberal inclusion criterion for substance use in our sample (<10 standard drinks <2 days per week for <1 year). Sixteen patients (1 female) took part in the study. Of the 16 patients, eight were involved in motor vehicle accidents, two were pedestrians hit by cars, one was a bicyclist hit by a car, two fell several meters, one was in a horse-riding accident, and one was assaulted.

Sixteen neurologically healthy controls (two female) were recruited, either through the patients' family or friends, or from the local community. All participants were Caucasian and had English as a first language. Many studies of TBI screen out patients with comorbid psychological problems or substance abuse to examine only TBI-specific factors. While there are many research advantages to this approach, substance use and mental health concerns are common in people with TBI (Graham & Cardon, 2008; Parry-Jones et al., 2006; Ponsford et al., 2008; Ponsford, Whelan-Goodinson, & Bahar-Fuchs, 2007). Therefore, we chose instead to balance these factors across our two samples. Because three of the patients reported premorbid difficulties with depression and one with anxiety, for which they had not received medical treatment, three controls with similar histories of untreated depression, and one with a history of untreated anxiety were recruited. Substance use histories were also balanced across groups (see Table 1).

Patients' medical charts revealed a mean time since injury of 6.2 years ($SD = 8.8$), loss of consciousness of 33.5 days ($SD = 40.8$), post-traumatic amnesia of 50.3 days ($SD = 37.9$), and Glasgow Coma Score on admission of 6.2 ($SD = 2.8$). Radiologists' readings of computerized axial tomography scans in the charts contained variable amounts of detail. Eight of the scan readings confirmed frontal lobe injury, four referred to widespread contusions or hematomas possibly including the frontal lobes, and one scan reading reported no visible cerebral damage, although the patient was comatose. Scan information was unavailable for three participants. Premorbid IQ was estimated by administering the Wechsler Test of Adult Reading (WTAR), using norms for the UK (The Psychological Corporation, 2001).

Analyses were performed by the Statistical Package for the Social Sciences (SPSS/IBM, 2007). Patients' and controls' ages (patients: $M = 39.9$, $SD = 12.2$; controls: $M = 35.5$, $SD = 12.0$) and years of education (patients: $M = 12.2$, $SD = 1.67$; controls: $M = 12.9$, $SD = 2.67$) were normally distributed, and were not significantly different, as determined by two-tailed, independent-samples t-tests age: $t(30) = 1.00$, $p = .325$; education: $t(30) = -0.874$, $p = .390$.

Procedure

Participants were asked to have a friend or family member complete the GIST before the testing session. Participants then completed the ASC, Social Interpretations, and Awareness of Interoception tasks with one or two experimenters present in a quiet room. The GIST for one control participant was not returned, leaving 15 controls in the final analysis of the GIST. Hardware failure caused two patients' Awareness of Interoception data sets to be lost, leaving 14 patients and 16 controls in the final analysis for that task.

All participants completed brief psychometric testing, including Digit Span and Digit Symbol Coding from the Wechsler Adult Intelligence Scale (WAIS) (The Psychological Corporation, 1997); phonemic (FAS) and semantic fluency (Animals) from the Verbal Fluency Test (Benton & Hamsher, 1989); and the Trail Making Test, Parts A and B (Reitan, 1958). All tests were converted to standard scores for age, and education where appropriate, and percentile scores are reported. We used norms for the Australian adaptation of the WAIS (The Psychological Corporation, 1997). Norms for the Verbal Fluency and the Trail Making Test were taken from Spreen and Strauss (1998). For one of the patients, there was insufficient time to complete the psychometric battery due to slow performance. Trail B was discontinued for another patient, due to his inability to complete the task after 5 min. Animals for one patient, and both FAS and Animals for one control, were spoiled by noise or interruptions during administration. These cognitive test data are summarized in Table 2.

TABLE 1

Levels of substance use among patients with nonpenetrating traumatic brain injuries and controls. Binge drinking = >4 standard drinks in one sitting; heavy drinking = >4 standard drinks per day NHMRC, 2009, cannabis = >0.25 g per month; polydrug = consumption of two or more of the following substances more than three times during lifetime: cannabis, amphetamines, MDMA, LSD, cocaine, heroin

	Patients	*Controls*
Alcohol use	2 premorbid heavy drinking, 3 premorbid binge drinking, 1 current binge drinking	1 past binge drinking, 7 current binge drinking
Other substance use	1 premorbid cannabis, 1 post-injury cannabis, 6 premorbid polydrug	2 current cannabis, 2 past cannabis, 2 past polydrug

TABLE 2

Mean scores and results from MANOVA on neuropsychological test performance for patients with nonpenetrating TBI and controls. Test results are expressed in mean percentile scores, with *SD* in parentheses. We used the conservative Bonferroni corrected threshold of 0.05/11 = 0.0045 for all comparisons due to sample size limitations (11 = 7 cognitive measures + 4 social/emotional measures)

Neuropsychological test	Patients	Controls	MANOVA
Digit span	54.1 (32.7)	50.5 (24.4)	$F(1, 31) = 0.480, p = .494$
Digit symbol coding	33.2 (27.5)	49.1 (25.4)	$F(1, 30) = 1.87, p = .183$
Phonemic fluency	22.2 (27.6)	53.5 (25.3)	$F(1, 29) = 12.3, p < .0045**$
Semantic fluency	39.9 (31.0)	66.8 (27.5)	$F(1, 28) = 6.15, p < .05$
Trails A	46.6 (30.5)	63.3 (20.4)	$F(1, 30) = 2.74, p = .110$
Trails B	35.8 (39.3)	55.6 (24.4)	$F(1, 29) = 3.00, p = .094$
WTAR	48.0 (21.5)	64.4 (19.0)	$F(1, 30) = 4.75, p < .05$

Notes: **represents significance below .0045.

RESULTS

Data checking

Distributions were checked for normality. The GIST for controls was found to have high levels of kurtosis, indicating the presence of a participant who was an outlier, which was confirmed by inspecting the distribution. Removal of the outlier did not alter the interpretation of the data; thus, the outlier remained in the final analyses. All other variables were normally distributed.

We conducted two multivariate analyses of variance (MANOVA), one for all the standard neuropsychological measures, and one for all the novel social/emotional competence measures, with the presence of TBI (Patient vs. Control) as a grouping variable. Because of the small sample, we opted for two MANOVAs to avoid having almost as many variables as participants in each cell of the MANOVA (7 cognitive + 4 social/emotional = 11 variables vs. 13 participants per cell, due to missing data for some tasks), and used conservative Bonferroni adjustments for *p* levels in order to control for multiple comparisons.

In a MANOVA with GIST, ASC, Social Interpretations, and Awareness of Interoception scores entered as dependent variables and Patients versus Controls as fixed factors, people with TBI scored significantly lower than controls on ASC, Awareness of Interoception, and informant-rated social skill (GIST) (Table 3).

Within the ASC, a separate analysis showed that all three variables, Intention, Emotion, and Attitude, were significantly different between TBI and control groups, Intention: $F(3, 31) = 14.78, p < .005$; Emotion: $F(3, 31) = 18.86, p < .0001$; Attitude: $F(3, 31) = 13.20, p < .005$. Furthermore, briefer versions of the task are possible. A post hoc *t*-test showed that when only 5 of the 10 clip pairs were used, patients' and controls' total scores were still significantly different, $t(29) = -5.41, p < .0001$, and still correlated with the GIST, $\rho(31) = .940, p < .0001$.

The second MANOVA revealed that phonemic fluency (FAS) was significantly different between the patient and control groups (Table 2).

Pearson's correlations among each social competence measure revealed significant correlations as predicted between ASC and GIST scores, $r(31) = .655, p < .0001$, and between Social Interpretations and GIST scores, $r(31) = .460, p < .0001$. The GIST had high internal consistency ($\alpha = .97$), pooled across all 69 items. No other correlations were significant.

TABLE 3

Results from MANOVA of social/emotional competence measures for patients with nonpenetrating TBI and controls. Score ranges are −207 to +207 for the GIST, 0–3 for Assessment of Social Context, and 0–1 for Social Interpretations and Awareness of Interoception. *SD* in parentheses. We used the conservative Bonferroni corrected threshold of 0.05/11 = 0.0045 for all comparisons due to sample size limitations (11 = 7 cognitive measures + 4 social/emotional measures)

Social/emotional competence measures	Patients	Controls	MANOVA
GIST	67.0 (46.5)	126.4 (46.7)	$F(1, 29) = 19.6, p < .0001†$
Assessment of Social Context Task	1.76 (.404)	2.34 (.292)	$F(1, 30) = 19.5, p < .0001†$
Social Interpretations Task	0.862 (.112)	0.930 (.062)	$F(1, 30) = 3.48, p = .073$
Awareness of Interoception Task	0.461 (.153)	0.688 (.202)	$F(1, 28) = 9.69, p < .0045**$

Notes: **represents significance below .0045; †represents significance below .0001.

TABLE 4
Self-reported post-injury emotional changes for each participant with nonpenetrating traumatic brain injury, rank-ordered by percent accuracy score on the heartbeat-detection task

Patient	Awareness of Interoception Task (% Accuracy)	Self-reported emotional changes
01	9	More labile
02	29	More emotional blunting
03	36	More emotional blunting
04	36	More labile
05	43	More labile
06	43	More labile
07	50	More labile
08	50	More labile
09	50	More labile
10	57	More labile
11	57	More labile
12	57	More labile
13	57	No changes reported
14	71	Increased sadness*

Notes: *increased sadness reported in the context of marital breakup.

The emotional changes reported by patients and their performance on Awareness of Interoception are presented in Table 4. Only one patient reported no emotional changes post-injury and scored similarly on the task to others who reported changes (57% accuracy). One patient reported increased sadness, attributed to his marital breakup rather than the TBI.

DISCUSSION

We described four novel measures of social and emotional competence to supplement currently available tests for the assessment of social and emotional deficits after TBI. Overall, we found some evidence that these social/emotional measures may be sensitive to the presence of TBI: Three of our measures were significantly different between people with TBI and controls, even in this small sample.

The GIST fills a gap in the available neuropsychology toolkit by enabling patients' caregivers to identify specific social traits and behaviors that may be sources of difficulty. It was validated separately in a group of 38 neurologically healthy participants, and was sensitive to social skill variation in that group, with good convergent validity: Ratings on the GIST correlated highly with how well liked a person was (ratings of fondness), and multiple raters' ratings of an individual's social skill converged. In a separate sample, informant ratings of people with TBI were significantly lower than those of neurologically healthy controls, indicating that global social skills, as perceived by close others, are significantly impaired by TBI.

The ASC task examines patients' comprehension of social contextual information including, intentions, feelings, and attitudes during a videotaped dyadic interaction. The ASC is relatively brief to administer, and shorter versions of the task are possible (10–15 min). There were no apparent ceiling effects in this sample: Control participants' average performance was below 90% accuracy. Our sample with moderate to severe TBI scored significantly lower than controls. Furthermore, scores were significantly correlated with other-rated social skill (GIST).

Contrary to predictions, the Social Interpretations Task was not significantly different between TBI patients and controls, although we did find a correlation between Social Interpretations scores and other-rated social skill (GIST). The multiple-choice responses may have provided too much structure on this task, and future research with this task should investigate open-ended responses requiring participants to interpret the actions of each animation (cf. Scheibel et al., 2011 this issue, for similar results with a similar task).

The Awareness of Interoception Task measures patients' sensitivity to their own heartbeat, which may underlie awareness of one's emotions (Barrett et al., 2004; Wiens et al., 2000). This study represents the first demonstration that interoception of heartbeat is poorer among people with TBI than neurologically healthy controls. Qualitative information about emotional changes in our sample suggests that task performance may be relevant to post-injury emotional changes, especially emotional lability and blunting. These preliminary data indicate a potential use for this task as a brief, noninvasive, clinical screening measure for changes in emotional awareness following TBI. The data we collected on emotional changes were not quantifiable, as our goal in this study was first to investigate whether such a task would be different between people with TBI and controls. In future research, including a structured questionnaire investigating emotional experience, alexithymia, and emotion-regulation will help clarify the relationship between this task and subjective emotional changes in TBI.

The purpose of introducing these new tests was to expand the range of social/emotional competencies that can be measured, and to begin to evaluate these tests' utility as neuropsychological assessments by investigating psychometric properties, relationships among measures, and the sensitivity of these tasks to the presence of moderate to severe TBI. Future research with larger-sized and more representative samples is necessary to continue their validation and collect norms. This first step suggests that future research is warranted.

Limitations of current study

The sample size in this study was modest and heavily biased in terms of gender, with only 16 participants with TBI, and 16 controls, limiting generalizability of these results primarily to men with TBI. Future research should use samples that reflect the gender distribution of TBI more accurately (the males to females ratio is 2:1; Burns & Hauser, 2003). The TBI sample had relatively severe injuries, so generalization is also limited to those with moderate to severe TBI. Valuable comparison groups to add in future research would be a matched sample of people with mild TBIs, to investigate whether injury severity contributes significantly to deficits in emotional and social competence, and orthopedic injury controls, to control for the emotional effects of suffering a traumatic injury.

The TBI group was heterogeneous in time since injury, duration of loss of consciousness, depth of coma, and post-traumatic amnesia. There were also reports of premorbid anxiety, depression, and substance misuse among the patients. Although control participants with similar features were deliberately recruited, it would be ideal to collect a large enough

sample to conduct a covariance analysis of these non-TBI factors. Future research replicating these findings in a larger sample of people with TBI, with additional factors such as injury severity and education included as covariates, will make the current findings more robust.

The data available in patients' medical charts about lesion localization included only radiologists' readings of structural scans, usually CT scans, not the scans themselves. This precluded any detailed analysis linking results to lesion location. To examine brain areas necessary for these tasks, particularly orbitofrontal contributions, volumetric analysis of structural MRI scans would be necessary. Correlation with diffusion tensor imaging analyses of white matter damage, particularly connections to orbitofrontal regions, could be used to examine how axonal damage is related to these social/emotional competencies.

As in much psychology research, these measures were, in large part, pretested and validated in samples of undergraduate students, who are more likely to be young, highly educated, and used to regular assessments than the general population. Among people with TBI, the ratio of males to females is 2:1, young adults and the elderly are especially common, and lower socioeconomic status is more prevalent (Burns & Hauser, 2003). People with TBI may be unaccustomed to formal testing and may find neuropsychological testing sessions unusual, novel, and possibly intimidating and stressful. Socioeconomic status affects what is considered socially appropriate, and thus it is imperative that future validation research use populations with similar characteristics to people with TBI. This was our aim in recruiting the control group for the brain-injured participants, balancing the groups on age and education.

Finally, further validation of our social measures will need to examine relationships between these measures and specific, real-world outcomes. More research linking performance on each of our measures to measures of real-world social functioning in people with TBI would strengthen evidence for construct validity. Because our measures were designed to tap into areas of social/emotional competence not covered by previously developed measures, we would not necessarily expect our measures to correlate with all existing measures, such as TASIT. Just as not all cognitive variables are intercorrelated in TBI, not all social/emotional measures would be expected to intercorrelate. Nevertheless, to further investigate convergent validity of these tests, it would be useful to know their correlations with existing measures: Low or absent correlations would mean that different aspects of social cognition are being measured, whereas strong correlations with a measure would mean it is not worth administering both measures. In particular, correlations with real-world social variables, such as expert ratings of videotaped interactions or the size of people's social networks, would be valuable. Future research should also examine the relationship between each of our measures and real-world outcomes such as life satisfaction and employment.

Conclusions

This study represents a first step in expanding available measures of social/emotional competence, and more work lies ahead before these measures will be ready to use in the clinic. Sophisticated test interpretation requires normative data; the magnitude of healthy variation on these tests must be defined before an individual's scores are interpretable. Given the complexity of social competence, a large body of normative data will need to be collected in men and women of various ages, education levels, socioeconomic levels, and cultural backgrounds. Research suggesting gender differences in emotion perception (e.g., Lee, Liu, Chan, Fang, & Gao, 2005) makes it important to collect gender norms. Education is clearly a relevant variable for cognitive abilities; more research is required to determine whether education affects social skills.

Finally, socioeconomic and cultural differences play a key role in what is considered socially competent. Different cultures have varied interpretations of social roles and social status, and different emphases on harmony during interactions, which may result in different responses on social tests. For example, some low-income African-Americans found the Faux Pas Recognition task odd, as many of the faux pas did not seem inappropriate to them (E. Heerey, personal communication, October 2009). As with any neuropsychological test, normative data collection must include a consideration of how to address such cultural concerns.

Currently, the clinicians' toolkit for objective, quantitative assessments is much better stocked with cognitive measures than with social and emotional measures. Social neuroscience is in a position to help TBI patients by providing objective methods to define and measure social/emotional competence. Several questionnaires to assess personality change are available with norms, but do not focus on social skill in particular, capturing instead a broader range of psychiatric and behavioral disturbances. Our questionnaire, the GIST, focuses on social skill, had good convergent validity in 38 neurologically healthy participants, and

seems sensitive to moderate to severe TBI. Two of our performance-based measures, the ASC Task and the Awareness of Interoception Task are quantitative measures that seem to be sensitive to moderate to severe TBI. These three measures thus may warrant further development into clinically useful tools that measure aspects of social/emotional competence not captured by other measures, such as TASIT and other sarcasm-detection tests, empathy scales, clinician-rating scales for conversational pragmatics, and tests of responsivity to emotional pictures. In the meantime, we hope that neuropsychologists will use the measures that are already available. We encourage social neuroscientists and clinicians to work together to produce and use more objective performance measures that tap into a wider range of specific social/emotional competencies. People with TBI, their families, and rehabilitation specialists would benefit enormously from such progress in social neuroscience research.

REFERENCES

Arango-Lasprilla, J. C., Ketchum, J. M., Dezfulian, T., Kreutzer, J. S., O'Neil-Pirozzi, T. M., Hammond, F., et al. (2008). Predictors of marital stability 2 years following traumatic brain injury. *Brain Injury, 22*(7–8), 565–574.

Baldo, J. V., Schwartz, S., Wilkins, D., & Dronkers, N. F. (2006). Role of frontal versus temporal cortex in verbal fluency as revealed by voxel-based lesion symptom mapping. *Journal of the International Neuropsychological Society, 12*(6), 896–900.

Barrash, J., Tranel, D., & Anderson, S. W. (2000). Acquired personality disturbances associated with bilateral damage to the ventromedial prefrontal region. *Developmental Neuropsychology, 18*(3), 355–381.

Barrett, L. F., Quigley, K. S., Bliss-Moreau, E., & Aronson, K. R. (2004). Interoceptive sensitivity and self-reports of emotional experience. *Journal of Personality and Social Psychology, 87*(5), 684–697.

Bayly, P. V., Cohen, T. S., Leister, E. P., Aj, D., Leuthardt, E. C., & Genin, G. M. (2005). Deformation of the human brain induced by mild acceleration. *Journal of Neurotrauma, 22*(8), 845–856.

Benton, A., & Hamsher, K. (1989). *Multilingual Aphasia Examination*. Iowa City, IA: AJA Associates.

Berlin, H. A., Rolls, E. T., & Kischka, U. (2004). Impulsivity, time perception, emotion and reinforcement sensitivity in patients with orbitofrontal cortex lesions. *Brain, 127*(5), 1108–1126.

Bibby, H., & McDonald, S. (2005). Theory of mind after traumatic brain injury. *Neuropsychologia, 43*(1), 99–114.

Blais, M. C., & Boisvert, J. M. (2007). Psychological adjustment and marital satisfaction following head injury. Which critical personal characteristics should both partners develop? *Brain Injury, 21*(4), 357–372.

Bond, F., & Godfrey, H. P. D. (1997). Conversation with traumatically brain-injured individuals: A controlled study of behavioral changes and their impact. *Brain Injury, 11*(5), 319–329.

Boone, K. B. (2007). A reconsideration of the Slick et al. criteria for malingered neurocognitive dysfunction. In K. B. Boone (Ed.), *Assessment of feigned cognitive impairment* (pp. 29–49). New York, NY: Guilford Press.

Burns, J., Jr., & Hauser, W. A. (2003). The epidemiology of traumatic brain injury: A review. *Epilepsia, 44*(Suppl. 10), 2–10.

Castelli, F., Frith, C., Happé, F., & Frith, U. (2002). Autism, Asperger syndrome and brain mechanisms for the attribution of mental states to animated shapes. *Brain, 125*(8), 1839–1849.

Channon, S., Pellijeff, A., & Rule, A. (2005). Social cognition after head injury: Sarcasm and theory of mind. *Brain & Language, 93*(2), 123–134.

Critchley, H. D., Wiens, S., Rotshtein, P., Ohman, A., & Dolan, R. J. (2004). Neural systems supporting interoceptive awareness. *Nature Neuroscience, 7*(2), 189–195.

Cummings, J. L., Mega, M., Gray, K., & Rosenberg-Thompson, S. (1994). The Neuropsychiatric Inventory: Comprehensive assessment of psychopathology in dementia. *Neurology, 44*(12), 2308–2314.

Dahlberg, C. A., Cusick, C. P., Hawley, L. A., Newman, J. K., Morey, C. E., Harrison-Felix, C. L., et al. (2007). Treatment efficacy of social communication skills training after traumatic brain injury: A randomized treatment and deferred treatment controlled trial. *Archives of Physical Medicine & Rehabilitation, 88*(12), 1561–1573.

de Sousa, A., McDonald, S., Rushby, J., Li, S., Dimoska, A., & James, C. (2010). Why don't you feel how I feel? Insight into the absence of empathy after severe traumatic brain injury. *Neuropsychologia, 48*(12), 3585–3595.

de Sousa, A., McDonald, S., Rushby, J., Li, S., Dimoska, A., & James, C. (2011). Understanding deficits in empathy after traumatic brain injury: The role of affective responsivity. *Cortex, 47*(5), 526–535.

Devinsky, O., & D'Esposito, M. (2004). *Neurology of cognitive and behavioral disorders* (vol. 68). New York, NY: Oxford University Press.

Drummond, S. S., & Boss, M. R. (2004). Functional communication screening in individuals with traumatic brain injury. *Brain Injury, 18*(1), 41–56.

Engberg, A. W., & Teasdale, T. W. (2004). Psychosocial outcome following traumatic brain injury in adults: A long-term population-based follow-up. *Brain Injury, 18*(6), 533–545.

Gallagher, H. L., Happé, F., Brunswick, N., Fletcher, P. C., Frith, U., & Frith, C. (2000). Reading the mind in cartoons and stories: An fMRI study of 'theory of mind' in verbal and nonverbal tasks. *Neuropsychologia, 38*(1), 11–21.

Garmoe, W., Newman, A. C., & O'Connell, M. (2005). Early self-awareness following traumatic brain injury: Comparison of brain injury and orthopedic inpatients using the Functional Self-Assessment Scale (FSAS). *Journal of Head Trauma Rehabilitation, 20*(4), 348–358.

Grace, J., & Malloy, P. F. (2001). *Frontal Systems Behavior Scale (FrSBe): Professional Manual.* Lutz, FL: Psychological Assessment Resources.

Graham, D. P., & Cardon, A. L. (2008). An update on substance use and treatment following traumatic brain injury. In G. R. Uhl (Ed.), *Addiction reviews 2008. Annals of the New York Academy of Sciences* (pp. 148–162). Malden, MA: Blackwell Publishing.

Gregory, C., Lough, S., Stone, V., Erzinclioglu, S., Martin, L., Baron-Cohen, S., et al. (2002). Theory of mind in patients with frontal variant frontotemporal dementia and Alzheimer's disease: Theoretical and practical implications. *Brain, 125*(4), 752–764.

Gronwall, D., Wrightson, P., & Waddell, P. (1998). *Head injury: The facts* (2nd ed.). Oxford, UK: Oxford University Press.

Hanks, R. A., Temkin, N., Machamer, J., & Dikmen, S. S. (1999). Emotional and behavioral adjustment after traumatic brain injury. *Archives of Physical Medicine and Rehabilitation, 80*(9), 991–997.

Happé, F. (1994). An advanced test of theory of mind: Understanding of story characters' thoughts and feelings by able autistic, mentally handicapped and normal children and adults. *Journal of Autism and Developmental Disorders, 24*(2), 129–154.

Happé, F., Brownell, H., & Winner, E. (1999). Acquired "theory of mind" impairments following stroke. *Cognition, 70*(3), 211–240.

Hart, T., Sherer, M., Whyte, J., Polansky, M., & Novack, T. A. (2004). Awareness of behavioral, cognitive, and physical deficits in acute traumatic brain injury. *Archives of Physical Medicine & Rehabilitation, 85*(9), 1450–1456.

Hawley, L. A., & Newman, J. K. (2010). Group interactive structured treatment (GIST): A social competence intervention for individuals with brain injury. *Brain Injury, 24*(11), 1292–1297.

Hawthorne, G., Gruen, R. L., & Kaye A. H. (2009). Traumatic brain injury and long-term quality of life: Findings from an Australian study. *Journal of Neurotrauma, 26*(10), 1623–1633.

Heberlein, A. S., & Adolphs, R. (2004). Impaired spontaneous anthropomorphizing despite intact perception and social knowledge. *Proceedings of the National Academy of Sciences of the United States of America, 101*(19), 7487–7491.

Heider, F., & Simmel, M. (1944). An experimental study of apparent behavior. *American Journal of Psychology, 57*, 243–259.

Jorge, R. E. (2005). Neuropsychiatric consequences of traumatic brain injury: A review of recent findings. *Current Opinion in Psychiatry, 18*(3), 289–299.

Kertesz, A., Davidson, W., & Fox, H. (1997). Frontal behavioral inventory: Diagnostic criteria for frontal lobe dementia. *Canadian Journal of Neurological Sciences, 24*(1), 29–36.

Kim, E. (2002). Agitation, aggression, and disinhibition syndromes after traumatic brain injury. *NeuroRehabilitation, 17*(4), 297–310.

Klin, A. (2000). Attributing social meaning to ambiguous visual stimuli in higher-functioning autism and Asperger syndrome: The Social Attribution Task. *Journal of Child Psychology and Psychiatry, 41*(7), 831–846.

Knox, L., & Douglas, J. (2009). Long-term ability to interpret facial expression after traumatic brain injury and its relation to social integration. *Brain and Cognition, 69*(2), 442–449.

Kreutzer, J. S. (1996). Validation of a neurobehavioral functioning inventory for adults with traumatic brain injury. *Archives of Physical Medicine and Rehabilitation, 77*(2), 116–124.

Kreutzer, J. S., Marwitz, J. H., Seel, R. T., & Devany Serio, C. (1996). Validation of a neurobehavioural functioning inventory for adults with traumatic brain injury. *Archives of Physical Medicine and Rehabilitation, 77*(2), 116–124.

Kringelbach, M., & Rolls, E. T. (2003). Neural correlates of rapid reversal learning in a simple model of human social interaction. *NeuroImage, 20*(2), 1371–1383.

Lang, P. J., Bradley, M. M., & Cuthbert, B. N. (2005). *International Affective Picture System (IAPS), Technical report A-6.* Gainesville, FL: University of Florida, Center for Research in Psychophysiology.

Lee, T. M. C., Liu, H. L., Chan, C. C. H., Fang, S. Y., & Gao, J. H. (2005). Neural activities associated with emotion recognition observed in men and women. *Molecular Psychiatry, 10*(5), 450–455.

Levenson, R. W. (1992). Autonomic nervous system differences among emotions. *Psychological Science, 3*(1), 23–27.

Levine, B., Kovacevic, N., Nica, E. I., Cheung, G., Gao, F., Schwartz, M. L., et al. (2008). The Toronto Traumatic Brain Injury Study: Injury severity and quantified MRI. *Neurology, 70*(10), 771–778.

Linscott, R., Knight, R., & Godfrey, H. (1996). The Profile of Functional Impairment in Communication (PFIC). *Brain Injury, 10*(6), 397–412.

Lowe, M. R., & Cautela, J. R. (1978). A self-report measure of social skill. *Behavior Therapy, 9*(4), 535–544.

McDonald, S., Bornhofen, C., & Hunt, C. (2009). Addressing deficits in emotion recognition after severe traumatic brain injury: The role of focused attention and mimicry. *Neuropsychological Rehabilitation, 19*(3), 321–339.

McDonald, S., Flanagan, S., Martin, I., & Saunders, C. (2004). The ecological validity of TASIT: A test of social perception. *Neuropsychological Rehabilitation, 14*(3), 285–302.

McDonald, S., Flanagan, S., & Rollins, J. (2002). *The Awareness of Social Inference Test.* San Antonio, TX: Pearson Assessments.

McDonald, S., Flanagan, S., Rollins, J., & Kinch, J. (2003). TASIT: A new clinical tool for assessing social perception after traumatic brain injury. *Journal of Head Trauma Rehabilitation, 18*(3), 219–238.

McDonald, S., & Pearce, S. (1996). Clinical insights into pragmatic theory: Frontal lobe deficits and sarcasm. *Brain & Language, 53*(1), 81–104.

McGann, W., Werven, G., & Douglas, M. (1997). Social competence and head injury: A practical approach. *Brain Injury, 11*(9), 621–628.

Milders, M., Fuchs, S., & Crawford, J. R. (2003). Neuropsychological impairments and changes in emotional and social behavior following severe traumatic brain injury. *Journal of Clinical and Experimental Neuropsychology, 25*(2), 157–172.

Milders, M., Ietswaart, M., Crawford, J. R., & Currie, D. (2006). Impairments in theory of mind shortly after traumatic brain injury and at 1-year follow-up. *Neuropsychology, 20*(4), 400–408.

Milders, M., Ietswaart, M., Crawford, J. R., & Currie, D. (2008). Social behavior following traumatic brain injury and its association with emotion recognition, understanding of intentions, and cognitive flexibility. *Journal of the International Neuropsychological Society, 14*(2), 318–326.

Muller, F., Simion, A., Reviriego, E., Galera, C., Mazaux, J. M., Barat, M., et al. (2010). Exploring theory of mind after severe traumatic brain injury. *Cortex, 46*(9), 1088–1099.

Nelson, L. D., Drebing C., Satz P., & Uchiyama C. (1998). Personality change in head trauma: A validity study of the Neuropsychology Behavior and Affect Profile. *Archives of Clinical Neuropsychology, 13*(6), 549–560.

NHMRC. (2009). *Australian guidelines to reduce health risks from drinking.* Canberra, Australia: National Health and Medical Research Council of Australia.

Nyhus, E., & Barceló, F. (2009). The Wisconsin Card Sorting Test and the cognitive assessment of prefrontal executive functions: A critical update. *Brain and Cognition, 71*(3), 437–51.

Ownsworth, T. L., & McKenna, K. (2004). Investigation of factors related to employment outcome following traumatic brain injury: A critical review and conceptual model. *Disability and Rehabilitation, 26*(13), 765–784.

Pagulayan, K. F., Hoffman, J. M., Temkin, N. R., Machamer, J. E., & Dikmen, S. S. (2008). Functional limitations and depression after traumatic brain injury: Examination of the temporal relationship. *Archives of Physical Medicine and Rehabilitation, 89*(10), 1887–1892.

Parry-Jones, B. L., Vaughan, F. L., & Cox, W. M. (2006). Traumatic brain injury and substance misuse: A systematic review of prevalence and outcomes research (1994–2004). *Neuropsychological Rehabilitation, 16*(5), 537–560.

Ponsford, J., Draper, K., & Schonberger, M. (2008). Functional outcome 10 years after traumatic brain injury: Its relationship with demographic, injury severity, and cognitive and emotional status. *Journal of the International Neuropsychological Society, 14*(2), 233–242.

Ponsford, J., Whelan-Goodinson, R., & Bahar-Fuchs, A. (2007). Alcohol and drug use following traumatic brain injury: A prospective study. *Brain Injury, 21*(13–14), 1385–1392.

Prigatano, G. P. (2005). Disturbances of self-awareness and rehabilitation of patients with traumatic brain injury: A 20-year perspective. *Journal of Head Trauma Rehabilitation, 20*(1), 19–29.

Reitan, R. (1958). Validity of the Trail Making Test as an indicator of organic brain damage. *Perceptual and Motor Skills, 8,* 271–276.

Rochat, L., Ammann, J., Mayer, E., Annoni, J. M., & Van der Linden, M. (2009). Executive disorders and perceived socio-emotional changes after traumatic brain injury. *Journal of Neuropsychology, 3*(2), 213–227.

Saunders, J. C., McDonald, S., & Richardson, R. (2006). Loss of emotional experience after traumatic brain injury: Findings with the startle probe procedure. *Neuropsychology, 20*(2), 224–231.

Shamay-Tsoory, S. G., & Aharon-Peretz, J. (2007). Dissociable prefrontal networks for cognitive and affective theory of mind: A lesion study. *Neuropsychologia, 45*(13), 3054–3067.

Shamay-Tsoory, S. G., Tomer, R., Berger, B. G., & Aharon-Peretz, J. (2003). Characterization of empathy deficits following prefrontal brain damage: The role of the right ventromedial prefrontal cortex. *Journal of Cognitive Neuroscience, 15*(3), 324–337.

Shamay-Tsoory, S. G., Tomer, R., Berger, B. D., Goldsher, D., & Aharon-Peretz, J. (2005). Impaired "affective theory of mind" is associated with right ventromedial prefrontal damage. *Cognitive and Behavioral Neurology*, *18*(1), 55–67.

Spreen, O., & Strauss, E. (1998). *A compendium of neuropsychological tests* (3rd. ed.). New York, NY: Oxford University Press.

SPSS/IBM. (2007). Statistical Package for the Social Sciences [computer software]. Armonk, NY: SPSS/IBM.

Stone, V. E., Baron-Cohen, S., & Knight, R. T. (1998). Frontal lobe contributions to theory of mind. *Journal of Cognitive Neuroscience*, *10*(5), 640–656.

Stone, V. E., & Hynes, C. A. (2011). Real-world consequences of social deficits: Executive functions, social competencies, and theory of mind in patients with ventral frontal damage and traumatic brain injury. In J. Decety & J. Cacciopo (Eds.), *The Oxford handbook of social neuroscience* (pp. 455–476). New York, NY: Oxford University Press. Retrieved February 28, 2011, from http://www.assesscompetency.com/articles.html

Struchen, M. A., Clark, A. N., Sander, A. M., Mills, M. R., Evans, G., & Kurtz, D. (2008). Relation of executive functioning and social communication measures to functional outcomes following TBI. *NeuroRehabilitation*, *23*(2), 185–198.

Temkin, N. R., Corrigan, J. D., Dikmen, S. S., & Machamer, J. (2009). Social functioning after traumatic brain injury. *Journal of Head Trauma Rehabilitation*, *24*(6), 460–467.

The Psychological Corporation. (1997). *Wechsler Adult Intelligence Scale 3rd Edition: Administration and Scoring Manual. (Autralian Adaptation)*. San Antonio, TX: The Psychological Corporation.

The Psychological Corporation. (2001). *Wechsler Test of Adult Reading*. San Antonio, TX: The Psychological Corporation.

Wang, J. Y., Bakhadirov, K., Devous, M. D., Sr., Abdi, H., McColl, R., et al. (2008). Diffusion tensor tractography of traumatic diffuse axonal injury. *Archives of Neurology*, *65*(5), 619–626.

Wiens, S., Mezzacappa, E. S., & Katkin, E. (2000). Heartbeat detection and the experience of emotions. *Cognition and Emotion*, *14*(3), 417–427.

Williams, C., & Wood, R. L. (2010). Alexithymia and emotional empathy following traumatic brain injury. *Journal of Clinical and Experimental Neuropsychology*, *32*(3), 259–267.

Yamasato, M., Satoh, S., Ikejima, C., Kotani, I., Senzaki, A., & Asada, T. (2007). Reliability and validity of questionnaire for neurobehavioral disability following traumatic brain injury. *Psychiatry and Clinical Neurosciences*, *61*(6), 658–664.

Transcultural differences in brain activation patterns during theory of mind (ToM) task performance in Japanese and Caucasian participants

Katja Koelkebeck[1], Kazuyuki Hirao[2], Ryousaku Kawada[2], Jun Miyata[2], Teruyasu Saze[2], Shiho Ubukata[2], Shoji Itakura[3], Yasuhiro Kanakogi[3], Patricia Ohrmann[1], Jochen Bauer[1], Anya Pedersen[1], Nobukatsu Sawamoto[4], Hidenao Fukuyama[4], Hidehiko Takahashi[2], and Toshiya Murai[2]

[1]Department of Psychiatry, School of Medicine, University of Münster, Münster, Germany
[2]Department of Neuropsychiatry, Graduate School of Medicine, Kyoto University, Kyoto, Japan
[3]Department of Psychology, Graduate School of Letters, Kyoto University, Kyoto, Japan
[4]Human Brain Research Center, Graduate School of Medicine, Kyoto University, Kyoto, Japan

Background: Theory of mind (ToM) functioning develops during certain phases of childhood. Factors such as language development and educational style seem to influence its development. Some studies that have focused on transcultural aspects of ToM development have found differences between Asian and Western cultures. To date, however, little is known about transcultural differences in neural activation patterns as they relate to ToM functioning.

Experimental methods: The aim of our study was to observe ToM functioning and differences in brain activation patterns, as assessed by functional magnetic resonance imaging (fMRI). This study included a sample of 18 healthy Japanese and 15 healthy Caucasian subjects living in Japan. We presented a ToM task depicting geometrical shapes moving in social patterns. We also administered questionnaires to examine empathy abilities and cultural background factors.

Results: Behavioral data showed no significant group differences in the subjects' post-scan descriptions of the movies. The imaging results displayed stronger activation in the medial prefrontal cortex (MPFC) in the Caucasian sample during the presentation of ToM videos. Furthermore, the task-associated activation of the MPFC was positively correlated with autistic and alexithymic features in the Japanese sample.

Discussion: In summary, our results showed evidence of culturally dependent sociobehavioral trait patterns, which suggests that they have an impact on brain activation patterns during information processing involving ToM.

Keywords: Transcultural; fMRI; Theory of mind; MPFC; Sociocultural background.

Correspondence should be addressed to: Katja Koelkebeck, Department of Psychiatry, School of Medicine, University of Münster, Albert-Schweitzer-Strasse 11, 48149 Münster, Germany. E-mail: katja.koelkebeck@ukmuenster.de

The study was supported by a research fellowship from the Japan Society for the Promotion of Science (JSPS) (PE 07550 to K.K.); a grant for the initiation of bilateral cooperation from the DFG (German Research Foundation) (KO 4038/1-1 to K.K.); grants-in-aid for scientific research from the Japan Society for the Promotion of Science (21890119 to J.M.) and the Ministry of Education, Culture, Sports, Science and Technology, Japan (20691401 to T.M.); a grant from the Ministry of Health, Labor and Welfare, Japan (20E-3 to T.M.); a research grant from the Research Group for Schizophrenia sponsored by Astellas Pharma, Inc.; and a research grant from the Mitsubishi Pharma Research Foundation. We are grateful to all of the subjects for their participation in this study and to the staff involved in data acquisition. We would also like to thank Christoph Paulus of the University of Saarbrücken, Germany, for his helpful suggestions on the German version of the IRI.

www.psypress.com/socialneuroscience   http://dx.doi.org/10.1080/17470919.2011.620763

Theory of mind (ToM) is the tendency to explain one's own and others' actions in terms of beliefs, desires, and goals (Castelli, Frith, Happé, & Frith, 2002). This aspect of social cognition is especially crucial for active engagement in a culture (Duffy, Toriyama, Itakura, & Kitayama, 2009, p. 358), because mentalizing abilities are necessary to react adequately to the environment in a socially accepted way (Bruene & Bruene-Cohrs, 2006). ToM abilities have been found to rely on individuals' cultural backgrounds. A cross-cultural study examining children and adults from North America and China showed that educational styles have an effect on the time course of ToM development (Wellman, Fang, Liu, Zhu, & Liu, 2006). Naito and Koyama (2006) found that Japanese children developed false-belief task skills later than Western culture children but had a tendency to interpret situations by using implicit social information. The authors hypothesized that Japanese children's tendency to infer social meanings might be due to early social behavior training, including expressing and understanding emotions and thoughts implicitly (see also Hendry, 1986). Furthermore, Tardif and Wellman (2000) observed different uses of ToM-related language in Chinese- and English-speaking children (Li & Rao, 2000). According to Naito and Koyama (2006), who investigated Japanese and Caucasian children, as well as Lee and colleagues (Lee, Olson, & Torrance, 1999), who observed Chinese children, this might be due to cultural-specific differences in language use, such as the use of the verb "to think (falsely)" in Asian cultures. In their study of adults, Matsuda and Nisbett (2001) found that the Japanese made more statements about contextual information and relationships while performing a picture-viewing task than Americans did. Thus, it seems that Asians pay more attention to the general emotional implication of a situation than to the details (Shweder et al., 1998). Taken together, these results indicate that the development and application of ToM abilities (i.e., the use of ToM-related language and application strategies) may be culturally dependent.

In imaging studies, the following specific brain networks have been suggested to be important for ToM abilities: the temporal and prefrontal cortical areas—(medial (MPFC) and dorsolateral prefrontal cortex (DLPFC)—and the amygdala (see Castelli, Happé, Frith, & Frith, 2000; Voellm et al., 2006). Because there are cultural differences in the mode and development of ToM, one might expect different ethnocultural groups to exhibit different brain activation patterns when performing ToM tasks. To date, however, only two studies have explicitly addressed cultural differences in ToM-related brain activation

in adults. A functional magnetic resonance imaging (fMRI) study that compared American monolinguals and Japanese bilinguals proposed a differential effect of language education on ToM-related brain activation (Kobayashi, Glover, & Temple, 2006). While the ventromedial prefrontal cortex and precuneus were activated in both groups, the inferior frontal gyrus and the temporoparietal junction (TPJ) were activated in a culture-dependent manner during ToM task-performance. Additionally, a study comparing Japanese and American subjects using an Asian and an American version of the "eyes task" (Baron-Cohen, Jolliffe, Mortimore, & Robertson, 1997) found activation in the superior temporal sulcus during same-versus other-culture mental state decoding in both groups but did not hint at a differential recruitment of brain areas (Adams et al., 2010).

In the present study, we investigated ToM abilities in a sample of native Japanese and a group of Caucasian individuals living in Japan. Activation patterns were examined by fMRI in implementing a sophisticated ToM task depicting moving geometrical shapes acting in social patterns (Abell, Happé, & Frith, 2000). This paradigm has been validated for the assessment of ToM abilities in autism spectrum disorders (Castelli et al., 2002), schizophrenia (Russell, Reynaud, Herba, Morris, & Corcoran, 2006), brain damage (Weed, McGregor, Feldbaek Nielsen, Roepstorff, & Frith, 2010), and alexithymia (Moriguchi et al., 2006). Using this paradigm, this study aimed to examine culturally dependent brain activation patterns associated with ToM abilities.

Our first aim was to assess behavioral performance on the task. We expected that the appropriateness of oral descriptions of the videos and the use of ToM-related language by Japanese participants would differ from that used by Caucasian participants. Because Japanese individuals reportedly evaluate information more contextually and less emotionally (Matsuda & Nisbett, 2001; Moriguchi et al., 2005), we expected the Western participants to express more ToM-related vocabulary and the Japanese participants to describe the videos more accurately in a social context. The differences between Caucasians and Japanese were expected to be more prominent for descriptions of movements that include social patterns than for goal-directed or random movements. Moreover, language differences concerning mentalizing terms might occur due to differences in styles of describing social interactions.

Our second aim was to analyze brain activation patterns by fMRI. Looking at differential activation patterns in two cultural groups would likely increase our knowledge about biological and neuropsychological factors related to mentalizing abilities, because cultural similarities and differences in basic brain functions or brain activations could exist. Consistent with previous findings related to general social cognition tasks (Chiao et al., 2009, 2010; Moriguchi et al., 2005; Zhu, Zhang, Fan, & Han, 2007), we expected the activation of the amygdala and the MPFC to be more distinct in Caucasian participants presented with a ToM task.

If brain activation differences between the study groups were found, our third aim was to analyze the correlations between the activation patterns and the participants' transcultural backgrounds to target reasons for ToM-performance differences and differential activation patterns. Thus, we administered questionnaires assessing the participants' level of acculturation. The questionnaires also assessed alexithymic traits, empathy abilities, and autistic traits, as these have been shown to correlate with reduced ToM abilities (Castelli et al., 2002; Moriguchi at al., 2006; Rogers, Dziobek, Hassenstab, Wolf, & Convit, 2007).

This is one of the first studies to observe transcultural differences in ToM performance by a sophisticated paradigm and a functional imaging approach. This study is the first to apply wide-ranging questionnaires to examine the possible impact of sociobehavioral traits on ToM functioning.

EXPERIMENTAL METHODS

Participants

The sample consisted of 18 healthy Japanese (mean age = 34.8, SD = 9.8; 9 women; 9 men) and 15 healthy Caucasian (mean age = 30.9, SD = 8.1; 8 women; 7 men) participants. Movement-provoked mental state attribution was examined by implementing a ToM paradigm using moving geometrical shapes. All Japanese participants were native speakers and were recruited personally. The Caucasian participants were recruited via advertisements at the international centers of Kyoto language institutes and Kyoto University. All of the Caucasian participants were native speakers of German and English or had excellent English abilities (nationalities: 5 German, 4 American/Canadian, 1 French, 2 Dutch, 1 Russian, 1 Danish, 1 Polish), and their mean duration of stay in Japan was 28.1 months (SD = 20.6). None of the participants had a history of psychiatric disorders, according to the SCID I, and none were receiving psychotropic medication or had neurological disorder. All had normal vision and were right-handed. The study protocol was approved by the University of Kyoto ethics committee according to the Declaration of Helsinki (1975, revised 1984). Written, informed consent was obtained from all participants before they were enrolled in the study.

Stimuli and procedure

The "moving shapes" paradigm, first used by Abell and colleagues (2000), was presented to the participants. We presented nine of the 12 original silent, animated videos (see http://sites.google.com/site/utafrith/research for examples); each had a big, red triangle and a small, blue triangle moving in a framed, white screen. Because a behavioral study with a clinical study group (Koelkebeck et al., 2010) had found that three videos were less reliable for discriminating between patients and control subjects than the other videos, these videos were excluded. The remaining nine videos were provided in a shorter version (24 s each) without diminishing the original meaning. Three types of animations were displayed: (1) random movement sequences (RM), in which the triangles purposelessly moved around (e.g., bouncing off the walls); (2) goal-directed movement sequences (GD), in which one triangle acted and the other one reacted (e.g., fighting) with no indication of one reading the other's mind; and (3) ToM sequences, in which the triangles interacted as if they read each other's mind (e.g., seducing). Before the experiment, the participants were informed about the session procedure and the different conditions without explicitly being told the aims of the study.

The nine videos were presented in a blocked design. The participants saw the videos in 30-s blocks, consisting of 24 s of video presented in a pseudorandomized sequence, followed by a 6-s question period. The order of blocks was counterbalanced across participants, and there were three counterbalanced orders of presentation. The overall presentation time was 4.5 min. The participants were told that they would see videos and should pay attention to them. The experiment used the Presentation software package (by Neurobehavioral Systems Inc., Albany, CA, USA). Each participant's head position was stabilized with a vacuum head cushion. To ensure vigilance, the participants were asked whether they thought the triangles' movements were (1) related to each other, (2) random, or (3) expressing feelings and emotions; they responded by pressing a button after each video was presented. A forced-choice paradigm with three response categories (corresponding to the video types) was used.

The "moving shapes" videos were presented a second time, after the fMRI experiment, in a quiet room free of auditory and visual distractions. The computer monitor was placed directly in front of the participants. The participants were asked what they thought the triangles were doing. Their answers were recorded and later evaluated by the scoring criteria provided by Abell and colleagues (2000). The answers were evaluated by experienced raters according to three dimensions: intentionality (degree of mental state attribution, amount of ToM-related language use: 0–5 points), appropriateness of the answers (degree of correctness: 0–2 points), and length of answers (0–4 points).

Questionnaires and statistical analysis

We administered several questionnaires in a Japanese, German, or English version. An adapted version of the Suinn-Lew Asian Self-Identity Acculturation Scale (SL-Asia) (Suinn, Ahuna, & Khoo, 1992) was given to the participants. It measures the level of identification with one's own culture and a foreign culture. In this study, a score of 1 indicated high acculturation with Japanese culture, whereas a score of 5 indicated high acculturation to Western culture. We also administered the American Asian Multidimensional Acculturation Scale (AAMAS) (Chung, Kim, & Abreu, 2004) in English and in translated German and Japanese versions. The AAMAS measures acculturation level and consists of two scores: acculturation to one's own culture and acculturation to the foreign culture. The scale can be subdivided into several subscales, including food consumption habits, cultural knowledge, language knowledge, and identification with their own and the foreign culture. We also administered the Individualism/Collectivism Scale (IND/COL) by Triandis (1994). The foreign participants also completed a questionnaire about living circumstances, adapted from Tanaka (2000).

Participants were assessed with a validated English, German, or Japanese version of the Autism Spectrum Questionnaire (AQ) by Baron-Cohen (Baron-Cohen, Wheelwright, Skinner, Martin, & Clubley, 2001; Wakabayashi, Tojo, Baron-Cohen, & Wheelwright, 2004). The AQ assesses autistic traits in a normal population. Additionally, the 20-item version of the Toronto Alexithymia Scale (TAS) (Komaki et al., 2003; Taylor, Ryan, & Bagby, 1985) was administered to measure alexithymic traits, such as the ability to judge and describe one's own feelings. The Japanese version was used with the permission of Gen Komaki. We also administered the Interpersonal Reactivity Index (IRI) by Davis (Aketa, 1999; Davis, 1983; Paulus, 2006), which measures empathy abilities. Handedness was assessed with the Edinburgh Handedness Questionnaire (Oldfield, 1971). Socioeconomic status (SES) was assessed with the Hollingshead scale (Hollingshead, 1975).

Statistical analyses were performed with SPSS (SPSS 18.0 for Windows, SPSS, Inc., Chicago, IL, USA). Normal distribution was confirmed with the Kolmogorov-Smirnov test. For between-group comparisons, Student's t-test and chi-squared tests were applied as appropriate. Correlations were determined with the Pearson correlation coefficient. The significance level for all analyses was set at $p < .05$ (two-tailed).

Behavioral data were first analyzed via repeated-measures analyses of variance (ANOVA) for each rating type (intentionality, appropriateness, and length), with one between-subjects factor (group: Caucasian or Japanese participants) and one within-subjects factor (video type: ToM, GD, and RM sequences). Correlational analyses were also conducted. Furthermore, we used analyses of covariance (ANCOVA) to control for possible effects of neuropsychological functioning on ToM performance.

fMRI procedure and analysis

All participants received MRI scans with a 3-T scanner equipped with an 8-channel, phased-array head coil (Trio, Siemens, Erlangen, Germany). Functional images were obtained in a T2*-weighted gradient echo-planar imaging sequence, with the following sequence: TE = 30 ms; TR = 2500 ms; flip angle = 90°; FOV = 192 × 192 mm; matrix = 64 × 64; 40 interleaved axial slices of 3-mm thickness without gaps; resolution = 3-mm cubic voxels. The first two volumes were discarded for signal stabilization; the total number of volumes was 114. Functional imaging data were motion-corrected by a set of six rigid body transformations determined for each image. The images were spatially normalized to standard Montreal Neurological Institute (MNI) space and smoothed (Gaussian kernel, 6-mm, full-width at half maximum), using statistical parametric mapping (SPM8, Wellcome Department of Imaging Neuroscience, London, UK). A statistical analysis was performed by modeling the different types of video presentation conditions as variables within the context of the general linear model (convolved with a standard hemodynamic response function). The analyses in the present study focused on the contrast between ToM–RM video conditions, as suggested by Moriguchi and colleagues (2006). On the single-subject level, we extracted the contrast values for ToM-RM conditions. First, one-sample t-tests were performed on activation data for both groups separately. Random-effects analyses (t-tests for independent samples) were then performed to examine brain activation differences between the two groups. Age and gender were included as nuisance covariates.

Anatomical labels of reported coordinates (transformed from MNI to Talairach space, Mathworks, Natick, MA, USA; "mni2tal" MATLAB script available from http://eeg.sourceforge.net/) for peak voxel clusters were retrieved from the Talairach Daemon database (Lancaster et al., 2000) within a 5-mm cubical search range or from the SPM anatomy toolbox. The statistical significance level was set at $p < .001$, uncorrected.

RESULTS

Behavioral results

Demographic data, neuropsychology, and questionnaire results

There were no significant differences between the age, gender, or SES of the participants. None of the participants met pathology criteria for autism or alexithymia. The Japanese participants displayed higher TAS scores than the Caucasian participants; differences between groups were significant for the TAS total score and the second and third TAS subscales, indicating difficulties in describing emotions and a tendency of individuals to focus their attention externally. However, there was no significant difference between groups on the first subscale, which describes difficulties in identifying emotions. Furthermore, although the Japanese participants displayed higher scores on the AQ, this difference was not significant. There were significant differences on all AAMAS subscale scores except for "food consumption." According to the AAMAS, the Caucasian participants showed a lower identification with their own culture compared with the Japanese participants but were more knowledgeable when asked about a foreign culture. On the IND/COL, the Caucasian participants displayed higher levels of individualism, and the Japanese participants showed higher levels of collectivism, but these differences were not significant. No significant differences on any of the subscales of the IRI were observed. Acculturation levels for the two groups, as measured by the SL Asia, were comparable: Japanese: 1.9 (0.1), Caucasians: 3.9 (0.3) (for results of questionnaires, see Table 1).

Rating of videos

To analyze the rating results, we conducted three (video type) × 2 (group) ANOVAs. In the first ANOVA, rating type intentionality was the dependent variable. In this analysis, the main effect of video type, $F(1, 30) = 291.52, p < .001$, was significant. The main effect of group, $F(1, 30) = .64, p = .43$, and the interaction between group and video type, $F(1, 30) = 1.16$, $p = .29$, were not significant, indicating no response differences regarding the three conditions between the two groups. Planned post-hoc t-tests showed significantly reduced use of ToM vocabulary (intentionality) when describing the GD sequences, $t(30) = -2.028$, $p < .05$, in Caucasian participants compared with the Japanese participants. For the ToM and RM sequences, the differences failed to reach significance.

In the second ANOVA, the appropriateness of the answers was the dependent variable. In this analysis, neither the main effect of video type, $F(1, 30) = .21, p = .65$, or group, $F(1, 30) = .21$, p = .89; nor the interaction between group and video type, $F(1, 30) = 1.96, p = .17$, was significant.

The third ANOVA used rating type length as the dependent variable. In this analysis, the main effects for video type, $F(1, 30) = 63.19, p < .001$, and group, $F(1, 30) = 9.80, p = .004$, were significant, but the interaction between group and video type, $F(1, 30) = 3.66, p = .07$, was not. Planned post-hoc t-tests showed significantly reduced length when describing the ToM, $t(30) = -4.261, p < .001$, and GD sequences, $t(30) = -3.151, p < .005$, in Japanese participants compared with Caucasians. For the RM sequences, no difference was found.

fMRI

Within-group activations in response to ToM videos

For the ToM-RM contrast for Caucasian participants, differences were significant in the MPFC bilaterally—Brodmann's area (BA) 9, the left middle temporal gyrus (BA 21), the right temporal lobe (BA 40), the left parahippocampal gyrus (BA 36), and the right thalamus (see Table 1 and Figure 1)—at a statistical threshold of $p < .001$, uncorrected. No significant differences were found for Caucasian participants on the RM-ToM contrast. For the Japanese participants, no significant differences were observed for either the ToM-RM contrast or the RM-ToM contrast (see Table 2).

Between-group differences in response to ToM videos

A group comparison at a statistical threshold of $p < .001$, uncorrected, revealed significantly greater activation in the Caucasian participants compared with the Japanese participants in the MPFC (BA 8). At a threshold of $p < .005$, uncorrected, the Caucasians

TABLE 1
Characteristics and results of the study groups' questionnaires

	Japanese participants (n = 18)	Caucasian participants (n = 15)	Statistics
Age	34.8 (9.8)	30.9 (8.1)	$t(df = 31) = 1.21, p = 0.24$
Gender	♀9, ♂9	♀8, ♂7	$\chi^2(1) = 0.030, p = .86$
SES	49.3 (13.3)	45.7 (6.5)	$t(df = 30) = 0.96, p = 0.34$
TAS total subscale	45.4 (8.8)	38.3 (9.6)	$t(df = 31) = 2.58, p = .015$
(1) Difficulties in identifying emotions	12.1 (4.3)	10.8 (4.1)	$t(df = 31) = 0.95, p = .35$
(2) Difficulties in describing emotions	12.8 (4.3)	8.6 (2.7)	$t(df = 31) = 3.57, p = .001$
(3) Externally oriented thinking	20.5 (2.9)	18.9 (4.4)	$t(df = 31) = 2.11, p = .047$
AQ	15.3 (6.4)	12.8 (5.8)	$t(df = 30) = 1.24, p = .23$
AAMAS	73.4 (7.0)	58.4 (9.9)	$t(df = 31) = 4.50, p < .001$
Identification of own culture ~ foreign culture	41.2 (8.2)	73.5 (7.2)	$t(df = 31) = -12.04, p < .001$
Language	10.56 (2.71)	22.87 (1.25)	$t(df = 31) = -16.22, p < .001$
Cultural identity	16.0 (3.45)	27.89 (4.45)	$t(df = 31) = -8.63, p < .001$
Cultural knowledge	6.72 (2.6)	12.93 (1.8)	$t(df = 31) = -7.81, p < .001$
Food consumption	8.11 (2.11)	9.80 (2.65)	$t(df = 31) = -2.04, p < .050$
IND/COL			
Collectivism	101.6 (29.0)	99.7 (11.5)	$t(df = 30) = -0.05, p = .96$
Individualism	83.3 (25.1)	92.1 (13.2)	$t(df = 31) = -1.46, p = .16$
IRI subscale			
(1) Fantasy	11.7 (3.8)	17.3 (16.5)	$t(df = 29) = -1.33, p = .22$
(2) Perspective taking	12.5 (2.1)	12.9 (7.2)	$t(df = 29) = -0.23, p = .83$
(3) Empathic concern	9.1 (2.3)	12.5 (9.3)	$t(df = 29) = -1.36, p = .19$
(4) Personal distress	12.7 (4.3)	21.7 (24.9)	$t(df = 29) = -1.38, p = .19$

Notes: AAMAS: Asian-American Multidimensional Acculturation Scale; AQ: Autism Spectrum Quotient; IND/COL.; Individualism/Collectivism Scale; IRI: Interpersonal Reactivity Index; SES: Socioeconomic Status; Suinn-Lew Asian Self-Identity Acculturation Scale; TAS: Toronto Alexithymia Scale.

displayed areas of greater activation on the ToM-RM contrast in the right superior frontal gyrus (BA 7), the right claustrum and the right parahippocampal gyrus (BA 35). The Japanese participants did not show any greater brain activation for this contrast at a threshold of p < .001 or at a threshold of $p < .005$, uncorrected (see Table 3). In Figure 2, activation differences at a threshold of $p < .005$ are shown, so that the distribution of group differences is more evident.

Correlational analyses

For the correlational analyses, we extracted the averaged eigenvariates to obtain the fMRI signal intensity time series from the largest cluster of significant ToM-RM contrast group differences at a statistical threshold of $p < .005$, uncorrected, for both groups (MPFC;

coordinates: 12; 48; 34). We then performed a partial correlational analysis in SPSS between the MPFC signal intensity time series and the questionnaire results, using gender and age as covariates. MPFC activation was found to be correlated with the first subscale (difficulties in identifying emotions) of the TAS ($r = .57$, $p = .027$) (see Figure 3A) and with scores from the AQ ($r = .52$, $p = .048$) (see Figure 3B) in the Japanese participants. For the Caucasian participants, no significant correlations were demonstrated between MPFC activation and the questionnaire scores.

DISCUSSION

The aim of our study was to observe intercultural differences in ToM in two groups of participants from different cultural/ethnic backgrounds. Therefore, we assessed behavioral and functional imaging data in a

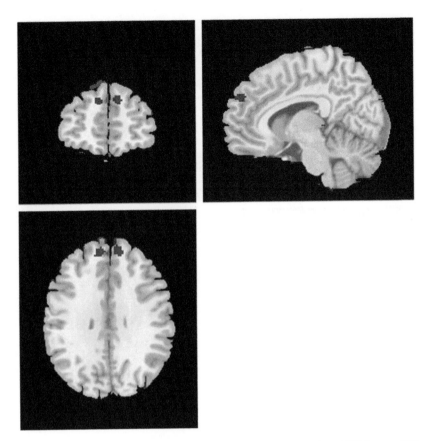

Figure 1. fMRI images showing activation for Caucasian participants during the presentation of ToM versus RM sequences (*p* < .001, uncorrected).

sample of Caucasian and Japanese participants, using a sophisticated ToM paradigm, and correlated these findings with data from transcultural and social-behavioral questionnaires.

We did not find any significant behavioral differences in the use of ToM-related vocabulary or the appropriateness of descriptions in the groups studied. A lack of significant behavioral ToM findings is consistent with previous data that show differences in ToM abilities at a younger age but similar levels across cultures in adulthood (for an account of ToM development, see Wellman, Cross, & Watson, 2001). These results indicate that, on the behavioral level, the paradigm we used might be relatively independent of transcultural performance differences. As the movies used as stimuli were silent, confounding variables such as language comprehension were reduced during task presentation.

On the other hand, fMRI results indicated a cultural background effect on brain activation. We found a higher level of MPFC activation in Caucasian participants compared with Japanese controls as well as higher activation of temporal parts of the brain on the ToM-RM contrast. In a previous ToM imaging study, Kobayashi and colleagues (2006) found a

culturally dependent activation in the frontal parts of the brain among American monolinguals and Japanese bilinguals completing a second-order false-belief task.

We believe the above-mentioned group comparisons indicate that Japanese participants activate the MPFC to a lesser extent because they have been taught from early childhood to "read the air" (*kuuki wo yomu*), or to be attuned to unspoken social signals all around and to react in a socially accepted way. Naito and Koyama (2006) argued that Japanese individuals have a delay in ToM development compared with Western children but that they are able to understand social implications without explicit information. Thus, even though Japanese children seem to develop ToM abilities later than Western children, their performance might be more sophisticated, and they may mentalize with a lower level of ToM network activation.

This interpretation is partially consistent with a previous study by Chiao and colleagues (2009), who showed that during a self-estimation task, Westerners activated the MPFC more than Asian controls. They associated this finding with individualistic traits. However, they interpreted that individualistic Caucasians overactivate the MPFC because they constantly need to distinguish between themselves, others,

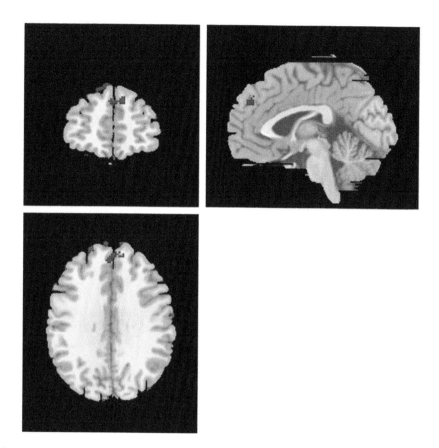

Figure 2. fMRI images comparing activation in Caucasian and Japanese participants during ToM video presentation in fMRI ($p < .005$, uncorrected).

and their surroundings, citing findings by Kitayama and colleagues (Kitayama, Duffy, Kawamura, & Larsen, 2003). They found that even when judging external objects, North Americans tend to relate this information to themselves, while Asians attend more to the social context. We could not simply reduce our findings to cultural differences because we could not find any significant between-group differences on the IND/COL. Therefore, this discrepancy should be addressed in further studies.

Our imaging results are consistent with previous studies; however, our investigation is one of the first not only to show transcultural differences but also to reveal the impact of transcultural sociobehavioral traits on brain activation during the presentation of a ToM task. We could not find any association between acculturation level (e.g., language abilities and cultural knowledge) or empathy, as assessed through the IRI; however, we did find differences between the TAS subscale score "difficulties in identifying feelings" and the AQ score and MPFC activation pattern in the Japanese participants. These results suggest that, within the Japanese group, those with more prominent alexithymic or autistic traits have to activate the ToM network more due to difficulty in describing

emotions and feelings and their externally oriented thinking style. Thus, our attempt to explain the correlation between MPFC activation and AQ and a TAS subscore assumes that the stronger MPFC activation plays a compensatory role. This idea parallels one proposed by Marjoram and colleagues (Marjoram et al., 2006), who performed a functional imaging study on a sample of relatives of schizophrenia patients at high risk of psychosis.

Finally, a study on Chinese parental style showed that impulse control is especially valued in Asian countries (Chen et al., 1998). A similar manner of educational style might thus have an effect on the reaction to emotional stimuli (Moriguchi et al., 2005) and could result in a weaker activation of the MPFC. An alternative hypothesis that we cannot exclude might be that the Japanese participants did not perceive RM sequences as genuinely random and tried to find goals and intentions in the triangles' behavior. This might account for the lack of differences in the contrast of ToM and RM sequences.

Taking all this together, we hypothesize that, due to their cultural background and education, Japanese individuals need to activate specialized brain areas less intensively than Caucasians. This might be due

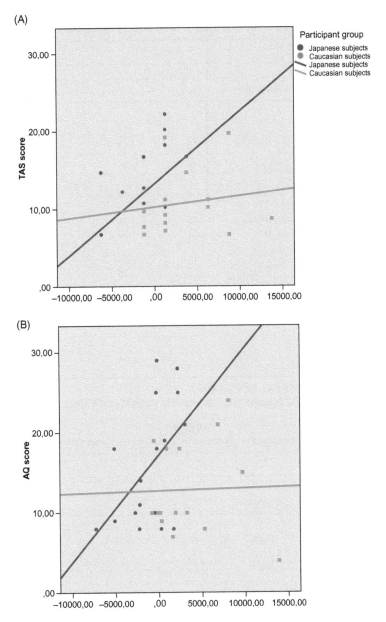

Figure 3. Correlations between the eigenvariates of MPFC activation of ToM-RM and the TAS subscale 1 (A) and the AQ score (B) in Caucasian and Japanese participants.

to better mentalizing abilities or a different manner of utilizing the brain areas that participate in the ToM network. Caucasians are thought to constantly monitor themselves and their surroundings; thus, their ToM network is activated more frequently. However, when Japanese participants display autistic or alexithymic sociobehavioral traits, their activation pattern equals that of Caucasian participants, probably due to compensation for reduced mentalizing abilities.

There are some limitations to our study. First, we do not have neuropsychological data for both study groups. Consequently, intelligence as a confounder

cannot be ruled out; however, because the socioeconomic data of both groups did not differ significantly, we assumed that there were no substantial differences in intelligence levels. Second, the sample size in this study was small, and our Caucasian sample was quite heterogeneous regarding linguistic background and nationality. Group comparison may have been compromised by in-group differences among Caucasians. Few studies have focused on European–European or European–American comparisons in ToM. Lillard occupied herself with the question of a "European-American ToM" (1998). She

TABLE 2
Activations in response to videos in Japanese ($n = 18$) and Caucasian participants ($n = 15$)[a]

Talairach and Tournoux coordinates

Group, contrast and brain region	Brodmann's area	Hemisphere	x	y	z	Number of activated voxels	z[b]
Caucasian participants							
Theory of Mind vs. random movement							
Superior frontal gyrus (MPFC)	9	Right	12	52	30	70	3.66
Middle temporal gyrus	21	Left	−58	−12	−12	78	3.58
Superior frontal gyrus (MPFC)	9	Left	−8	50	30	52	3.54
Temporal lobe	40	Right	62	−54	22	59	3.44
Parahippocampal gyrus	36	Left	−22	−30	−20	9	3.36
Thalamus		Right	12	−30	8	10	3.36
Thalamus		Right	20	−28	10	8	3.28
Random movement vs. Theory of mind sequences							
−	−	−	−	−	−	−	−
Japanese participants							
Theory of mind vs. random movement sequences							
−	−	−	−	−	−	−	−
Random movement vs. Theory of mind sequences							
−	−	−	−	−	−	−	−

Notes: [a]In one-sample *t*-tests, $p < .001$, uncorrected, for all results.
[b]Expressed as the maximum within each area; local maxima are separated by a minimum of 8 mm.

TABLE 3
Differences in the ToM-RM contrast in Japanese ($n = 18$) and Caucasian participants ($n = 15$)[a]

Talairach and Tournoux coordinates

Group, contrast and brain region	Brodmann's area	Hemisphere	x	y	z	Number of activated voxels	z[b]
Caucasian > Japanese participants							
Superior frontal gyrus (MPFC)	8	Right	4	40	54	52	3.09
Superior frontal gyrus (MPFC)	9	Right	12	48	34	92	3.07
Claustrum		Right	30	−4	18	27	3.20
Parahippocampal gyrus	28	Right	22	−24	−12	22	2.89
Temporal lobe	37	Right	44	−46	−10	26	2.87
Japanese > Caucasian participants							
−	−	−	−	−	−	−	−

Notes: [a]$p < .005$, uncorrected, for all results.
[b]Expressed as the maximum within each area; local maxima are separated by a minimum of 8 mm.

referred to middle- and upper-class Americans with European origins, but assumed that many elements of their ToM are shared with Europeans as well. On the other hand, Lecce and Hughes (2010) recently found that British children outperformed Italian children on ToM tasks. It must be assumed that there is at least a small intragroup difference in our Caucasian sample, adding to interindividual differences. As a whole, we think the differences between Caucasian and Japanese participants far outweigh those; however, this needs to be proven with larger sample sizes.

Furthermore, we have to consider an effect of test location, as all tests were conducted in Japan. We also have to assume a selection bias on our results, as the participants all chose to live in Japan and might thus display specific personality traits. On average, the subjects stayed in Japan for approximately 28 months, and some acculturation effects have to be considered.

However, those are not likely to outweigh the effects of an individual's acculturation to his or her culture of origin. This issue should be addressed in future studies.

Finally, during the scan, the subjects were asked to assign each video sequence to a category to ensure their attention. This kind of instruction might have constituted a cue that may have influenced activation results.

Especially concerning imaging studies with clinical samples, our results suggest that attention should be paid to the participants' cultural backgrounds. Differences in the activation patterns of clinical samples are only relevant to those investigated in healthy populations of the same culture and may not apply when individuals of different cultures are compared.

In summary, we found a culturally dependent manner of activation in Japanese and Caucasian participants that seemed to be independent of factors such as acculturation or IND/COL but relied on personality traits such as autistic or alexithymic features. Research using a larger number of subjects is needed to further investigate specific cultural, socioeconomic, and personality aspects in different cultures.

REFERENCES

Abell, F., Happé, F. G., & Frith, U. (2000). Do triangles play tricks? Attribution of mental states to animated shapes in normal and abnormal development. *Cognitive Development, 15*, 1–16.

Adams, R. B., Jr., Rule, N. O., Franklin, R. G., Jr., Wang, E., Stevenson, M. T., Yoshikawa, S., et al. (2010). Cross-cultural reading the mind in the eyes: An fMRI investigation. *Journal of Cognitive Neuroscience, 22*(1), 97–108.

Aketa, H. (1999). Structure and measurement of empathy: Japanese version of Davis's Interpersonal Reactivity Index (IRI-J). *Psychological Reports of Sophia University, 23*, 19–31.

Baron-Cohen, S., Jolliffe, T., Mortimore, C., & Robertson, M. (1997). Another advanced test of theory of mind: Evidence from very high functioning adults with autism or Asperger syndrome. *Journal of Child Psychology and Psychiatry, 38*(7), 813–822.

Baron-Cohen, S., Wheelwright, S., Skinner, R., Martin, J., & Clubley, E. (2001). The autism-spectrum quotient (AQ): Evidence from Asperger syndrome/high-functioning autism, males and females, scientists and mathematicians. *Journal of Autism and Developmental Disorders, 31*(1), 5–17.

Bruene, M., & Bruene-Cohrs, U. (2006). Theory of Mind—evolution, ontogeny, brain mechanisms and psychopathology. *Neuroscience and Biobehavioral Reviews, 30*(4), 437–455.

Castelli, F., Frith, C. D., Happé, F. G., & Frith, U. (2002). Autism, Asperger syndrome and brain mechanisms for the attribution of mental states to animated shapes. *Brain, 125*(Pt 8), 1839–1849.

Castelli, F., Happé, F. G., Frith, U., & Frith, C. D. (2000). Movement and mind: A functional imaging study of perception and interpretation of complex intentional movement patterns. *NeuroImage, 12*(3), 314–325.

Chen, X., Hastings, P. D., Rubin, K. H., Chen, H., Cen, G., & Stewart, S. L. (1998). Child-rearing attitudes and behavioral inhibition in Chinese and Canadian toddlers: A cross-cultural study. *Developmental Psychology, 34*(4), 677–686.

Chiao, J. Y., Harada, T., Komeda, H., Li, Z., Mano, Y., Saito, D., et al. (2009). Neural basis of individualistic and collectivistic views of self. *Human Brain Mapping, 30*(9), 2813–2820.

Chiao, J. Y., Harada, T., Komeda, H., Li, Z., Mano, Y., Saito, D., et al. (2010). Dynamic cultural influences on neural representations of the self. *Journal of Cognitive Neurosciences, 22*(1), 1–11.

Chung, R. H. G., Kim, B. S. K., & Abreu, J. M. (2004). Asian American Multidimensional Acculturation Scale: Development, factor analysis, reliability and validity. *Cultural Diversity and Ethnic Minority Psychology, 10*, 66–80.

Davis, M. H. (1983). Measuring individual differences in empathy: Evidence for a multidimensional approach. *Journal of Personality and Social Psychology, 44*, 113–126.

Duffy, S., Toriyama, R., Itakura, S., & Kitayama, S. (2009). Development of cultural strategies of attention in North American and Japanese children. *Journal of Experimental Child Psychology, 102*(3), 351–359.

First, M. B., Spitzer, R. L., Gibbon, M., & Williams, J. B. W. (1996). *Structured Clinical Interview for DSM-IV Axis I disorders (SCID)*. New York, NY: New York State Psychiatric Institute, Biometrics Research.

Hendry, J. (1986). *Becoming Japanese: The world of the pre-school child*. Manchester,UK: Manchester University Press.

Hollingshead, A. B. (1975). *A four-factor classification of social status*. New Haven, CT: Yale University Press.

Kitayama, S., Duffy, S., Kawamura, T., & Larsen, J. T. (2003). Perceiving an object and its context in different cultures: A cultural look at new look. *Psychological Science, 14*(3), 201–206.

Kobayashi, F. C., Glover, G. H., & Temple, E. (2006). Cultural and linguistic influence on neural bases of 'theory of mind': An fMRI study with Japanese bilinguals. *Brain and Language, 98*(2), 210–220.

Koelkebeck, K., Pedersen, A., Suslow, T., Kueppers, K. A., Arolt, V., & Ohrmann, P. (2010). Theory of mind in first-episode schizophrenia patients: Correlations with cognition and personality traits. *Schizophrenia Research, 119*(1–3), 115–123.

Komaki, G., Maeda, M., Arimura, T., Nakata, A., Shinoda, H., Ogata, I., et al. (2003). The reliability and factorial validity of the Japanese version of the 20-item Toronto Alexithymia Scale. *Journal of Psychosomatic Research, 55*(2), 143.

Lancaster, J. L., Woldorff, M. G., Parsons, L. M., Liotti, M., Freitas, C. S., Rainey, L., et al. (2000). Automated Talairach atlas labels for functional brain mapping. *Human Brain Mapping, 10*(3), 120–131.

Lecce, S., & Hughes, C. (2010). The Italian job?: Comparing theory of mind performance in British and Italian children. *British Journal of Developmental Psychology, 28*(Pt 4), 747–766.

Lee, K., Olson, D. R., & Torrance, N. (1999). Chinese children's understanding of false beliefs: The role of language. *Journal of Child Language, 26*(1), 1–21.

Li, H., & Rao, N. (2000). Parental influences on Chinese literacy development: A comparison of preschoolers in Beijing, Hong Kong and Singapore. *International Journal of Behavioral Development, 24*, 82–90.

Lillard, A. (1998). Ethnopsychologies: Cultural variations in theories of mind. *Psychological Bulletin, 123*(1), 3–32.

Marjoram, D., Job, D. E., Whalley, H. C., Gountouna, V. E., McIntosh, A. M., Simonotto, E., et al. (2006). A visual joke fMRI investigation into theory of mind and enhanced risk of schizophrenia. *NeuroImage, 31*(4), 1850–1858.

Matsuda, T., & Nisbett, R. E. (2001). Attending holistically versus analytically: comparing the context sensitivity of Japanese and Americans. *Journal of Personality and Social Psychology, 81*(5), 922–934.

Moriguchi, Y., Ohnishi, T., Kawachi, T., Mori, T., Hirakata, M., Yamada, M., et al. (2005). Specific brain activation in Japanese and Caucasian people to fearful faces. *Neuroreport, 16*(2), 133–136.

Moriguchi, Y., Ohnishi, T., Lane, R. D., Maeda, M., Mori, T., Nemoto, K., et al. (2006). Impaired self-awareness and theory of mind: An fMRI study of mentalizing in alexithymia. *NeuroImage, 32*(3), 1472–1482.

Naito, M., & Koyama, K. (2006). The development of false-belief understanding in Japanese children: Delay and difference? *International Journal of Behavioral Development, 30*(4), 290–304.

Oldfield, R. C. (1971). The assessment and analysis of handedness: The Edinburgh inventory. *Neuropsychologia, 9*(1), 97–113.

Paulus, C. (2006). *Saarbruecker Persoenlichkeits-Fragebogen (SPF)*. Saarbruecken, DE: Saarland University, Department of Education Science.

Rogers, K., Dziobek, I., Hassenstab, J., Wolf, O. T., & Convit, A. (2007). Who cares? Revisiting empathy in Asperger syndrome. *Journal of Autism and Developmental Disorders, 37*(4), 709–715.

Russell, T. A., Reynaud, E., Herba, C., Morris, R., & Corcoran, R. (2006). Do you see what I see? Interpretations of intentional movement in schizophrenia. *Schizophrenia Research, 81*(1), 101–111.

Shweder, R. A., Goodnow, J., Hatano, G., LeVine, R. A., Markus, H., & Miller, P. (1998). The cultural psychology of development: One mind, many mentalities. In W. Damon & R. M. Lerner (Eds.), *Handbook of child psychology: Theoretical models of human development*. New York, NY: Wiley.

Suinn, R. M., Ahuna, C., & Khoo, G. (1992). The Suinn-Lew Asian Self-Identity Acculturation Scale: Concurrent and factorial validation. *Educational and Psychological Measurement, 52*(4), 1041–1046.

Tanaka, K. (2000). *Ryugakusei no sosharu nettowaku to sosharu sukiru* [*Social network and social skill in foreign students*]. Tokyo, Japan: Nakanishiya.

Tardif, T., & Wellman, H. M. (2000). Acquisition of mental state language in Mandarin- and Cantonese-speaking children. *Developmental Psychology, 36*(1), 25–43.

Taylor, G. J., Ryan, D., & Bagby, R. M. (1985). Toward the development of a new self-report alexithymia scale. *Psychotherapy and Psychosomatics, 44*(4), 191–199.

Triandis, H. C. (1994). *Culture and social behavior*. New York, NY: McGraw-Hill.

Voellm, B. A., Taylor, A. N., Richardson, P., Corcoran, R., Stirling, J., McKie, S., et al. (2006). Neuronal correlates of theory of mind and empathy: A functional magnetic resonance imaging study in a nonverbal task. *NeuroImage, 29*(1), 90–98.

Wakabayashi, A., Tojo, Y., Baron-Cohen, S., & Wheelwright, S. (2004). [The Autism-Spectrum Quotient (AQ) Japanese version: Evidence from high-functioning clinical group and normal adults]. *Shinrigaku Kenkyu, 75*(1), 78–84.

Weed, E., McGregor, W., Feldbaek Nielsen, J., Roepstorff, A., & Frith, U. (2010). Theory of mind in adults with right hemisphere damage: What's the story? *Brain and Language, 113*(2), 65–72.

Wellman, H. M., Cross, D., & Watson, J. (2001). Meta-analysis of theory-of-mind development: The truth about false belief. *Child Development, 72*(3), 655–684.

Wellman, H. M., Fang, F., Liu, D., Zhu, L., & Liu, G. (2006). Scaling of theory-of-mind understandings in Chinese children. *Psychological Science, 17*(12), 1075–1081.

Zhu, Y., Zhang, L., Fan, J., & Han, S. (2007). Neural basis of cultural influence on self-representation. *NeuroImage, 34*(3), 1310–1316.

Identification of psychopathic individuals using pattern classification of MRI images

João R. Sato[1,3], Ricardo de Oliveira-Souza[2], Carlos E. Thomaz[4], Rodrigo Basílio[2], Ivanei E. Bramati[2], Edson Amaro Jr[3], Fernanda Tovar-Moll[2,5], Robert D. Hare[6], and Jorge Moll[2]

[1]Center for Mathematics, Computation, and Cognition, Universidade Federal do ABC, Santo André, Brazil
[2]Cognitive and Behavioral Neuroscience Unit, D'Or Institute for Research and Education, Rio de Janeiro, Brazil
[3]NIF/LIM44, Departamento de Radiologia da Faculdade de Medicina da Universidade de São Paulo, São Paulo, Brazil
[4]Department of Electrical Engineering, Centro Universitário da FEI, São Bernardo do Campo, Brazil
[5]Biomedical Sciences Institute, Federal University of Rio de Janeiro, Rio de Janeiro, Brazil
[6]Department of Psychology, University of British Columbia, and Darkstone Research Group, Vancouver, Canada

Background: Psychopathy is a disorder of personality characterized by severe impairments of social conduct, emotional experience, and interpersonal behavior. Psychopaths consistently violate social norms and bring considerable financial, emotional, or physical harm to others and to society as a whole. Recent developments in analysis methods of magnetic resonance imaging (MRI), such as voxel-based-morphometry (VBM), have become major tools to understand the anatomical correlates of this disorder. Nevertheless, the identification of psychopathy by neuroimaging or other neurobiological tools (e.g., genetic testing) remains elusive.

Methods/Principal findings: The main aim of this study was to develop an approach to distinguish psychopaths from healthy controls, based on the integration between pattern recognition methods and gray matter quantification. We employed support vector machines (SVM) and maximum uncertainty linear discrimination analysis (MLDA), with a feature-selection algorithm. Imaging data from 15 healthy controls and 15 psychopathic individuals (7 women in each group) were analyzed with SPM2 and the optimized VBM preprocessing routines. Participants were scanned with a 1.5 Tesla MRI system. Both SVM and MLDA achieved an overall leave-one-out accuracy of 80%, but SVM mapping was sparser than using MLDA. The superior temporal sulcus/gyrus (bilaterally) was identified as a region containing the most relevant information to separate the two groups.

Conclusion/significance: These results indicate that gray matter quantitative measures contain robust information to predict high psychopathy scores in individual subjects. The methods employed herein might prove useful as an adjunct to the established clinical and neuropsychological measures in patient screening and diagnostic accuracy.

Keywords: Psychopathy; Antisocial; Voxel-based morphometry; Moral; Machine learning.

Correspondence should be addressed to: João Ricardo Sato, Center of Mathematics, Computation, and Cognition, Universidade Federal do ABC, Avenida Atlântica, 420, Valparaíso, CEP: 09060-000, Santo André, SP, Brazil. E-mail: joao.sato@ufabc.edu.br or to Jorge Moll, IDOR, R. Diniz Cordeiro 30, 3 andar, CEP:22281-100, Rio de Janeiro, RJ, Brazil. E-mail: jorge.moll@idor.org

This research was supported by FAPERJ (Pronex and INNT grants), FAPESP, CAPES, and CNPq, as well as by intramural grants from IDOR, Brazil.

www.psypress.com/socialneuroscience   http://dx.doi.org/10.1080/17470919.2011.562687

Psychopathy, the first personality disorder recognized in psychiatry (Millon, Simonsen, Davis, & Birket-Smith, 2002), is defined by a cluster of interpersonal, affective, lifestyle, and antisocial traits and behaviors that include grandiosity, egocentricity, deceptiveness, lack of empathy or remorse, irresponsibility, impulsivity, and a tendency to violate social norms (Hare & Neumann, 2005). Psychopathy can be assessed in forensic settings by the Psychopathy Checklist–Revised (PCL–R) and in nonforensic contexts by the Psychopathy Checklist: Screening Version (PCL–SV), each supported by extensive evidence for their reliability and validity (Hare, 2006). As a stand-alone instrument for assessing psychopathy in civil psychiatric and community populations (de Oliveira-Souza, Ignácio, Moll, & Hare, 2008a; Guy & Douglas, 2006), the PCL–SV is strongly related to the PCL–R both conceptually and empirically (Cooke, Michie, Hart, & Hare, 1999). The widespread adoption of the PCL scales as a common metric for psychopathy has led to a dramatic increase in theoretical and empirical work, paving the way for research on its neurobiological substrates (Glenn & Raine, 2008). In particular, recent advances in magnetic resonance imaging (MRI) acquisition and processing, such as the methods for quantitative and automated assessment of brain structure (Good et al., 2001) have opened up new possibilities. Converging evidence indicates that the core features of the psychopathic personality are related to discrete volumetric changes in a set of frontotemporal and subcortical brain regions (De Brito et al., 2009; de Oliveira-Souza et al., 2008b; Müller et al., 2008; Tiihonen et al., 2008) that underlie moral cognition and behavior (Moll et al., 2005). These recent, quantitative, voxel-based studies have confirmed and extended the findings of earlier ones, which employed radioisotope and volumetric MRI manual tracing techniques (e.g., Soderstrom et al., 2002; Yang et al., 2005). Important progress has also been made on the more proximate causes of psychopathy; evidence suggests that the brain differences in psychopathic individuals are neurodevelopmental in nature, and arise from genetic and environmental factors (e.g., physical abuse) and their interactions (Bezdjian et al., 2011; Gao et al., 2010). A detailed account of these aspects is outside the scope of the current paper, and can be found in recent authoritative reviews (Gao, Glenn, Schug, Yang, & Raine, 2009).

Exciting as these neuroanatomical findings may be, their clinical utility remains elusive. Imaging studies on patient populations generally provide results on a group level, but their use for diagnostic classification of individual patients has yet to be established. The development of reliable and specific neuroanatomical biomarkers for psychiatric disorders—including psychopathy—would obviously be important. Typically, group studies using brain imaging rely on massive, voxel-by-voxel application of the general linear model (GLM) (Friston et al., 1995) or its particular cases, such as t-tests and ANOVA. This model is useful to provide statistical hypothesis testing for group differences or linear associations among variables. In neuroimaging, GLM is applied independently to all intracranial voxels, providing p values for group comparisons at each voxel. These p values are corrected for multiple comparisons, using the false discovery rate (FDR) (Benjamini & Hochberg, 1995) or random fields theory (RFT) (Worsley, 1995), and then compared to a prespecified significance level. This procedure is the core of univariate statistical analysis for brain mapping, by far the most frequently used approach. However, univariate analysis may not be the most suitable approach in clinical neuroimaging for two reasons: (1) the brain is organized in several highly structured networks and univariate, voxel-by-voxel analysis does not take into account this property, treating brain voxels independently; and (2) statistically significant differences do not necessarily mean cognitive or clinically relevant differences.

The interconnected structure of the brain implies that regions may influence one another, both structurally and functionally, and thus multivariate approaches may be more suitable than univariate ones (Lukic, Wernick, & Strother, 2002). Furthermore, groups or conditions may be characterized by the topology or changes of these relationships (Sato et al., 2008a). In addition, statistical hypothesis testing for group differences is an inferential procedure that compares parameters between two populations based on measures calculated over the samples. Therefore, assuming that an adequate test is applied, any difference in a parameter of interest between two populations will be detected, provided the sample sizes are large enough. Thus, complementary to statistical tests, one major point of concern is whether or not differences between groups can be used to allocate each subject to a particular group based on *a priori* defined individual variables (e.g., symptom clusters, diagnostic categories, genetic markers). This is, in fact, an ultimate challenge of clinical diagnosis research, with vast implications for diagnosis and treatment. In the realm of imaging techniques, functional and structural MRI findings have been increasingly employed as "intermediate phenotypes" or "endophenotypes," capturing at a meso- or macroscopic level features that reflect complex and often subtle factors, including environmental, genetic, and epigenetic ones (Meyer-Lindenberg & Weinberger, 2006). The potential of this approach is reflected by the rapidly growing number of studies using quantitative techniques such as diffusion tensor imaging and voxel-based morphometry (VBM) in this context (Bertisch et al., 2010; Bradley et al., 2009; Camchong, Lim, Sponheim, & Macdonald, 2009; Honea et al., 2008).

Statistical learning or pattern recognition methods have become attractive approaches in computer-aided diagnosis, mostly because they can be applied in a multivariate fashion and they provide classification rules for predicting the group membership of a *new* subject. In the last decade, multivariate pattern recognition providing a joint analysis of all voxels was proposed for neuroimaging analysis (Fan, Shen, & Davatzikos, 2005; Golland, Grimson, Shenton, & Kikinis, 2000; Golland et al., 2002; Lao et al., 2004; Lukic et al., 2002; Sato et al., 2008b, 2009; Thomaz et al., 2007a, 2007b). Nevertheless, only a few studies have explored the predictive power of these approaches in neurological or neuropsychiatric disorders. Emblem et al. (2008) applied support vector machines (SVM) to predict glioma grades, using perfusion images. Gerardin et al. (2009) used hippocampal shape to classify Alzheimer's disease, mild cognitive impairment, and controls with an accuracy of 94%. A classification method for primary progressive aphasia was developed by Wilson et al. (2009), showing good accuracy and generalization power. Ecker et al. (2010) evaluated the predictive power of SVM for whole-brain structural images in autism (gray matter VBM), and also demonstrated good discriminative power between patients and normal controls (specificity of 86% and sensitivity of 88%). Davatzikos, Bhatt, Shaw, Batmanghelich, & Trojanowski, (2010) have shown that it is possible to predict the conversion of mild cognitive impaired patients to Alzheimer's disease, based on a combined analysis between VBM and spatial patterns of abnormalities. Recently, Koutsouleris et al. (2010) studied the prediction of vulnerability and transition to psychosis by using support vector regression.

Despite the clinical and societal relevance of psychopathy, pattern classification of neuroimaging data has not yet been employed in this condition. In this paper, we evaluate the application of pattern recognition methods to gray-matter images, focusing on distinguishing individuals with psychopathy from normal controls. Two classification methods were investigated: SVM and maximum uncertainty linear discriminant analysis (MLDA). Furthermore, we introduced an approach for feature selection to improve classification rates, a useful tool for general subject/group classification when dealing with relatively small samples of brain imaging data. The main goal of the present study was to investigate whether classifiers could discriminate between subject groups by multivariate pattern analysis of whole-brain gray matter voxels, which is essentially distinct from attempting to map which voxels show statistical differences between subject samples. From previous studies (De Brito et al., 2009; de Oliveira-Souza et al., 2008b; Moll, Zahn, de Oliveira-Souza, Krueger, & Grafman, 2005; Müller et al., 2008; Tiihonen et al., 2008), we had, nonetheless, some *a priori* expectations about which brain regions would likely shelter discriminant voxels. These regions included the anterior and ventral sectors of the prefrontal cortex and the superior temporal sulcus region, which were consistently activated in several functional MRI studies on control subjects (see Moll et al., 2005, for a review), and were found to be structurally abnormal in psychopaths (de Oliveira-Souza et al., 2008b; Müller et al., 2008; Tiihonen et al., 2008).

MATERIAL AND METHODS

All participants provided written informed consent before entering the study, which was approved by the D'Or Institutional Review Board (Rio de Janeiro, Brazil). The 15 patients (8 men, 7 women) who agreed to undergo MRI scanning were part of a larger group of 50 patients with neurological and/or neuropsychiatric disorders who were brought to consultation by relatives or acquaintances for a variety of emotional and behavioral problems (de Oliveira-Souza et al., 2008b). Each patient fulfilled the DSM-IV adult criteria for antisocial personality disorder (American Psychiatric Association, 1994) and was assessed with the PCL–SV as the primary measure of interest (Hart, Cox, & Hare, 1995). Their occupational history was erratic and unstable. They lived in the community, but eventually came to medical attention due to chronic and recurrent misbehaviors, which did not result in criminal prosecution. The control group included 15 normal volunteers matched on gender, age, and education, and without a history of neurological or psychiatric disorders or serious misconduct. Further details on participants' characteristics and behavioral results can be found elsewhere (de Oliveira-Souza et al., 2008b). Briefly, there were no significant differences between groups in gender, age, education, handedness, global cognitive status, and executive performance (Table 1).

TABLE 1
Demographic and neuropsychological information of the
sample evaluated in the current study

	Controls	Patients
Education (years)	11 ± 2	11 ± 2
Age (years)	32 ± 13	32 ± 14
MMSE (0–30)	28.9 ± 1.1	28.7 ± 1.9
Handedness (R/L)	13/2	14/1
PCL–SV (0–24)	0.4 ± 1.0	17.8 ± 3.8
WCST		
Categories completed (0–6)	5.4 ± 1.4	5.0 ± 1.6
Perseverative errors (0–127)	14 ± 12	18 ± 11
Set failures (0–22)	0.80 ± 1.1	1.2 ± 1.0

Notes: MMSE: Mini Mental Status Examination; PCL–SV:
Psychopathy Checklist, Screening Version; WCST: Wisconsin Card
Sorting Test. Patients and controls were matched in all scores,
except for the PCL–SV (t-test, $p < .05$).

Image acquisition

MRI scans were acquired at the Department of
Radiology at Barra D'Or Hospital, using a 1.5 Tesla
MR System (Siemens Medical Systems, Erlangen,
Germany), with a standard quadrature head coil.
For each volunteer, a high-resolution, T1-weighted,
3D structural volume was obtained (MPRAGE pulse
sequence, $TR = 9.7$ ms, $TE = 4$ ms, $TI = 300$ ms,
flip angle $= 12°$, field of view $= 256$ mm, slice thickness $= 1.25$ mm, matrix size $= 256 \times 256$, 128 sagittal
slices, in-plane resolution of 1 mm $\times$ 1 mm).

Image preprocessing

Automated preprocessing of structural images was
carried out using the package SPM2 (Wellcome
Department of Imaging Neuroscience, London, UK;
http://www.fil.ion.ucl.ac.uk/spm). The optimized
VBM protocol was used (for details on the preprocessing steps, see the Methods section in de
Oliveira-Souza et al. (2008b). For the purpose of the
present investigation, the normalized, smoothed (Full
width at half maximum $= 12$ mm) unmodulated gray
matter (GM) images (corresponding to GM "concentration," or GMC) were employed for all subsequent
pattern classification analyses. The statistical steps
of VBM processing and the resulting topographical
maps previously reported in (de Oliveira-Souza
et al., 2008b) were disregarded for the purposes of
the present study, and did not affect the statistical
inferences of reported herein.

SVM and maximum uncertainty linear discrimination

The typical task of the statistical learning (or classification) methods is to use the features provided by
the previous stages in this work the preprocessed VBM
gray matter values to assign the object of interest to a
specific group or class. This assignment can be done
directly, using risk minimization-based approaches
such as SVM (Vapnik, 1998), or indirectly as in the
spectral multivariate analysis of the data with the
Linear discriminant analysis (LDA)-based approaches
(Devijver & Kittler, 1982; Fukunaga, 1990). Both
linear kernel SVM and LDA are discriminant methods that seek to find a classification boundary that
separates data into different groups with maximum
precision. Recent studies have suggested that these two
approaches can be successfully applied to neuroimaging data sets (Mourão-Miranda, Bokde, Born, Hampel,
& Stetter, 2005; Sato et al., 2008b; Thomaz et al.,
2007b). This is because, by being linear methods,
they allow the quantification of discriminative information contained at each predictor variable (voxel),
which can be directly obtained from the separating
hyperplane coefficients. There are, however, important
differences between these two approaches on extracting and classifying discriminating information from
data.

The primary purpose of SVM is to maximize the
width of the margin between two distinct sample
classes (Vapnik, 1998). Given a training set that con-
sists of N pairs of $(x_1, y_1), (x_2, y_2), \ldots, (x_N, y_N)$, where
x_i denotes the K-dimensional training observations
and $y_i \in \{-1, +1\}$ the corresponding classification
labels, the SVM method seeks to find the hyperplane
defined by

$$f(x) = (x \cdot w) + b = 0,$$

which separates positive and negative observations
with the maximum margin. It can be shown that
the solution vector w_{svm} (hyperplane coefficients) is
defined in terms of a linear combination of the training
observations; that is,

$$w_{svm} = \sum_{i=1}^{N} \alpha_i y_i x_i,$$

where α_i are non-negative coefficients obtained by
solving a quadratic optimization problem with linear
inequality constraints. Those training observations x_i
with non-zero α_i lie on the boundary of the margin
and are called support vectors (Vapnik, 1998). In the

present study, only the linear kernel was applied, and the cost parameter was set to 1. The description of the SVM solution does not make any assumption about the distribution of the data, focusing on the observations that lie close to the opposite class; that is, on the observations that most count for classification (Hastie, Tibshirani, & Friedman, 2001).

The LDA solution, on the other hand, is a spectral matrix analysis of the data and is based on the assumption that each class can be represented by its distribution of data; that is, the corresponding mean vector (or class prototype) and covariance matrix (or spread of the sample group) (Hastie et al., 2001). In other words, LDA depends on all of the data, even points far away from the separating hyperplane; its main objective is to find a projection matrix W_{lda} that maximizes Fisher's criterion (Fukunaga, 1990):

$$\frac{\left|W^T S_b W\right|}{W^T S_w W},$$

where S_b and S_w are respectively the between- and within-class scatter matrices. Fisher's criterion is maximized when the projection matrix W_{lda} is composed of the eigenvectors of $S_w^{-1} S_b$ with at most number of classes − 1 non-zero eigenvalues (Devijver & Kittler, 1982). In the case of a two-class problem, the LDA projection matrix is, in fact, the leading eigenvector w_{lda} of $S_w^{-1} S_b$, assuming that S_w is invertible. However, in limited sample and high dimensional problems, such as the one under investigation, S_w is either singular or mathematically unstable, and the standard LDA cannot be used for the classification task. To avoid these critical issues, we have calculated the leading eigenvector w_{lda} (hyperplane coefficients) by a MLDA that considers the issue of stabilizing the S_w estimate with a multiple of the identity matrix (Thomaz, Kitani, & Gillies, 2006).

Classification and brain mapping

As described in the previous section, SVM and MLDA are classifiers based on finding a discriminative hyperplane, where the decision to which group a subject belongs is achieved by projecting the gray matter maps onto this hyperplane. For each voxel, there is an associated coefficient defining the discriminative hyperplane, which is fully specified by the set of coefficients of all voxels. The absolute value of this coefficient is a measure of how the related voxel predicts the grouping of the subjects; that is, it is an index of the amount of discriminative information contained in this brain region.

A well-known obstacle to the application of classification methods in neuroimaging is the huge dimensionality of the data. In most VBM studies, the number of subjects is in the order of tens, while the number of voxels is in the order of hundreds of thousands. Since we are interested in using the gray matter values at each voxel to predict the class of a subject, this means that we have thousands of variables to predict the class of tens of subjects. This obstacle, known as the "curse of dimensionality," may lead to overfitting; that is, the classifiers are excellent for predicting the subjects used to define the discriminative hyperplane, but they may perform poorly in predicting the class for a new subject. In other words, if a subject is used to train the classifier, the classifier would provide an accurate prediction for this particular subject, but the generalization power for a new individual is not guaranteed. In keeping with this limitation, all accuracies presented in this paper will be based on a leave-one-subject-out cross-validation procedure. This approach consists in leaving one subject out of the training set, training the classifier with the remaining subjects, and then evaluating the generalization power of the classifier by testing the excluded subject.

Despite the fact that both SVM and MLDA are pattern recognition approaches developed to deal with high-dimensional data, the inclusion of confounding variables with little relevant information, (i.e., noise) as predictors may lead to low accuracy rates. A feature (in this case, a voxel) selection step may be useful to improve the accuracy rates. The elimination of voxels that did not contain discriminative information to differentiate the groups may also be useful for brain mapping, since the relevant voxels are identified. Thus, the feature selection step is important to avoid overfitting and is also suitable for brain mapping.

In this paper, image analyses and processing were carried out in the following steps (Figure 1):

1. Process the data with the VBM pipeline to obtain unmodulated gray matter concentration maps (GMC) for each subject. This pipeline includes image intensity normalization, spatial normalization, and spatial smoothing and segmentation for cerebrospinal fluid, and gray and white matter.
2. Mask the volumes for considering only intracranial and gray matter high-probability voxels. This step is important to select only voxels with relevant and interpretative gray matter coefficients.
3. Leave one subject out of the sample.
4. Build a feature matrix X, where the columns correspond to voxels (K features), and each row contains the data of each subject (N individuals).

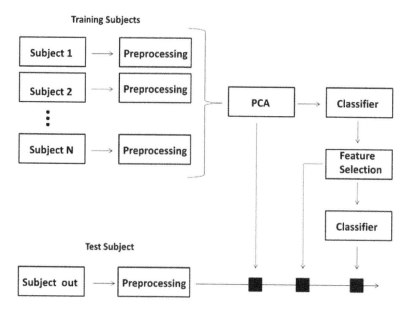

Figure 1. Diagram illustrating the steps for leave-one-out classification and feature selection implemented in this study. The subject left out in each iteration of the leave-one-out procedure is used to evaluate the prediction accuracy. PCA, principal component analysis.

5. Normalize each column of matrix X to have mean equal to zero and variance equal to one, obtaining a feature matrix Z.

6. Train the classifier (SVM or MLDA), using the feature matrix Z and the label (groups) of the subjects included in this matrix.

7. Rank each voxel (feature) by its level of discriminative information measured by the absolute coefficients in hyperplane vector w. Apply the feature selection, keeping the subset ($Q\%$) of more informative voxels set. Note that this step also produces a mask of voxels containing the discriminant information.

8. Retrain the classifiers for this subset of "most informative" voxels.

9. Process the data of the subject-out for normalization (step 4) and feature selection (step 8), and predict his or her group, using the classifier trained at Step 8.

10. Return to Step 3 until all subjects have been processed.

It is important to emphasize that in this procedure, the feature selection is applied for each subject-out. This implies that the voxels used by the classifiers in each leave-one-out may differ. Thus, we propose a brain-mapping strategy based on the proportion of overlap of informative voxels masks (obtained at step 7) between all leave-one-out loops. Note that, in practice, the solution will be sparse, since only a small number of voxels contain discriminative information. Furthermore, the overlap proportion at each voxel

allows the evaluation of the procedure's robustness. In this study, these steps were carried out for different values of Q (0.01%, 0.05%, 0.10%, and 1%), in order to evaluate the performance of the nested subset of voxels. In addition, it is important to emphasize that is not necessary to define regions of interest (ROI) *a priori*, and that the automated identification of relevant voxels is not based on mass-univariate statistical tests; instead, relevant voxels are identified in a multivariate fashion from the voxel maps obtained in step 7. The most discriminant regions were based on feature ranking. Furthermore, the discrimination maps were used solely to make sure that the results were not driven by spurious signals (e.g., border effects, CSF), but from brain regions relevant to social cognition and behavior.

RESULTS

The estimated classification rates for $Q\% = \{0.01\%, 0.05\%, 0.10\%,$ and $1\%\}$ were $\{70\%, 80\%, 73\%, 60\%\}$ and $\{67\%, 77\%, 80\%, 70\%\}$ for SVM and MLDA, respectively. Note that the classifiers achieved an overall accuracy of 80%, but to provide this rate, MLDA uses the 0.1% highest discriminative features, while SVM is more parsimonious, requiring only 0.05%. Figure 2 highlights the brain regions containing the discriminative information used by the classifiers, after the feature selection step. These maps point out that the unmodulated gray matter coefficient at the left superior temporal gyrus/sulcus (STG/STS)

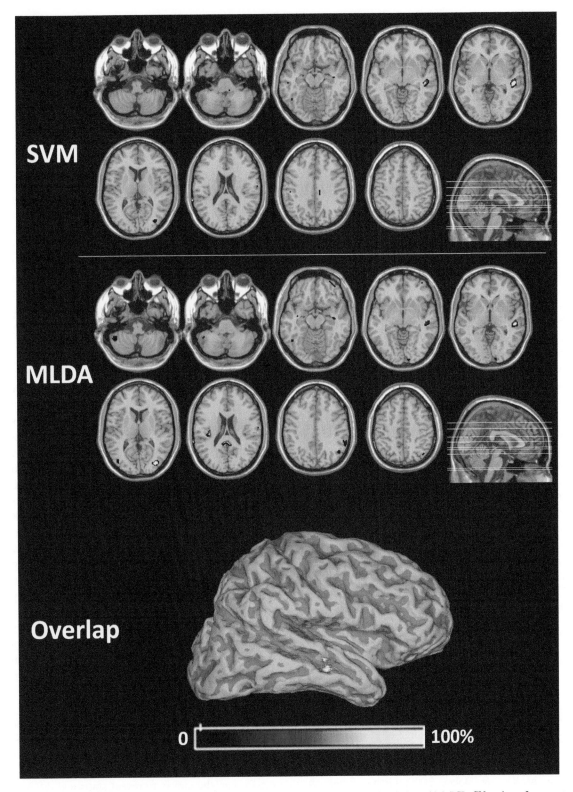

Figure 2. Brain mapping of regions containing the information used by the classifiers to discriminate high PCL–SV patients from controls. The color scale describes the proportion of leave-one-out interactions that selected the respective voxel at feature selection step (consistency).

TABLE 2
Statistical information at local maxima from discriminative clusters with overlap proportion across subjects greater than 70%

	X	Y	Z	Side	Overlap	Size (voxels)	BA	Area
SVM	48	−28	−4	R	100%	121	22	Superior temporal sulcus
	−68	−38	18	L	100%	21	22	Superior temporal gyrus
	6	−40	22	R	100%	143	26	Isthmus of cingulate gyrus
MLDA	36	−82	8	R	100%	96	19	Occipital peristriate cortex
	52	−28	−2	R	100%	88	22	Superior temporal sulcus

is an important predictor for both classifiers. On the other hand, since MLDA uses more discriminant regions, it also indicates that the coefficients at the right STG/STS, left occipital cortex, and posterior cingulate gyrus contain relevant information to predict the classes. The statistical information about the clusters of discriminative regions is presented in Table 2.

For each classifier, the leave-one-out projections of each subject onto the discriminative hyperplane space are shown in Figure 3. This figure indicates that the variability within each group is approximately the same at this space. Subject 13 stood out as a "control group" outlier for both classifiers, since his projection was far from the decision boundary and at the control group space set. Finally, note that the misclassified subjects are not exactly the same for the two classifiers, but most of them fall close to the decision boundary.

In addition, in order to explore the sensitivity and specificity of MLDA and SVM, Figure 4 describes the Receiver-operator-characteristic (ROC) curves built by ranking the leave-one-out decision values of each classifier. Note that in some points of the curve the SVM may achieve a specificity and sensitivity of 80% and 86.7%, and MLDA achieved 86.7% and 80%, respectively.

DISCUSSION

The study of the neural basis of psychopathy is of great relevance to the understanding of this severe disorder. This is reflected in the recent surge of electrophysiological, structural, and functional MRI investigations (Blair, Peschardt, Budhani, Mitchell, Pine, & 2006; Fullam, McKie, & Dolan, 2009; Glenn, Raine, & Schug, 2009; Glenn, Raine, Yaralian, & Yang, 2010; Kiehl, Bates, Laurens, Hare, & Liddle, 2006; Müller et al., 2008; Rilling et al., 2007; Veit et al., 2010). In this study, we explored the applicability of multivariate machine learning techniques to psychopathy diagnosis based on gray matter indexes resulting from VBM data preprocessing. Importantly, no group statistical inferences were made by VBM procedures; instead, linear SVM and MLDA classifiers were used to discriminate between patients and controls based only on their GM images ("GM concentration"). In addition, we showed that these approaches may also provide anatomical information on the brain mapping of regions containing relevant information used to discriminate between the two classes.

From the neurobehavioral and anatomical perspective, the STS region identified by SVM and MLDA has figured prominently in several studies of social cognition and emotion, being implicated in social feature representation (e.g., emotional faces and body posture), intentionality inferences, empathy, and moral sentiments such as guilt, compassion, and embarrassment (Decety, Chaminade, Grèzes, & Meltzoff, 2002; de Gelder, 2006; Grèzes, Pichon, & de Gelder, 2007; Kalbe et al., 2009; Leibenluft, Gobbini, Harrison, Haxby, & 2004; Materna, Dicke, & Thier, 2008). Although acquired damage to this region has not so far been associated with the development of severe antisocial behaviors, it has been suggested that this region may critically work in concert with other frontal, temporal, and subcortical regions to enable complex social and emotional abilities, such as interpersonal feelings. Despite these converging lines of evidence, the problem of why the STS showed up more prominently in the present study, and the neuroanatomical implications of this finding for the diagnosis of psychopathy, are matters of careful analyses, as discussed below.

The goals of conventional brain-mapping methods, such as independent samples group comparisons of means, using VBM or ROI volumetry and pattern classification methods, are essentially distinct. In the former, the aim is to reveal statistical differences between groups on a voxel or ROI basis, independently of how well this finding may help categorize a given subject in a diagnostic group. The pattern recognition methods focus less on how much information a given region carries individually, and more

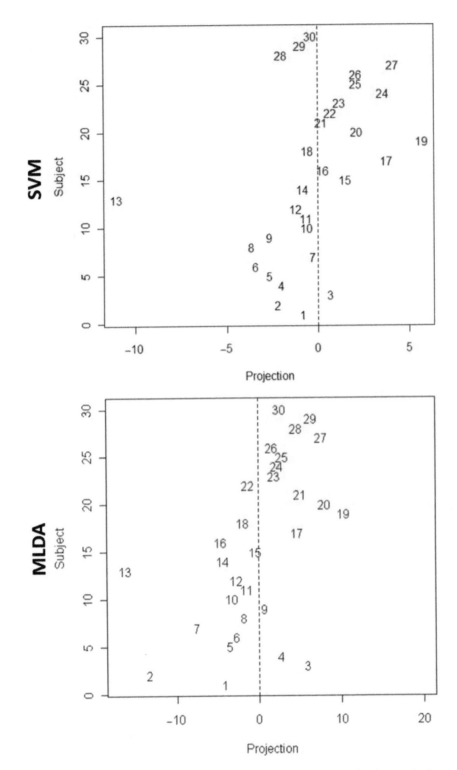

Figure 3. Leave-one-out projections of subjects onto discriminative hyperplane. The decision boundary is at zero in the *x*-axis. The black and red points describe the control and patient groups, respectively.

on how well a combination of different brain regions contributes to the correct classification of a given individual into diagnostic categories. The present results therefore show that the gray matter concentration in the right STS, and to a lesser degree in a few other frontoparieto-occipital regions (including the left

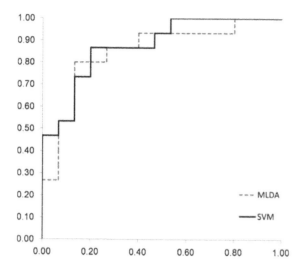

Figure 4. ROC curves for MLDA and SVM obtained by ranking the leave-one-out decision values of each classifier.

STS), when classified according to SVM and MLDA, can reach moderate-to-high accuracies in discriminating patients with high PCL– SV from normal individuals. Thus, the present results emphasize that classification methods are promising in predicting diagnostic categories based on subtle differences in brain structure.

Another advantage of pattern classification methods over more classical approaches is that they are not intrinsically dependent on contiguous spatial relationships of voxels. For example, the influence of anatomical gyral variations is less critical for pattern classification than for methods dependent on contiguous variations (i.e., clustering, Gaussian fields theory). It follows that these techniques can be useful in alterations involving subtle changes spread across the cortical mantle, regardless of spatial distribution. Thus, the alterations can be uniformly distributed, grouped in clusters, or any combination of these— and all could be picked by pattern classification methods.

Although, the potential clinical applicability of this method is straightforward, a number of technical shortcomings and ethical implications must be addressed. Potentially, pattern classification of brain data can increase the diagnostic accuracy of psychopathy, and even help to determine pathological subtypes—but the practical value of this method has yet to be established. One possibility would be to use the method, in conjunction with other neurobehavioral tests, as a supplement to clinical rating tools, such as PCL–R or PCL–SV. Similarly, it is possible that the addition of pattern classifications of brain data will add to the utility of current procedures used to assess the risk of

crime and violence, and to select treatment options for psychopathy. Medical diagnostic imaging in general might benefit from pattern classification methods, since interpretation of single-case images is a challenging task, exacerbated by technological advances that may lead to increased diagnostic sensitivity at the expense of specificity.

The small sample size is a limitation of the current study. Leave-one-out and feature selection were implemented in order to avoid results that arise from overfitting. However, leave-one-out accuracy rates were statistically greater than chance ($p < .05$, for binomial distribution with probability of success equal to 50%). It should also be emphasized that discriminating subtle cortical changes in individuals with high levels of psychopathy from healthy controls is less challenging than discriminating psychopathy from other psychiatric conditions associated with a variety of cortical abnormalities, such as borderline personality, drug abuse, attention deficit-hyperactivity disorder, and other neuropsychiatric disorders (Georgopoulos et al., 2010; Nardo et al., 2010; Schaufelberger et al., 2007; Schlaepfer et al., 2006; Zhu et al., 2005). Critically, reduced gray matter volume in the STS region, which had the greatest discriminative power in this study, is found not only in psychopathy (de Oliveira-Souza et al., 2008b; Müller et al., 2008), but also in schizotypal and borderline personality disorders (Goldstein et al., 2009). Moreover, abnormalities in several frontotemporo-limbic regions are implicated in both psychopathy and other psychiatric disorders (Benetti et al., 2010; Brunner et al., 2010; Soloff, Nutche, Goradia, & Diwadkar, 2008; Zou et al., 2010). Clearly, much more research is needed to determine the ability of classification algorithms to discriminate among different conditions, especially given the fact that patients may show multiaxial patterns of psychopathology. Future studies should investigate how the anatomical distributions of cortical anomalies relate to the diagnostic accuracies in discriminating among disorders, not only in offender and patient samples but also in various other settings, such as the corporate world.

Another caveat when interpreting the anatomical results emerging from pattern classification, as well as from other anatomical and functional imaging data, relates to the fact that the pathogenesis of psychopathy are multifactorial, emerging from a range of neurological, genetic, and environmental ingredients (Weber, Habel, Amunts, & Schneider, 2008). Combining data from genetic biomarkers, neuropsychology, and multimodal imaging may be extremely helpful in furthering our understanding of complex disorders such as psychopathy. It is likely that classifiers using these different types of data will prove to be even more useful and powerful than imaging data alone.

In summary, the results of the current study suggest that it is possible to discriminate patients with psychopathy from healthy controls, with relatively high sensitivity and specificity rates, solely by measuring the gray matter MRI features. Future studies may extend the present approach by employing multivariate approaches to predict the psychopathy scores in larger patient samples. In addition, pattern classification of functional MRI experiments addressing moral judgments and feelings in psychopathy (Glenn et al., 2009; Moll et al., 2002, 2007; Veit et al., 2010) may reveal how functional impairments of specific frontotemporo-limbic networks can facilitate the identification of individuals with this severe disorder.

REFERENCES

American Psychiatric Association (1994). *Diagnostic and statistical manual of mental disorders* (4th ed.). Washington, DC: Author.

Benetti, S., McCrory, E., Arulanantham, S., De Sanctis, T., McGuire, P., & Mechelli, A. (2010). Attachment style, affective loss and gray matter volume: A voxel-based morphometry study. *Human Brain Mapping, 1*(10), 1482–1489.

Benjamini, Y., & Hochberg, Y. (1995). Controlling the false discovery rate: A practical and powerful approach to multiple testing. *Journal of the Royal Statistical Society. Series B (Methodological), 57*, 289–300.

Bertisch, H., Li, D., Hoptman, M. J., & Delisi, L. E. (2010). Heritability estimates for cognitive factors and brain white matter integrity as markers of schizophrenia. *American Journal of Medical Genetics. Part B, Neuropsychiatric Genetics, 153B*(4), 885–894.

Bezdjian, S., Raine, A., Baker, L. A., & Lynam, D. R. (2011). Psychopathic personality in children: Genetic and environmental contributions. *Psychological Medicine, 41*, 589–600.

Blair, R. J. R., Peschardt, K. S., Budhani, S., Mitchell, D. G. V., & Pine, D. S. (2006). The development of psychopathy. *Journal of Child Psychology and Psychiatry, 47*, 262–276.

Bradley, D., Whelan, R., Walsh, R., Reilly, R. B., Hutchinson, S., Molloy, F., et al. (2009). Temporal discrimination threshold: VBM evidence for an endophenotype in adult onset primary torsion dystonia. *Brain, 132*(9), 2327–2335.

Brunner, R., Henze, R., Parzer, P., Kramer, J., Feigl, N., Lutz, K., et al. (2010). Reduced prefrontal and orbitofrontal gray matter in female adolescents with borderline personality disorder: Is it disorder specific? *NeuroImage, 49*, 114–120.

Camchong, J., Lim, K. O., Sponheim, S. R., & Macdonald, A. W. (2009). Frontal white matter integrity as an endophenotype for schizophrenia: Diffusion tensor imaging in monozygotic twins and patients' nonpsychotic relatives. *Frontiers in Human Neuroscience, 3*, 35.

Cooke, D. J., Michie, C., Hart, S. D., & Hare, R. D. (1999). Evaluating the screening version of the Hare Psychopathy Checklist–Revised (PCL:SV): An item response theory analysis. *Psychological Assessment, 11*(1), 3–13.

Davatzikos, C., Bhatt, P., Shaw, L. M., Batmanghelich, K. N., & Trojanowski, J. Q. (2010). Prediction of MCI to AD conversion, via MRI, CSF biomarkers, and pattern classification. *Neurobiology of Aging*. Advance online publication. DOI: 10.1016/j.neurobiolaging.2010.05.023

De Brito, S. A., Mechelli, A., Wilke, M., Laurens, K. R., Jones, A. P., Barker, G. J., et al. (2009). Size matters: Increased grey matter in boys with conduct problems and callous-unemotional traits. *Brain, 132*(4), 843–852.

Decety, J., Chaminade, T., Grèzes, J., & Meltzoff, A. N. (2002). A PET exploration of the neural mechanisms involved in reciprocal imitation. *NeuroImage, 15*(1), 265–272.

de Gelder, B. (2006). Towards the neurobiology of emotional body language. *Nature Reviews Neuroscience, 7*(3), 242–249.

de Oliveira-Souza, R., Ignácio, F. A., Moll, J., & Hare, R. D. (2008a). Psychopathy in a civil psychiatric outpatient sample. *Criminal Justice and Behavior, 35*(4), 427–437.

de Oliveira-Souza, R., Hare, R.D., Bramati, I. E., Garrido, G. J., Ignácio, F. A., Tovar-Moll, F., et al. (2008b). Psychopathy as a disorder of the moral brain: Fronto-temporo-limbic gray matter reductions demonstrated by voxel-based morphometry. *NeuroImage, 40*(3), 1202–1213.

Devijver, P. A., & Kittler, J. (1982). *Pattern classification: A statistical approach*. Englewood Cliffs, NJ: Prentice-Hall.

Ecker, C., Rocha-Rego, V., Johnston, P., Mourão-Miranda, J., Marquand, A., Daly, E. M., et al. (2010). Investigating the predictive value of whole-brain structural MR scans in autism: A pattern classification approach. *NeuroImage, 49*, 44–56.

Emblem, K. E., Zoellner, F. G., Tennoe, B., Nedregaard, B., Nome, T., Due-Tonnessen, P., et al. (2008). Predictive modeling in glioma grading from MR perfusion images using support vector machines. *Magnetic Resonance in Medicine, 60*, 945–952.

Fan, Y., Shen, D., & Davatzikos, C. (2005). Classification of structural images via high-dimensional image warping, robust feature extraction, and SVM. *Medical Image Computing and Computer-Assisted Intervention, 8*, 1–8.

Friston, K. J., Holmes, A. P., Worsley, K. J., Poline, J. B., Frith, C., & Frackowiak, R. S. J. (1995). Statistical parametric maps in functional imaging: A general linear approach. *Human Brain Mapping, 2*, 189–210.

Fukunaga, K. (1990). *Introduction to statistical pattern recognition* (2nd ed.). Boston, MA: Academic Press.

Fullam, R. S., McKie, S., & Dolan, M. C. (2009). Psychopathic traits and deception: Functional magnetic resonance imaging study. *British Journal of Psychiatry, 194*, 229–235.

Gao, Y., Glenn, A. L., Schug, R. A., Yang, Y., & Raine, A. (2009). The neurobiology of psychopathy: A neurodevelopmental perspective. *Canadian Journal of Psychiatry (Revue Canadienne de Psychiatrie), 54*(12), 813–823.

Gao, Y., Raine, A., Chan, F., Venables, P. H., & Mednick, S. A. (2010). Early maternal and paternal bonding, childhood physical abuse and adult psychopathic personality. *Psychological Medicine, 40*(6), 1007–1016.

Georgopoulos, A. P., Tan, H. M., Lewis, S. M., Leuthold, A. C., Winskowski, A. M., Lynch, J. K., et al. (2010). The synchronous neural interactions test as a functional neuromarker for post-traumatic stress disorder (PTSD): A robust classification method based on the bootstrap. *Journal of Neural Engineering*, 7, 16011.

Gerardin, E., Chételat, G., Chupin, M., Cuingnet, R., Desgranges, B., Kim, H. S., et al. (2009). Alzheimer's disease neuroimaging initiative. Multidimensional classification of hippocampal shape features discriminates Alzheimer's disease and mild cognitive impairment from normal aging. *NeuroImage*, 47, 1476–1486.

Glenn, A. L., & Raine, A. (2008). The neurobiology of psychopathy. *Psychiatric Clinics of North America*, 31, 463–475.

Glenn, A. L., Raine, A., & Schug, R. A. (2009). The neural correlates of moral decision-making in psychopathy. *Molecular Psychiatry*, 14, 5–6.

Glenn, A. L., Raine, A., Yaralian, P. S., & Yang, Y. (2010). Increased volume of the striatum in psychopathic individuals. *Biological Psychiatry*, 67, 52–58.

Goldstein, K. E., Hazlett, E. A., New, A. S., Haznedar, M. M., Newmark, R. E., Zelmanova, Y., et al. (2009). Smaller superior temporal gyrus volume specificity in schizotypal personality disorder. *Schizophrenia Research*, 112, 14–23.

Golland, P., Fischl, B., Spiridon, M., Kanwisher, N., Buckner, R. L., Shenton, M. E., et al. (2002). Discriminative analysis for image-based studies. In *Proceedings of the 5th International Conference on Medical Image Computing and Computer-Assisted Intervention*, LNCS 2488, pp. 508–515, Tokyo, Japan, September 25–28.

Golland, P., Grimson, W., Shenton, M., & Kikinis, R. (2000). Small sample size learning for shape analysis of anatomical structures. In *Proceedings of the 3rd International Conference on Medical Image Computing and Computer-Assisted Intervention*, LNCS 1935, pp. 72–82, Pittsburgh, PA, USA, October 11–14.

Good, C. D., Johnsrude, I. S., Ashburner, J., Henson, R. N., Friston, K. J., & Frackowiak, R. S. (2001). A voxel-based morphometric study of ageing in 465 normal adult human brains. *NeuroImage*, 14(1), 21–36.

Grèzes, J., Pichon, S., & de Gelder, B. (2007). Perceiving fear in dynamic body expressions. *NeuroImage*, 35(2), 959–967.

Guy, L. S., & Douglas, K. S. (2006). Examining the utility of the PCL–SV as a screening measure using competing factor models of psychopathy. *Psychological Assessment*, 18(2), 225–230.

Hare, R., & Neumann, C. (2005). Structural models of psychopathy. *Current Psychiatry Reports*, 7(1), 57–64.

Hare, R. D. (2006). Psychopathy: A clinical and forensic overview. *Psychiatric Clinics of North America*, 29(3), 709–724.

Hart, S., Cox, D., & Hare, R. (1995). *Manual for the Psychopathy Checklist: Screening Version (PCL:SV)*. Toronto, Canada: Multi-Health Systems.

Hastie, T., Tibshirani, R., & Friedman, J. H. (2001). *The elements of statistical learning: Data mining, inference, and prediction*. New York, NY: Springer.

Honea, R. A., Meyer-Lindenberg, A., Hobbs, K. B., Pezawas, L., Mattay, V. S., Egan, M. F., et al. (2008). Is gray matter volume an intermediate phenotype for schizophrenia? A voxel-based morphometry study of patients with schizophrenia and their healthy siblings. *Biological Psychiatry*, 63(5), 465–474.

Kalbe, E., Nowak, D. A., Brand, M., Kessler, J., Dafotakis, M., Bangard, C., et al. (2010). Dissociating cognitive from affective theory of mind: A TMS study. *Cortex*, 46(6), 769–780.

Kalbe, E., Schlegel, M., Sack, A. T., Nowak, D. A., Dafotakis, M., Bangard, C., et al. (2010). Dissociating cognitive from affective theory of mind: A TMS study. *Cortex*, 46(6), 769–780.

Kiehl, K. A., Bates, A. T., Laurens, K. R., Hare, R. D., & Liddle, P. F. (2006). Brain potentials implicate temporal lobe abnormalities in criminal psychopaths. *Journal of Abnormal Psychology*, 115, 443–453.

Koutsouleris, N., Gaser, C., Bottlender, R., Davatzikos, C., Decker, P., Jäger, M., et al. (2010). Use of neuroanatomical pattern regression to predict the structural brain dynamics of vulnerability and transition to psychosis. *Schizophrenia Research*, 123(2–3), 175–187.

Lao, Z., Shen, D., Xue, Z., Karacali, B., Resnick, S., & Davatzikos, C. (2004). Morphological classication of brains via high-dimensional shape transformations and machine learning methods. *NeuroImage*, 21, 46–57.

Leibenluft, E., Gobbini, M. I., Harrison, T., & Haxby, J. V. (2004). Mothers' neural activation in response to pictures of their children and other children. *Biological Psychiatry*, 56(4), 225–232.

Lukic, A. S., Wernick, M. N., & Strother, S. C. (2002). An evaluation of methods for detecting brain activations from functional neuroimages. *Artificial Intelligence in Medicine*, 25, 69–88.

Materna, S., Dicke, P. W., & Thier, P. (2008). The posterior superior temporal sulcus is involved in social communication not specific for the eyes. *Neuropsychologia*, 46(11), 2759–2765.

Meyer-Lindenberg, A., & Weinberger, D. R. (2006). Intermediate phenotypes and genetic mechanisms of psychiatric disorders [Review]. *Nature Reviews Neuroscience*, 7(10), 818–827.

Millon, T., Simonsen, E., Davis, R. D., & Birket-Smith, M. (2002). *Psychopathy: Antisocial, criminal, and violent behavior*. New York, NY: Guilford Press.

Moll, J., de Oliveira-Souza, R., Eslinger, P. J., Bramati, I. E., Mourao-Miranda, J., Andreiuolo, P. A., et al. (2002). The neural correlates of moral sensitivity: A functional magnetic resonance imaging investigation of basic and moral emotions. *Journal of Neuroscience*, 22, 2730–2736.

Moll, J., de Oliveira-Souza, R., Garrido, G. J., Bramati, I. E., Caparelli-Daquer, E. M. A., Paiva, M. L. M. F., et al. (2007). The self as a moral agent: Linking the neural bases of social agency and moral sensitivity. *Social Neuroscience*, 2, 336–352.

Moll, J., Zahn, R., de Oliveira-Souza, R., Krueger, F., & Grafman, J. (2005). The neural basis of human moral cognition. *Nature Reviews Neuroscience*, 6(10), 799–809.

Mourão-Miranda, J., Bokde, A. L.W., Born, C., Hampel, H., & Stetter, S. (2005). Classifying brain states and determining the discriminating activation patterns: Support vector machine on functional MRI data. *NeuroImage*, 28(4), 980–995.

Müller, J. L., Gänssbauer, S., Sommer, M., Döhnel, K., Weber, T., Schmidt-Wilcke, T., et al. (2008). Gray matter changes in right superior temporal gyrus in criminal psychopaths. Evidence from voxel-based morphometry. *Psychiatry Research, 163*, 213–222.

Nardo, D., Högberg, G., Looi, J. C. L., Larsson, S., Hällström, T., & Pagani, M. (2010). Gray matter density in limbic and paralimbic cortices is associated with trauma load and EMDR outcome in PTSD patients. *Journal of Psychiatric Research, 44*(7),477–785.

Rilling, J. K., Glenn, A. L., Jairam, M. R., Pagnoni, G., Goldsmith, D. R., Elfenbein, H. A., et al. (2007). Neural correlates of social cooperation and non-cooperation as a function of psychopathy. *Biological Psychiatry, 61*, 1260–1271.

Sato, J. R., da Graça Morais Martin, M., Fujita, A., Mourão-Miranda, J., Brammer, M. J., & Amaro, E. Jr. (2009). An fMRI normative database for connectivity networks using one-class support vector machines. *Human Brain Mapping, 30*, 1068–1076.

Sato, J. R., Mourão-Miranda, J., Morais Martin Mda, G., Amaro, E. Jr., Morettin, P. A., & Brammer, M. J. (2008a). The impact of functional connectivity changes on support vector machines mapping of fMRI data. *Journal of Neuroscience Methods, 172*, 94–104.

Sato, J. R., Thomaz, C. E., Cardoso, E. F., Fujita, A., Martin Mda, G., & Amaro, E. Jr. (2008b). Hyperplane navigation: A method to set individual scores in fMRI group datasets. *Neuroimage, 42*, 1473–1480.

Schaufelberger, M. S., Duran, F. L. S., Lappin, J. M., Scazufca, M., Amaro, E., Leite, C. C., et al. (2007). Gray matter abnormalities in Brazilians with first-episode psychosis. *British Journal of Psychiatry, Suppl 51*, s117–s122.

Schlaepfer, T. E., Lancaster, E., Heidbreder, R., Strain, E. C., Kosel, M., Fisch, H., et al. (2006). Decreased frontal white-matter volume in chronic substance abuse. *International Journal of Neuropsychopharmacology, 9*, 147–153.

Soderstrom, H., Hultin, L., Tullberg, M., Wikkelso, C., Ekholm, S., & Forsman, A. (2002). Reduced frontotemporal perfusion in psychopathic personality. *Psychiatry Research, 114*(2), 81–94.

Soloff, P., Nutche, J., Goradia, D., & Diwadkar, V. (2008). Structural brain abnormalities in borderline personality disorder: A voxel-based morphometry study. *Psychiatry Research, 164*, 223–236.

Thomaz, C. E., Boardman, J. P., Counsell, S., Hill, D. L. G., Hajnal, J. V., Edwards, A. D., et al. (2007a). A multivariate statistical analysis of the developing human brain in preterm infants. *Image and Vision Computing, 25*(6), 981–994.

Thomaz, C. E., Duran, F. L. S., Busatto, G. F., Gillies, D. F., & Rueckert, D. (2007b). Multivariate statistical differences of MRI samples of the human brain. *Journal of Mathematical Imaging and Vision, 29*(2–3),95–106.

Thomaz, C. E., Kitani, E. C., & Gillies, D. F. (2006). A maximum uncertainty LDA-based approach for limited sample size problems – with application to face recognition. *Journal of the Brazilian Computer Society, 12*(2), 7–18.

Tiihonen, J., Rossi, R., Laakso, M. P., Hodgins, S., Testa, C., Perez, J., et al. (2008). Brain anatomy of persistent violent offenders: More rather than less. *Psychiatry Research, 163*(3), 201–212.

Vapnik, V. N. (1998). *Statistical learning theory.* New York, NY: Wiley.

Veit, R., Lotze, M., Sewing, S., Missenhardt, H., Gaber, T., & Birbaumer, N. (2010). Aberrant social and cerebral responding in a competitive reaction time paradigm in criminal psychopaths. *NeuroImage, 49*, 3365–3372.

Weber, S., Habel, U., Amunts, K., & Schneider, F. (2008). Structural brain abnormalities in psychopaths--a review. *Behavioral Sciences & the Law, 26*, 7–28.

Williams, K. M., Paulhus, D. L., & Hare, R. D. (2007). Capturing the four-factor structure of psychopathy in college students via self-report. *Journal of Personality Assessment, 88*(2), 205–219.

Wilson, S. M., Ogar, J. M., Laluz, V., Growdon, M., Jang, J., Glenn, S., et al. (2009). Automated MRI-based classification of primary progressive aphasia variants. *NeuroImage, 47*, 1558–1567.

Worsley, K. (1995). Estimating the number of peaks in a random field using the Hadwiger characteristic of excursion sets with applications to medical images. *Annals of Statistics, 23*, 640–669.

Zhu, C. Z., Zang, Y. F., Liang, M., Tian, L. X., He, Y., Li, X. B., et al. (2005). Discriminative analysis of brain function at resting-state for attention-deficit/hyperactivity disorder. *Medical Image Computing and Computer-Assisted Intervention, 8*, 468–475.

Zou, K., Deng, W., Li T., Zhang, B., Jiang, L., Huang, C., et al. (2010). Changes of brain morphometry in first-episode, drug-naïve, non-late-life adult patients with major depression: An optimized voxel-based morphometry study. *Biological Psychiatry, 67*, 186–188.

Yang, Y., Raine, A., Lencz, T., Bihrle, S., LaCasse, L., & Colletti, P. (2005). Volume reduction in prefrontal gray matter in unsuccessful criminal psychopaths. *Biological Psychiatry, 57*(10), 1103–1108.

A somatic marker perspective of immoral and corrupt behavior

Mona Sobhani[1] and Antoine Bechara[1,2,3]

[1]Brain and Creativity Institute, University of Southern California, Los Angeles, CA, USA
[2]Department of Psychology, University of Southern California, Los Angeles, CA, USA
[3]Psychiatry Department, McGill University, Montreal, Canada

Individuals who engage in corrupt and immoral behavior are in some ways similar to individuals with psychopathy. Normal people refrain from engaging in such behaviors because they tie together the moral value of society and the risk of punishment when they violate social rules. What is it, then, that allows these immoral individuals to behave in this manner, and in some situations even to prosper? When there is a dysfunction of somatic markers, specific disadvantageous impairments in decision-making arise, as in moral judgment, but, paradoxically, under some circumstances, the damage can cause the patient to make optimal financial investment decisions. Interestingly, individuals with psychopathy, a personality disorder, share many of the same behavioral characteristics seen in VMPFC and amygdala lesion patients, suggesting that defective somatic markers may serve as a neural framework for explaining immoral and corrupt behaviors. While these sociopathic behaviors of sometimes famous and powerful individuals have long been discussed, primarily within the realm of social science and psychology, here we offer a neurocognitive perspective on the possible neural roots of immoral and corrupt behaviors.

Keywords: Somatic marker hypothesis; Psychopathy; Corruption; Immorality; Decision-making.

We begin by giving a definition of corruption. According to *Merriam-Webster*, corruption is an "impairment of integrity, virtue, or moral principle; inducement to wrong by improper or unlawful means (as bribery)" (*Merriam-Webster Online English Dictionary*, 2010). Many people may fit this description; however, psychopaths are one psychiatric group whose social behaviors strongly match this definition. In one definition, a psychopath is "a self-centered, callous, and remorseless person profoundly lacking in empathy" (Hare, 1999, p. 2), and psychopaths are also considered "social predators who charm, manipulate, and ruthlessly plow their way through life . . . completely lacking in conscience and in feelings for others" (Hare, 1999, p. xi). They break the moral code of society through various antisocial acts, mostly for personal gain. How does one explain their behavior?

Though the term "psychopath" has long been used colloquially to describe those whose destructive and immoral behaviors do not fit neatly into the social thread of society, in the scientific literature psychopathy is considered a personality disorder in itself (Hare, 1996). What exactly characterizes one as a psychopath? From his clinical observations and interactions with institutionalized psychopaths, Hervey Cleckley wrote what many consider to be the definitive book on psychopathy, *The Mask of Sanity* (Cleckley, 1982). He became a pioneer of the field when he provided a clinical description of the disorder, including a list of 16 behavioral criteria (e.g., having superficial charm, lack of remorse, poor judgment, failure to follow any life plan, failure to learn by experience) that help characterize a person as a psychopath, and which would be later used by Robert Hare

Correspondence should be addressed to: Mona Sobhani, University of Southern California, Hedco Neuroscience Building, Room B-19, Mail Code 2520, 3641 Watt Way, Los Angeles, CA 90089-2520, USA. E-mail: msobhani@usc.edu

This research was supported by research NIDA grants R01 DA16708 and R01 DA022549 and NINDS grant P01 NS19632.

in making his own Psychopathy Checklist (PCL-R) (Hare, 1991). Cleckley emphasized that although psychopathy should be considered a mental illness, there was no delirium or delusion to be observed, and in fact, there seemed to be a total absence of "a lesion of the intellect" (Cleckley, 1982, p. 122). Within the field of psychopathy, a theoretical distinction between primary and secondary variants has been suggested. Karpman's classic theory is that primary psychopaths are "born" with the core interpersonal and affective features of the disorder, whereas secondary psychopaths develop similar traits in response to such adverse environmental experiences as parental rejection and abuse (Karpman, 1941). Trait anxiety is traditionally used to distinguish between the two subtypes. Many psychopaths can function normally in society, and they have been labeled as successful or "functional psychopaths" (Spencer, 2005, p. D1).

In this paper, we would like to offer a somatic marker perspective on corrupt and immoral behavior, using psychopathy as an example. The field of decision-making neuroscience is well developed, and, in parallel, extensive work on psychopathy has also become well established over the decades. The pioneering work of Adrian Raine (Raine, Lee, Yang, & Colletti, 2010) and James Blair (Blair, 2008), as well as that of Michael Koenigs and Joseph Newman (Koenigs, Kruepke, & Newman, 2010) and J. Moll and colleagues (Moll, Zahn, de Oliveira-Souza, Krueger, & Grafman, 2005), has made the link between psychopathy and abnormalities in the prefrontal cortex, the amygdala, and the septal region, all of which are neural regions implicated in decision-making, moral judgment, and the somatic marker hypothesis (SMH). While the literature on psychopathy (especially primary psychopathy) and its neural correlates is relatively rich, the current perspective capitalizes on this existing literature and expands it in order to offer a potential neural road map to investigate and understand the underpinnings of a commonly encountered but generally overlooked social behavior, namely immoral and corrupt behavior, which has been discussed and described in the literature as secondary psychopathy. Therefore, hard empirical evidence that causally connects brain mechanisms and psychopathic behavior is naturally lacking. However, the gathering of the available information from a variety of decision-making tasks, a careful analysis of the different types of decisions that may be engaged by different tasks, and the known neural correlates for these mechanisms of decisions provide a compelling rationale for the perspective presented here on the use of the somatic marker framework as a neural guide for future understanding of the complicated behavioral and neural

mechanisms associated with psychopathy and moral judgment, and their implications for immoral and corrupt behaviors.

STATEMENT OF THE PROBLEM

One key question relates to whether there is a neural basis for psychopathic behavior, and, more specifically, corrupt behavior. We propose that people do not normally engage in immoral and corrupt behavior, primarily because they tie together the moral value of society and the risk of punishment should they violate social rules (e.g., accepting bribes). However, there are two possible explanations of why some people may engage in corrupt behavior. One is that they have an abnormal ventromedial prefrontal cortex (VMPFC) function. Indeed, there are striking similarities between psychopaths and patients who have lesions of the VMPFC with respect to characteristics that include lack of empathy, irresponsibility, poor decision-making, inappropriate social behavior, failure to plan ahead, and diminished sense of guilt (Koenigs et al., 2010; Krajbich, Adolphs, Tranel, Denburg, & Camerer, 2009) . Studies examining the neural correlates of moral judgment reveal regions that overlap with the same VMPFC areas implicated in lesion patients (Moll et al., 2005). Like VMPFC patients, psychopaths can know and say "the right thing" but do "the wrong thing" (Cleckley, 1982).

Since a psychopath does not show an obvious lesion in the VMPFC, what brings about this putative VMPFC dysfunction? Genetic or environmental factors can lead to abnormal wiring of the prefrontal cortex and alterations in its function. For example, early-life stressors are known to cause alterations in the wiring of the frontostriatal neural circuitry, thereby leading to behaviors associated with frontal lobe dysfunction (Braun, Lange, Metzger, & Poeggel, 2000; Hanson et al., 2010). Additionally, variations in the serotonin-transporter-linked polymorphic region (5-HTTLPR) exert a great influence on decision-making under uncertainty (He et al., 2010), as well as affecting functional connectivity between the VMPFC and the amygdala (Heinz et al., 2005). In the disorder of psychopathy, these kinds of neurobiological aberrations may have varying degrees of abnormality that may lead to a wide range of psychopathic behaviors, from crimes and violence at one extreme (primary psychopathy) to antisocial behavior at the other extreme (secondary psychopathy). However, another possibility in both primary (e.g., belonging to a gang and shooting innocent people) and secondary (e.g., merely corrupt behavior) psychopathy is a faulty learning environment (i.e., learning that killing or corrupt behavior is good), which does not necessarily reflect an underlying brain problem as a precursor. The importance of distinguishing between psychopathic behaviors rooted in abnormalities in VMPFC function (and somatic marker activation) and psychopathic behaviors that are willful and controlled because they are learned in certain environmental contexts, is very crucial. The reason is that in the former case, neurological evidence suggests that individuals with decision-making impairments resulting from VMPFC damage never learn from repeated mistakes (Bechara & Damasio, 2005), and especially when the damage is suffered early in life (i.e., the earlier the damage, the worse the behavioral outcome in adulthood) (Anderson, Bechara, Damasio, Tranel, & Damasio, 1999). In contrast, in the latter case, the individuals have normal brains, and they are likely to adjust their behavior when the social and learning contingencies are changed, such as increasing the risk of negative consequences of their "corrupt" actions. Hence, our primary objectives in this paper are to show that damaged VMPFC leads (1) to impaired judgment and decision-making, and failure to learn from repeated mistakes, despite high intellect and explicit knowledge of the consequences of their decisions; (2) under certain circumstances, and paradoxically, to higher risk taking that results in making optimal financial investment decisions; and (3) to impairment in moral judgment. All these behaviors are characteristic of individuals with psychopathic traits, including those who engage in corrupt and immoral conducts. Although the neural circuitry underlying moral behavior has been explored previously by Moll and colleagues (Moll, de Oliveira-Souza, Bramati, & Grafman, 2002; Moll, de Oliveira-Souza, Eslinger et al., 2002), our perspective relies further on our understanding of the neural basis of these behaviors in neurological patients (e.g., those with VMPFC damage). Taking all this together, we propose a perspective on the neural basis of corrupt and immoral behavior, using the SMH (Bechara & Damasio, 2005; Damasio, 1994) as a theoretical guide.

IMPAIRED JUDGMENT AND DECISION-MAKING AFTER VMPFC DAMAGE

The SMH: overview

One of the first and most famous cases of the so-called frontal lobe syndrome was that of the patient Phineas Gage, described by Harlow (1848, 1868). Interestingly, the case of Phineas Gage and similar cases that were described after him received little attention for many years. The revival of interest in this case, and in various aspects of the frontal lobe syndrome, came from the patient described by Eslinger and Damasio (1985). Over the years, we have studied numerous patients with VMPFC lesions. Such patients develop severe impairments in personal and social decision-making, in spite of otherwise largely preserved intellectual abilities. These patients were intelligent and creative before their brain damage. After the damage, the actions they elect to pursue, often lead to losses of diverse kinds, such as financial losses, losses in social standing, and losses of family and friends. The choices they make are no longer advantageous, and are remarkably different from the kinds of choices they were known to make before their brain damage. These patients often decide against their best interests. They are unable to learn from previous mistakes, as reflected by repeated engagement in decisions that lead to negative consequences. In striking contrast to this real-life decision-making impairment, the patients perform normally in most laboratory tests of problem solving. Their intellect remains normal, as measured by conventional clinical neuropsychological tests.

Damasio proposed the SMH (Damasio, 1994), which posits that the neural basis of the decision-making impairment characteristic of patients with VMPFC damage is defective activation of somatic states (emotional signals) that attach value to given options and scenarios. These emotional signals (which are perceived by specific neural regions in the brain) function as covert, or overt, biases for guiding decisions. Deprived of these emotional signals, patients may resort to deciding based on the immediate reward of an option. The failure to enact somatic states, and consequently to decide advantageously, results from dysfunction in a neural system in which the VMPFC is a critical component. However, the VMPFC is not the only region. Other neural regions, including the amygdala, insula and somatosensory cortices, dorsolateral prefrontal cortex, and hippocampus, are also components of this same neural system, although the different regions may provide different contributions to the overall process of decision-making (Bechara & Damasio, 2005) (Figure 1). Thus, the somatic marker framework articulated elsewhere (e.g., Bechara & Damasio, 2005) does not simply provide a list of brain structures involved in somatic marker activation; rather, the framework provides a detailed account of how different neural systems play different roles in the overall process of decision-making.

More specifically, the amygdala, as well as the VMPFC, is a critical structure in triggering somatic

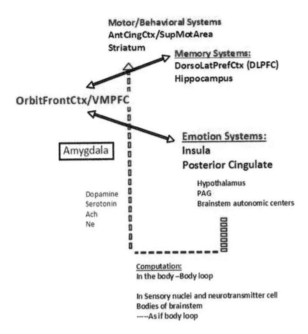

Figure 1. A schematic of all the brain regions involved in decision-making according to the somatic marker hypothesis.

states, but the amygdala seems more important in triggering somatic states from emotional events that occur in the environment (i.e., primary inducers), whereas the VMPFC region seems more important in triggering somatic states from memories, knowledge, and cognition (i.e., secondary inducers) (Bechara & Damasio, 2005). Decision-making is a complex process that relies on the integrity of at least two sets of neural systems: (1) one set is important for memory (e.g., the hippocampus), and especially working memory (e.g., the dorsolateral prefrontal cortex), in order to access knowledge and information used during the deliberation of a decision; (2) another set is important in triggering emotional responses. This set includes effector structures such as the hypothalamus and autonomic brainstem nuclei, which produce changes in internal milieu and visceral structures, along with other effector structures, such as the ventral striatum, periacqueductal gray, and other brainstem nuclei, which produce changes in facial expression and specific approach or withdrawal behaviors. It also includes cortical structures that receive afferent input from the viscera and internal milieu, such as the insular cortex, posterior cingulate gyrus, retrosplenial cortex, and cuneus region.

During the process of pondering decisions, the immediate prospects of an option may be driven by more subcortical mechanisms (e.g., via the amygdala) that do not require a prefrontal cortex. However, weighing the future consequences requires a prefrontal cortex for triggering somatic responses about possible future consequences. Specifically, in pondering the

decision, the immediate and future prospects of an option may trigger numerous somatic responses that conflict with each other (i.e., positive and negative somatic responses). The end result, though, is that an overall positive or negative signal emerges (a "go" or "stop" signal). There is a debate as to where this overall somatic state may be computed. We have argued that this computation occurs in the body proper (via the so-called body loop), but it can also occur in the brain itself, in areas that represent "body" states such as the dorsal tegmentum of the midbrain, or areas such as the insula and posterior cingulate (via the so-called as-if-body loop). The controversy surrounding the hypothesis has largely been in relation to the body loop, with certain investigators arguing that decision-making is not necessarily dependent on "somatic markers" expressed in the body (e.g., Dunn, Dalgleish, & Lawrence, 2006; Maia & McClelland, 2004); this, we admit, is the weakest link of the theory—but, for counterarguments, see also Bechara, Damasio, Tranel, and Damasio (2005) and Persaud, McLeod, and Cowey (2007). Irrespective of whether this computation occurs in the body itself, or within the brain, we have proposed that the emergence of this overall somatic state is consistent with the principle of natural selection. In other words, numerous and conflicting signals may be triggered simultaneously, but stronger ones gain selective advantage over weaker ones, until a winner-takes-all emerges, a positive or negative somatic state that consequently bias the decision one way or the other (Bechara et al., 2005).

In order for somatic signals to influence cognition and behavior, they must act on the appropriate neural systems. One target for somatic state action is the striatum. A large number of channels convey body information (that is, somatic signals) to the central nervous system (e.g., spinal cord, vagus nerve, and humoral signals). Evidence suggests that the vagal route is especially critical for relaying somatic signals (Martin, Denburg, Tranel, Granner, & Bechara, 2004). Further, it was proposed that the next link in this body–brain channel involves neurotransmitter systems (Bechara & Damasio, 2005; Damasio, 1996). Indeed, the cell bodies of the neurotransmitters dopamine, serotonin, noradrenaline, and acetylcholine are located in the brainstem; the axon terminals of these neurotransmitter neurons synapse on cells and/or terminals all over the cortex and striatum (Blessing, 1997). When somatic state signals are transmitted to the cell bodies of dopamine or serotonin neurons, for example, the signaling influences the pattern of dopamine or serotonin release at the terminals. In turn, changes in dopamine or serotonin release modulate synaptic activities of neurons subserving behavior and cognition within the cortex. Preliminary pharmacological studies indicate that learning to perform advantageously on the Iowa Gambling Task (IGT) is influenced by at least two neurotransmitter systems: dopamine and serotonin (Bechara, 2003; Sevy et al., 2007). This chain of neural mechanisms provides a way for somatic states to exert a biasing effect on decisions. At the cellular level, and more recently the functional neuroimaging level, the pioneering work of Schultz et al. (1997) has emphasized the role of dopamine in reward processing and error prediction (Schultz, Dayan, & Montague, 1997). While the cellular work of Schultz and colleagues focused solely on dopamine, and while the functional neuroimaging work cannot speak directly as to whether dopamine or serotonin is involved, our work and that of several others (Fineberg, 2010) suggest that both dopamine and serotonin contribute to the ability to learn from previous mistakes (e.g., improved learning on the IGT), a behavioral process that has been termed in more recent literature "reward prediction error". Thus, while our work with the IGT clearly demonstrates a reward prediction error curve (i.e., subjects adjust their next response based on the outcome of the previous trial), we did not use the same term. Given the fact that the dopamine mechanism addresses only one specific component of a larger neural network that is important for implementing decisions, it is quite possible that "somatic marker" and "dopamine reward prediction error signal" are two different terms that describe the same behavior.

Empirical tests of the SMH

For many years, VMPFC patients presented a puzzling defect, because it was difficult to explain their disturbance in terms of defects in knowledge pertinent to the situation or deficient general intellectual ability. Although the decision-making impairment was obvious in the real-world behavior life of these patients, there was no effective laboratory probe to detect and measure this impairment. Bechara's development of what became known as the IGT (Bechara, 1994) has enabled researchers, for the first time, to detect the decision-making impairment characteristic of patients with VMPFC lesions and investigate its possible causes. Such work using the IGT has provided the key empirical support for the proposal that somatic markers significantly influence decision-making (Bechara & Damasio, 2005). Why was the IGT successful in detecting the decision-making impairment in VMPFC patients, and why is it important for the study of the neurology of decision-making? Perhaps it was because the IGT mimics real-life decisions so closely. The task is carried out in real time, and it resembles real-world contingencies. It factors reward and punishment (i.e.,

winning and losing money) in such a way that it creates a conflict between an immediate, luring reward and a delayed, probabilistic punishment. Therefore, the task engages the subject in a quest to make advantageous choices. Each choice is full of uncertainty because a precise calculation or prediction of the outcome of a given choice is not possible.

Results of studies using the IGT revealed that the performance profile of patients with VMPFC lesions is comparable to their real-life inability to decide advantageously. This is especially true in personal and social matters, a domain for which in life, as in the task, an exact calculation of the future outcomes is not possible, and choices must be based on hunches and "gut feelings." Further studies addressed the question of whether the behavioral decision-making impairment in VMPFC lesion patients is linked to a failure in somatic (emotional) signaling (Bechara, Tranel, Damasio, & Damasio, 1996). We studied IGT performance of two groups, normal subjects and VMPFC lesion patients, while we recorded their electrodermal activity as skin conductance responses (SCRs), providing an indirect measure of the emotion experienced by the subject. Both normal subjects and VMPFC patients generated SCRs after they had picked a card and were told that they won or lost money. The most important difference, however, was that normal subjects, as they became experienced with the task, began to generate SCRs prior to the selection of any cards; that is, during the time when they were pondering from which pack of cards to choose. These anticipatory SCRs were more pronounced before picking a card from the disadvantageous choices (risky packs) when compared to the advantageous choices (the safe packs). In other words, these anticipatory SCRs were like "gut feelings" that warned the subject against picking from the bad packs. VMPFC patients failed to generate such SCRs before picking a card. This failure to generate anticipatory SCRs before picking cards from the bad packs correlates with their failure to avoid these bad packs and choose advantageously in this task (Figure 2). These results provide strong support for the notion that decision-making is guided by emotional signals ("gut feelings") that are generated in anticipation of future events.

Further experiments revealed that these biasing somatic signals ("gut feelings") do not need to be perceived consciously. We carried out an experiment, similar to the previous one, in which we tested normal subjects and VMPFC patients on the IGT while recording their SCRs. However, every time the subject picked 10 cards from the packs, we would stop the game briefly, and ask the subject to declare whatever they knew about what was going on in the game (Bechara, Damasio, Tranel, & Damasio, 1997).

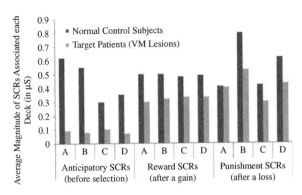

Figure 2. Results of skin conductance responses (SCRs) after selecting a card (Reward or Punishment SCRs) or before selecting a card (Anticipatory SCRs) in the Iowa Gambling Task from normal subjects and a group of patients with bilateral lesions of the VMPFC.

From the answers to the questions, we were able to distinguish four periods as subjects went from the first to the last trial in the task. The first was a pre-punishment period, when subjects sampled the packs, and before they had yet encountered any punishment. The second was a pre-hunch period, when subjects began to encounter punishment, but when asked about what was going on in the game, they had no clue. The third was a hunch period, when subjects began to express a hunch about which packs were riskier, but were not sure. The fourth was a conceptual period, when subjects knew very well the contingencies in the task, and which packs were the good ones, and which packs were the bad ones, and why this was so. When examining the anticipatory SCRs from each period, we found that, as expected, there was no significant activity during the pre-punishment period, because, at this stage, the subjects had not encountered any losses yet. Then there was a substantial rise in anticipatory responses during the pre-hunch period—that is, after encountering some money losses, but still before the subject had any clue about what was going on in the game. This SCR activity was sustained for the remaining periods—that is, during the hunch and then during the conceptual period. When examining the behavior during each period, we found that there was a preference for the high-paying packs (A and B) during the pre-punishment period. Then there was a hint of a shift in the pattern of card selection, away from the bad packs, even in the pre-hunch period. This shift in preference for the good packs became more pronounced during the hunch and conceptual periods. The VMPFC patients, on the other hand, never reported a hunch about which of the packs were good or bad. Furthermore, they never developed anticipatory SCRs, and they continued to choose more cards from the bad packs relative to the good packs. An especially intriguing observation was that not all the normal control subjects were able to figure out the

task, explicitly, in the sense that they did not reach the conceptual period. Only 70% of them were able to do so. Although 30% of controls did not reach the conceptual period, they still performed advantageously. On the other hand, 50% of the VMPFC patients were able to reach the conceptual period and state explicitly which packs were good and which ones were bad and why. Although 50% of the VMPFC patients did reach the conceptual period, they still performed disadvantageously. After the experiment, when these VMPFC patients were confronted with the question, why did you continue to pick from the packs you thought were bad?, these patients would resort to excuses such as "I knew that my luck was going to change and I would win."

These results show that VMPFC patients continue to choose disadvantageously in the gambling task, even after realizing explicitly the consequences of their action. This suggests that the anticipatory SCRs represent unconscious biases derived from prior experiences with reward and punishment. These biases help deter the normal subject from pursuing a course of action that is disadvantageous in the future. This occurs even before subjects become aware of the goodness or badness of the choice they are about to make. Without these biases, the knowledge of what is right and what is wrong may still become available. However, by itself, this knowledge is not sufficient to ensure an advantageous behavior. Therefore, although VMPFC patients may manifest declarative knowledge of what is right and what is wrong, they fail to act accordingly and may "say" the right thing but "do" the wrong thing. Thus, "knowledge" without "emotion/somatic signaling" leads to dissociation between what one knows or says and how one decides to act. This dissociation is not restricted to neurological patients, but it also applies to neuropsychiatric conditions with suspected disorder of the VMPFC or other components of the neural circuitry that process emotion. Psychopathy is one such an example, where the psychopaths can be fully aware of the consequences of their action, but they may fail to inhibit that action.

THE SMH, WORKING MEMORY, AND PSYCHOPATHY

The SMH describes working memory (and other executive processes of working memory such as response inhibition and reversal learning) as a key process in decision-making. Consequently, damage to neural structures that impair working memory, such as the dorsolateral prefrontal cortex (DLPFC), also lead to impaired decision-making. Nonetheless, some criticisms of the theory were made on the basis that deficits in decision-making as measured by the IGT may not be specific to the VMPFC (Manes et al., 2002), or they may be explained by deficits in other processes, such as reversal learning (Fellows & Farah, 2003). However, research has demonstrated that the relationship between decision-making, on the one hand, and working memory or reversal learning, on the other hand, are asymmetrical in nature (Bechara & Damasio, 2005). In other words, working memory and/or reversal learning are not dependent on the intactness of decision-making (that is, subjects can have normal working memory and normal reversal learning in the presence or absence of deficits in decision-making). Some patients with VMPFC lesions who were severely impaired in decision-making on the IGT had superior working memory, and are perfectly normal on simple reversal learning tasks. In contrast, decision-making seems to be influenced by the intactness or impairment of working memory and/or reversal learning. That is, decision-making is worse in the presence of abnormal working memory and/or poor reversal learning. Patients with right DLPFC lesions and severe working memory impairments showed low normal results on the IGT (Bechara, Damasio, Tranel, & Anderson, 1998). Patients with damage to the more posterior sector of the VMPFC (which includes the basal forebrain), such as the patients who were included in the study by Fellows and Farah (2003), showed impairments on reversal learning tasks, but similar patients with similar lesions also showed poor performance on the IGT (Bechara et al., 1998). Consistent with these neurological findings, psychopaths have been shown to have impaired response reversal (Budhani, Richell, & Blair, 2006). Therefore, perhaps the dysfunction in the VMPFC in these individuals extends to include this posterior region involved in response reversal, in addition to the somatic marker impairments, which may contribute to their disadvantageous decision-making. This possibility is corroborated by the work of Raine and colleagues, which confirms that the abnormalities in psychopaths lie far more posteriorly to include the septal nuclei (Raine et al., 2010).

FRONTAL LOBE DYSFUNCTION AS AN ADVANTAGE IN CERTAIN CIRCUMSTANCES: RELEVANCE TO PSYCHOPATHY

As indicated earlier, one of the peculiar aspects of psychopathy is that some individuals, the functional psychopaths specifically, can actually excel in certain domains of their life, such as the financial markets. For example, while people tend to be risk averse in a losing financial market, individuals who lack the emotion

(somatic) signal that fears risk (or at least individuals who can control their emotions) may fare better in an environment where risk taking is the rational thing to do. One study in neurological patients supported this very point (Shiv, Loewenstein, Bechara, Damasio, & Damasio, 2005), as we investigated how normal participants, patients with stable focal lesions in brain regions related to emotion (target patients), and patients with stable focal lesions in brain regions unrelated to emotion, such as orbital/VMPFC, insula, and amygdala (patient controls), made 20 rounds of investment decisions. We used a "risky decision-making task" closely modeled on a paradigm developed in previous economic research to demonstrate "myopic loss aversion." In each round, participants decide to invest $1 or not to invest. A coin is flipped, and they win $2.50 for heads and lose $1 for tails. It is clear that the most rational strategy is to keep investing in this task. The most intriguing results of this study were that target patients made more advantageous decisions and ultimately earned more money from their investments than the normal controls and patient controls. When normal controls and patient controls either won or lost money on an investment round, they adopted a conservative strategy and became more reluctant to invest on the subsequent round, suggesting that they were more affected than target patients by the outcomes of decisions made in the previous rounds (Shiv et al., 2005). This is an example of how the lack of a brain mechanism to trigger emotions (or somatic markers) can be advantageous in this particular context. Remember that, on the whole, VMPFC damage leads to impaired and disadvantageous decisions, but there are specific circumstances where this deficiency can be helpful. It may be these particular contexts that allow some psychopaths, not being hindered by emotional signals, to perform superiorly in certain financial situations and to lead successful lives (albeit this successful life can suddenly end in disaster, as in the American stock broker Bernie Madoff's case).

IMPAIRMENT IN MORAL JUDGMENT

Several studies have shown that VMPFC damage is associated with impairments in making moral judgments. For instance, patients with frontotemporal dementia, a disorder characterized by abnormal social behavior and potential sociopathy, have been shown to be impaired in making emotional moral judgments (Mendez, Anderson, & Shapira, 2005) and that these impairments may be related to impaired affective theory of mind (ToM), the ability to attribute mental states (Gleichgerrcht, Torralva, Roca, Pose, & Manes, 2011). This overlaps with the suggestion that the VMPFC is involved in empathy, specifically when using abstract visual information about another's affective state (Lamm, Decety, & Singer, 2011). Another example of moral judgment impairment is that of VMPFC patients who exhibit abnormal judgments when judging moral situations, using a more utilitarian approach than control subjects (Koenigs et al., 2007). As a result, they are more likely than control subjects and other brain-damaged patients to approve of harmful actions in situations they deem appropriate or reasonable (Young et al., 2010). Consequently, it was surmised that the VMPFC adds an emotional component to the decision-making process involved in moral judgments. When this component is absent, the person is left making a more pragmatic decision based on the facts of the situation, with a special emphasis on the outcome of the situation and less so on the inferred or abstract events (intentions) that came before it. To test this, VMPFC patients, other brain-damaged patients, and normal participants were given 24 scenarios where the task was to judge harmful intent (Young et al., 2010). The study used a 2 × 2 design: (1) the protagonist either intended to cause harm to another person (negative intent) or intended to cause no harm (neutral intent), and (2) the protagonist either caused harm to another person (negative outcome) or caused no harm (neutral outcome) (Young et al., 2010). The results were as predicted. Patients with VMPFC lesions judged attempted harms, including attempted murder, as more morally permissible, relative to controls. VMPFC patients showed neglect of negative intentions in moral judgment and instead focused on the action's neutral outcome. It may be that for an individual to judge an attempted harm as immoral, and therefore forbidden, the intent behind the attempted harm must elicit an emotional response (Valdesolo & DeSteno, 2006). It appears that VMPFC patients do not enjoy the benefit of this guiding emotional response and, therefore, instead focus on the outcome of a situation rather than the intent (Koenigs et al., 2007). Further, a review investigating the role of emotion in morality concluded that emotion is involved in moral judgments, particularly those mediated by the VMPFC (Young & Koenigs, 2007), and hence when it is damaged, impairments of this type of judgment arise.

In order to highlight the distinction between some of the behaviors of VMPFC lesion patients and sociopaths, the term "acquired sociopathy" has been previously used to describe these patients (Damasio, Tranel, & Damasio, 1990), thus reflecting the fact that the VMPFC lesion patients may not engage in the extreme immoral or corrupt acts (including criminal ones) that characterize developmental sociopaths or psychopaths. This distinction (acquired vs. developmental) may help explain some of the

ability of VMPFC lesion patients to use their lifetime's worth of social learning to avoid engaging in extreme immoral or corrupt acts. However these behavioral differences do not warrant fundamentally different brain mechanisms that mediate them. Indeed, some of our VMPFC patients did engage in financial decisions that involved unscrupulous people (e.g., (Damasio et al., 1990), but the handicap associated with their stroke (or tumor) may have helped curb the extent of their societal engagement, which could have led to more severe acts of immoral and corrupt behaviors. Furthermore, patients who acquire VMPFC damage earlier in life tend to grow up to commit more severe antisocial behavioral acts (Anderson, Barrash, Bechara, & Tranel, 2006), thus suggesting that if the onset of brain damage were at an earlier age, then the line between acquired and developmental sociopathy would become more blurred.

WHO AMONG US IS THE SOCIOPATH?: A NEUROCOGNITIVE PERSPECTIVE

The neural origin of psychopathy has been long debated (Blair, 2008; Glenn & Raine, 2008; Kiehl, 2006), the amygdala perhaps being implicated in the severe emotional aspects of psychopathy (e.g., Blair, 2008), and the VMPFC in the less severe aspects, albeit that a very posterior structure of this region, namely the septal region, has been linked to primary psychopathy as well (Raine et al., 2010). A recent study even highlights the possible structural differences between successful and unsuccessful psychopaths (based on criminal convictions, not cognitive or social functioning) in the greater prefrontal cortex and the amygdala, with unsuccessful psychopaths having reduced gray matter volume in these regions (Yang, Raine, Colletti, Toga, & Narr, 2010). In addition to the neuroanatomical evidence, behavioral data suggest that psychopaths show deficits in fear conditioning (Birbaumer et al., 2005), fearful facial expression recognition and processing (Blair, Colledge, Murray, & Mitchell, 2001), and augmentation of the startle reflex by visual threat primes (Levenston, Patrick, Bradley, & Lang, 2000), and show less interference by emotional distracters (Mitchell, Richell, Leonard, & Blair, 2006)–all of which are also symptoms of amygdala lesions and dysfunction. As indicated earlier, although the IGT has been used largely to detect impaired decision-making in patients with VMPFC lesions, the task is not specific to this region, and impaired performance can result from damage in other areas of the somatic marker circuitry, including the amygdala (Bechara & Damasio, 2005). However, this lack of specificity arises when other cognitive deficits are present, such as poor memory, and impaired conditioning learning. The poor IGT performance becomes specific to VMPFC damage when, and only when, all other cognitive deficits are ruled out. Thus, when normal individuals perform disadvantageously on the IGT, although this poor performance may implicate the VMPFC, abnormalities of other neural structures cannot be ruled out.

It is intriguing that when we test a large sample of the "normal" population on the IGT, there is a small subgroup that achieves scores that are comparable to those of patients with VMPFC lesions (Figure 3). Why does this small percentage of "normal subjects" perform like VMPFC patients on the IGT? As we indicated earlier, we suspect that poor performance arises from a potentially dysfunctional prefrontal cortex due to genetic and/or environmentally induced reasons. However, the key question is whether individuals with such low IGT scores also show signs of psychopathic behavior.

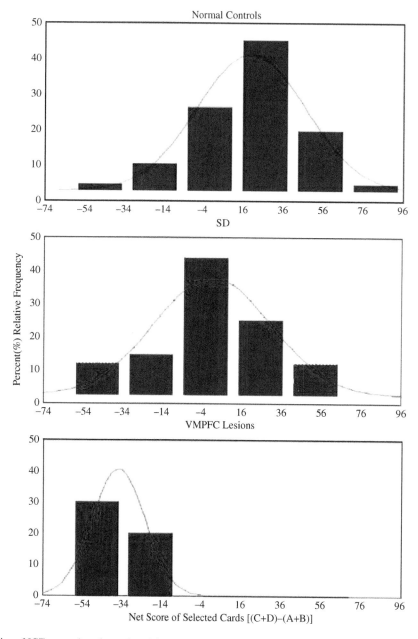

Figure 3. Distribution of IGT scores from 3 samples of the population: normal controls, substance dependent individuals (SD), and patients with VMPFC lesions. Negative scores reflect disadvantageous decisions, while positive scores reflect advantageous decisions. The Y-axis represents the percentage of the sample that achieves a particular score.

How do psychopaths fare on the IGT? Mitchell (2002) found poor performance on the IGT in psychopathic inmates as compared to non-psychopathic inmates (Mitchell, Colledge, Leonard, & Blair, 2002). Another study found that criminal psychopaths with low attention scores exhibited the expected disadvantageous decision-making performance on the IGT (Lösel & Schmucker, 2004). However, in another study that also used criminal psychopaths, Schmitt and colleagues administered the IGT and the Welsh Anxiety Scale (Schmitt, Brinkley, & Newman, 1999). The control group in this study was also a group of prisoners. However, this study was the only one that divided the psychopathy group by trait anxiety into primary (low-anxious) and secondary (high-anxious) psychopathy. They found that all groups performed poorly on the IGT, but levels of anxiety, and not psychopathy, predicted whether the subjects would learn to choose more advantageously (or become more risk-averse) over time. Moreover, this study did not find significant differences between the psychopathy groups and the control group, which also included incarcerated individuals, but who were considered non-psychopaths. The fact that all the psychopathy groups in this study performed poorly on the IGT is consistent with the somatic marker perspective argued here. The fact that the control group also performed poorly on the IGT is inconsistent with the earlier study of non-psychopathic inmates showing relatively more advantageous performance on the IGT (Mitchell et al., 2002). While it is clear that these non-psychopathic prisoners do not meet the criteria of primary psychopathy, it is not clear whether these individuals meet the criteria for secondary psychopathy, especially since the psychopathy instruments used to evaluate prisoners are more sensitive to detecting and measuring signs of primary psychopathy. Indeed, evidence suggests that an incarcerated population differs from a non-incarcerated population in important ways (e.g., risk aversion) (Raine, 1993). These differences in the ways of assessing psychopathy in incarcerated and non-incarcerated populations may explain these apparent inconsistencies. In support, other studies that looked at non-institutionalized populations of psychopaths found, in undergraduates, that high psychopathy traits—based on Levenson's Self-Report Psychopathy Scale: LSRP (Levenson, Kiehl, & Fitzpatrick, 1995)—performed significantly worse on the IGT (Mahmut, Homewood, & Stevenson, 2008). Another study found that boys with psychopathic tendencies also showed impaired performance on the IGT (Blair, Colledge, & Mitchell, 2001). Taken together, these results are consistent with our perspective that psychopathic behaviors (primary and to a milder extent secondary) are associated with

poor performance on the IGT, thus reflecting potential abnormalities in the activation of the somatic marker circuitry.

A more direct link between the VMPFC damage and psychopathy comes from a recent publication by Koenigs and colleagues (2010), who studied the responses of primary and secondary psychopathic prisoners on two different economic decision-making tasks (the Ultimatum Game and the Dictator Game), which were then analyzed and compared to those of VMPFC patients (Koenigs et al., 2010). The psychopathic subtypes were divided by trait anxiety scores and subsequently compared to each other and to criminal non-psychopaths, as well as to the VMPFC lesion patients. Primary psychopathy was associated with significantly lower acceptance rates of unfair Ultimatum offers and lower offer amounts in the Dictator Game when compared to secondary psychopaths and non-psychopaths (Koenigs et al., 2010). In addition, primary psychopaths were quantitatively more similar to the VMPFC lesion patients in their response patterns.

While the Ultimatum and Dictator games were not developed and discussed in the context of the SMH, studies have shown that performance on these tasks do indeed engage the "body loop" of the SMH circuitry (Hewig et al., 2011). Furthermore, patients with lesions of the VMPFC show abnormal performance on these tasks (i.e., higher rejection rates to unfair offers on the Ultimatum game and lower offers on the Dictator game) (Koenigs & Tranel, 2007; Krajbich et al., 2009) that is consistent with the type of "somatic marker" impairment described in these patients. For instance, it has been shown that while they appear apathetic in many social situations, their tendency to express anger and violent behavior is relatively more spared (Bechara, Damasio, & Damasio, 2003; Bechara, Tranel, & Damasio, 2002), thus explaining their intense reaction to "injustice" as reflected by their Ultimatum Game behavior. The recent study by Koenigs et al. (2010) provides additional support for a link between performance in these games and psychopathy. This supports further our perspective regarding a common neural link between psychopathy and the neural circuitry for somatic marker activation.

Another line of evidence that may link the poor performance of psychopaths on the Ultimatum and Dictator games to poor performance on the IGT and the somatic marker framework is the following: The IGT may be considered a measure of decisions under ambiguity (i.e., the outcome of the choice is completely unknown), whereas the Ultimatum and Dictator games, in which the contingencies are already known to the participant, may be considered as measures of decisions under risk. Lesion studies that examined the neural substrates underlying decisions under ambiguity versus risk (Weller, Levin, Shiv, & Bechara, 2007; Xue et al., 2009) did not reveal dissociations or fundamental differences, although a few functional neuroimaging studies have suggested partially separate neural substrates (Hsu, Bhatt, Adolphs, Tranel, & Camerer, 2005; Huettel, Stowe, Gordon, Warner, & Platt, 2006). Taken together, there is a considerable overlap in the neural circuits that subserve the different types of decision-making (ambiguity vs. risk) that are taxed by these different tasks.

CONCLUSION

The media has a penchant for sensational stories, and therefore when the general public hears the word "psychopath" it is usually the image of a killer or convict that comes to mind, though these are social stereotypes. Psychopaths, especially the functional or secondary type, can be successful entrepreneurs, politicians, and CEOs, or hold other respectable positions (Hare, 1999). Many psychopaths can function seemingly normally in society if they do not have official criminal records. These individuals commit crimes of another nature by using, manipulating, and hurting the people around them in order to enrich themselves. In the workforce, they are perfidious employees and untrustworthy businessmen who victimize the people around them (Hare, 1999). Perhaps it is time to use a neuroscientific perspective to revisit the underlying brain factors that lead to corruption and psychopathic behaviors, especially the non-criminal type, the "functional" one that is a part of our social realm.

REFERENCES

Anderson, S. W., Barrash, J., Bechara, A., & Tranel, D. (2006). Impairments of emotion and real-world complex behavior following childhood- or adult-onset damage to ventromedial prefrontal cortex. *Journal of International Neuropsychological Society, 12*, 224–235.

Anderson, S. W., Bechara, A., Damasio, H., Tranel, D., & Damasio, A. R. (1999). Impairment of social and moral behavior related to early damage in human prefrontal cortex. *Nature Neuroscience, 2*(11), 1032–1037.

Bechara, A. (2003). Risky business: Emotion, decision-making, and addiction. *Journal of Gambling Studies, 19*(1), 23–51.

Bechara, A., & Damasio, A. R. (2005). The somatic marker hypothesis: A neural theory of economic decision. *Games and Economic Behavior, 52*, 336–372.

Bechara, A., Damasio, H., & Damasio, A. R. (2003). Role of the amygdala in decision-making. *Annals of the New York Academy of Sciences, 985*, 356–369.

Bechara, A., Damasio, H., Tranel, D., & Anderson, S. W. (1998). Dissociation of working memory from decision making within the human prefrontal cortex. *Journal of Neuroscience, 18*(1), 428–437.

Bechara, A., Damasio, H., Tranel, D., & Damasio, A. R. (1997). Deciding advantageously before knowing the advantageous strategy. *Science, 275*(5304), 1293–1295.

Bechara, A., Damasio, H., Tranel, D., & Damasio, A. R. (2005). The Iowa Gambling Task and the somatic marker hypothesis: Some questions and answers. *Trends in Cognitive Sciences, 9*(4), 159–162; discussion 162–154.

Bechara, A., Tranel, D., & Damasio, A. R. (2002). The somatic-marker hypothesis and decision making. In F. Boller, J. Grafman, & G. Rizzolatti (Eds.), *Handbook of neuropsychology*, 2nd ed. (vol. 7, pp. 117–143). Amsterdam, Netherlands: Elsevier.

Bechara, A., Tranel, D., Damasio, H., & Damasio, A. R. (1996). Failure to respond autonomically to anticipated future outcomes following damage to prefrontal cortex. *Cerebral Cortex, 6*(2), 215–225.

Birbaumer, N., Veit, R., Lotze, M., Erb, M., Hermann, C., Grodd, W., et al. (2005). Deficient fear conditioning in psychopathy: A functional magnetic resonance imaging study. *Archives of General Psychiatry, 62*(7), 799–805.

Blair, R. J. (2008). The amygdala and ventromedial prefrontal cortex: Functional contributions and dysfunction in psychopathy. *Philosophical Transactions of the Royal Society of London. Series B: Biological Sciences, 363*(1503), 2557–2565.

Blair, R. J., Colledge, E., & Mitchell, D. G. (2001). Somatic markers and response reversal: Is there orbitofrontal cortex dysfunction in boys with psychopathic tendencies? *Journal of Abnormal Child Psychology, 29*(6), 499–511.

Blair, R. J., Colledge, E., Murray, L., & Mitchell, D. G. (2001). A selective impairment in the processing of sad and fearful expressions in children with psychopathic tendencies. *Journal of Abnormal Child Psychology, 29*(6), 491–498.

Blessing, W. W. (1997). Inadequate frameworks for understanding bodily homeostasis. *Trends in Neurosciences, 20*(6), 235–239.

Blumer, D., & Benson, D. F. (1975). Personality changes with frontal and temporal lobe lesions. In D. F. Benson & D. Blumer (Eds.), *Psychiatric aspects of neurological disease* (pp. 151–170). New York, NY: Grune and Stratton.

Braun, K., Lange, E., Metzger, M., & Poeggel, G. (2000). Maternal separation followed by early social deprivation affects the development of monoaminergic fiber systems in the medial prefrontal cortex of *Octodon degus*. *Neuroscience, 95*(1), 309–318.

Budhani, S., Richell, R. A., & Blair, R. J. (2006). Impaired reversal but intact acquisition: Probabilistic response reversal deficits in adult individuals with psychopathy. *Journal of Abnormal Psychology, 115*(3), 552–558.

Burgess, P. W., & Wood, R. L. (1990). Neuropsychology of behaviour disorders following brain injury. In R. L. Wood (Ed.), *Neurobehavioural sequelae of traumatic brain injury* (pp. 110–133). New York, NY: Taylor and Francis.

Cleckley, H. M. (1982). *The mask of sanity* (Rev. ed.). New York, NY: New American Library; Mosby.

Corruption (2010). In *Merriam-Webster online*. Retrieved from: www.merriam-webster.com/dictionary/corruption.

Damasio, A. R. (1994). *Descartes' error: Emotion, reason, and the human brain.* New York, NY: G. P. Putnam.

Damasio, A. R. (1996). The somatic marker hypothesis and the possible functions of the prefrontal cortex. *Philosophical Transactions of the Royal Society of London. Series B: Biological Sciences, 351*(1346), 1413–1420.

Damasio, A. R., Tranel, D., & Damasio, H. (1990). Individuals with sociopathic behavior caused by frontal damage fail to respond autonomically to social stimuli. *Behavioural Brain Research, 41*(2), 81–94.

Dunn, B. D., Dalgleish, T., & Lawrence, A. D. (2006). The somatic marker hypothesis: A critical evaluation. *Neuroscience & Biobehavioral Reviews, 30*(2), 239–271.

Eslinger, P. J., & Damasio, A. R. (1985). Severe disturbance of higher cognition after bilateral frontal lobe ablation: Patient EVR. *Neurology, 35*(12), 1731–1741.

Fellows, L. K., & Farah, M. J. (2003). Ventromedial frontal cortex mediates affective shifting in humans: Evidence from a reversal learning paradigm. *Brain, 126,* 1830–1837.

Fineberg, N. A., Potenza, M. N., Chamberlain, S. R., Berlin, H. A., Menzies, L., Bechara, A., et al. (2010). Probing compulsive and impulsive behaviors, from animal models to endophenotypes: A narrative review. *Neuropsychopharmacology, 35*(3), 591–604.

Gleichgerrcht, E., Torralva, T., Roca, M., Pose, M., & Manes, F. (2011). The role of social cognition in moral judgment in frontotemporal dementia. *Social Neuroscience, 6*(2), 113–122.

Glenn, A. L., & Raine, A. (2008). The neurobiology of psychopathy. *Psychiatric Clinics of North America, 31*(3), 463–475, vii.

Hanson, J. L., Chung, M. K., Avants, B. B., Shirtcliff, E. A., Gee, J. C., Davidson, R. J., et al. (2010). Early stress is associated with alterations in the orbitofrontal cortex: A tensor-based morphometry investigation of brain structure and behavioral risk. *Journal of Neuroscience, 30*(22), 7466–7472.

Hare, R. D. (1991). *Psychopathy Checklist: Revised.* New York, NY: Multi-Health Systems.

Hare, R. D. (1996). Psychopathy and antisocial personality disorder: A case of diagnostic confusion. *Psychiatric Times, 13*p. xi; p.2.

Hare, R. D. (1999). *Without conscience: The disturbing world of the psychopaths among us.* New York, NY: Guilford Press.

Harlow, J. M. (1848). Passage of an iron bar through the head. *Boston Medical and Surgical Journal, 39,* 389–393.

Harlow, J. M. (1868). Recovery from the passage of an iron bar through the head. *Publications of the Massachusetts Medical Society, 2,* 327–347.

He, Q., Xue, G., Chen, C., Lu, Z., Dong, Q., Lei, X., et al. (2010). Serotonin transporter gene-linked polymorphic region (5-HTTLPR) influences decision making under ambiguity and risk in a large Chinese sample. *Neuropharmacology, 59*(6), 518–526.

Heinz, A., Braus, D. F., Smolka, M. N., Wrase, J., Puls, I., Hermann, D., et al. (2005). Amygdala-prefrontal coupling depends on a genetic variation of the serotonin transporter. *Nature Neuroscience, 8*(1), 20–21.

Hewig, J., Kretschmer, N., Trippe, R. H., Hecht, H., Coles, M. G., Holroyd, C. B., et al. (2011). Why humans deviate from rational choice. *Psychophysiology, 48*(4), 507–514.

Hsu, M., Bhatt, M., Adolphs, R., Tranel, D., & Camerer, C. F. (2005). Neural systems responding to degrees of uncertainty in human decision-making. *Science, 310*(5754), 1680–1683.

Huettel, S. A., Stowe, C. J., Gordon, E. M., Warner, B. T., & Platt, M. L. (2006). Neural signatures of economic preferences for risk and ambiguity. *Neuron, 49*(5), 765–775.

Karpman, B. (1941). On the need of separating psychopathy into two distinct clinical types: The symptomatic and the idiopathic. *Journal of Criminal Psychopathology, 3,* 112–137.

Kiehl, K. A. (2006). A cognitive neuroscience perspective on psychopathy: Evidence for paralimbic system dysfunction. *Psychiatry Research, 142*(2–3), 107–128.

Koenigs, M., Kruepke, M., & Newman, J. P. (2010). Economic decision-making in psychopathy: A comparison with ventromedial prefrontal lesion patients. *Neuropsychologia, 48*(7), 2198–2204.

Koenigs, M., & Tranel, D. (2007). Irrational economic decision-making after ventromedial prefrontal damage: Evidence from the Ultimatum Game. *Journal of Neuroscience, 27*(4), 951–956.

Koenigs, M., Young, L., Adolphs, R., Tranel, D., Cushman, F., Hauser, M., et al. (2007). Damage to the prefrontal cortex increases utilitarian moral judgements. *Nature, 446*(7138), 908–911.

Krajbich, I., Adolphs, R., Tranel, D., Denburg, N. L., & Camerer, C. F. (2009). Economic games quantify diminished sense of guilt in patients with damage to the prefrontal cortex. *Journal of Neuroscience, 29*(7), 2188–2192.

Lamm, C., Decety, J., & Singer, T. (2011). Meta-analytic evidence for common and distinct neural networks associated with directly experienced pain and empathy for pain. *NeuroImage, 54*(3), 2492–2502.

Levenson, M. R., Kiehl, K. A., & Fitzpatrick, C. M. (1995). Assessing psychopathic attributes in a noninstitutionalized population. *Journal of Personality and Social Psychology, 68*(1), 151–158.

Levenston, G. K., Patrick, C. J., Bradley, M. M., & Lang, P. J. (2000). The psychopath as observer: Emotion and attention in picture processing. *Journal of Abnormal Psychology, 109*(3), 373–385.

Lösel, F., & Schmucker, M. (2004). Psychopathy, risk taking, and attention: A differentiated test of the somatic marker hypothesis. *Journal of Abnormal Psychology, 113*(4), 522–529.

Mahmut, M. K., Homewood, J., & Stevenson, R. J. (2008). The characteristics of non-criminals with high psychopathy traits: Are they similar to criminal psychopaths? *Journal of Research in Personality, 42,* 679–692.

Maia, T. V., & McClelland, J. L. (2004). A reexamination of the evidence for the somatic marker hypothesis: What participants really know in the Iowa Gambling Task. *Proceedings of the National Academy of Sciences of the United States of America, 101*(45), 16075–16080.

Manes, F., Sahakian, B., Clark, L., Rogers, R., Antoun, N., Aitken, M., et al. (2002). Decision-making processes following damage to the prefrontal cortex. *Brain, 125,* 624–639.

Martin, C. O., Denburg, N. L., Tranel, D., Granner, M. A., & Bechara, A. (2004). The effects of vagus nerve stimulation on decision-making. *Cortex, 40*(4–5), 605–612.

Mendez, M. F., Anderson, E., & Shapira, J. S. (2005). An investigation of moral judgement in frontotemporal dementia. *Cognitive and Behavioral Neurology, 18*(4), 193–197.

Mitchell, D. G., Colledge, E., Leonard, A., & Blair, R. J. (2002). Risky decisions and response reversal: Is there evidence of orbitofrontal cortex dysfunction in psychopathic individuals? *Neuropsychologia, 40*(12), 2013–2022.

Mitchell, D. G., Richell, R. A., Leonard, A., & Blair, R. J. (2006). Emotion at the expense of cognition: Psychopathic individuals outperform controls on an operant response task. *Journal of Abnormal Psychology, 115*(3), 559–566.

Moll, J., de Oliveira-Souza, R., Bramati, I. E., & Grafman, J. (2002). Functional networks in emotional moral and nonmoral social judgments. *NeuroImage, 16*(3 Pt 1), 696–703.

Moll, J., de Oliveira-Souza, R., Eslinger, P. J., Bramati, I. E., Mourão-Miranda, J., Andreiuolo, P. A., et al. (2002). The neural correlates of moral sensitivity: A functional magnetic resonance imaging investigation of basic and moral emotions. *Journal of Neuroscience, 22*(7), 2730–2736.

Moll, J., Zahn, R., de Oliveira-Souza, R., Krueger, F., & Grafman, J. (2005). Opinion: The neural basis of human moral cognition. *Nature Reviews Neuroscience, 6*(10), 799–1809.

Persaud, N., McLeod, P., & Cowey, A. (2007). Post-decision wagering objectively measures awareness. *Nature Neuroscience, 10*(2), 257–261.

Raine, A. (1993). *The psychopathology of crime*. New York, NY: Academic Press.

Raine, A., Lee, L., Yang, Y., & Colletti, P. (2010). Neurodevelopmental marker for limbic maldevelopment in antisocial personality disorder and psychopathy. *British Journal of Psychiatry, 197*(3), 186–192.

Schmitt, W. A., Brinkley, C. A., & Newman, J. P. (1999). Testing Damasio's somatic marker hypothesis with psychopathic individuals: Risk takers or risk averse? *Journal of Abnormal Psychology, 108*(3), 538–543.

Schultz, W., Dayan, P., & Montague, P. R. (1997). A neural substrate of prediction and reward. *Science, 275*(5306), 1593–1599.

Sevy, S., Burdick, K. E., Visweswaraiah, H., Abdelmessih, S., Lukin, M., Yechiam, E., et al. (2007). Iowa Gambling Task in schizophrenia: A review and new data in patients with schizophrenia and co-occurring cannabis use disorders. *Schizophrenia Research, 92*(1–3), 74–84.

Shiv, B., Loewenstein, G., Bechara, A., Damasio, H., & Damasio, A. R. (2005). Investment behavior and the negative side of emotion. *Psychological Science, 16*(6), 435–439.

Spencer, J. (2005). Lessons from the brain-damaged investor. Unusual study explores links between emotion and results; 'neuroeconomics' on Wall Street. *Wall Street Journal,* Eastern ed., 21 July, p. D1.

Valdesolo, P., & DeSteno, D. (2006). Manipulations of emotional context shape moral judgment. *Psychological Science, 17*(6), 476–477.

Weller, J. A., Levin, I. P., Shiv, B., & Bechara, A. (2007). Neural correlates of adaptive decision making for risky gains and losses. *Psychological Science, 18*(11), 958–964.

Xue, G., Lu, Z., Levin, I. P., Weller, J. A., Li, X., & Bechara, A. (2009). Functional dissociations of risk and reward processing in the medial prefrontal cortex. *Cerebral Cortex, 19*(5), 1019–1027.

Yang, Y., Raine, A., Colletti, P., Toga, A. W., & Narr, K. L. (2010). Morphological alterations in the prefrontal cortex and the amygdala in unsuccessful psychopaths. *Journal of Abnormal Psychology, 119*(3), 546–554.

Young, L., Bechara, A., Tranel, D., Damasio, H., Hauser, M., & Damasio, A. (2010). Damage to ventromedial prefrontal cortex impairs judgment of harmful intent. *Neuron, 65*(6), 845–851.

Young, L., & Koenigs, M. (2007). Investigating emotion in moral cognition: A review of evidence from functional neuroimaging and neuropsychology. *British Medical Bulletin, 84*, 69–79.

Apathy blunts neural response to money in Parkinson's disease

Andrew D. Lawrence[1], Ines K. Goerendt[2], and David J. Brooks[2]

[1]School of Psychology, Cardiff University, Cardiff, UK
[2]Centre for Neuroscience, Division of Experimental Medicine and MRC Clinical Sciences Centre, Faculty of Medicine, Hammersmith Hospital, Imperial College, London, UK

Apathy, defined as a primary deficit in motivation and manifested by the simultaneous diminution in the cognitive and emotional concomitants of goal-directed behavior, is a common and debilitating non-motor symptom of Parkinson's disease (PD). Despite the high prevalence and clinical significance of apathy, little is known about its pathophysiology, and in particular how apathy relates to alterations in the neural circuitry underpinning the cognitive and emotional components of goal-directed behavior. Here, we examined the neural coding of reward cues in patients with PD, with or without clinically significant levels of apathy, during performance of a spatial search task during $H_2{}^{15}O$ PET (positron emission tomography) functional neuroimaging. By manipulating search outcome (money reward vs valueless token), while keeping the actions of the participants constant, we examined the influence of apathy on the neural coding of money reward cues. We found that apathy was associated with a blunted response to money in the ventromedial prefrontal cortex, amygdala, striatum, and midbrain, all part of a distributed neural circuit integral to the representation of the reward value of stimuli and actions, and the influence of reward cues on behavior. Disruption of this circuitry potentially underpins the expression of the various manifestations of apathy in PD, including reduced cognitive, emotional, and behavioral facets of goal-directed behavior.

Keywords: Amygdala; Depression; Dopamine; Incentive salience; Ventromedial prefrontal cortex.

Once considered a positive quality (*apatheia* or freedom from emotion) (Sorabji, 2000), by the early nineteenth century apathy had come to be used in its current sense of blunted motivation (Berrios & Grli, 2007). In the modern clinical literature, debate focuses on a putative apathy syndrome, defined by Marin (1991) as a primary deficit in motivation, manifested by the "simultaneous diminution in the cognitive and emotional concomitants of goal-directed behaviour." Apathy is also a cardinal diagnostic symptom of major depression (Marin, Firinciogullari, & Biedrzycki, 1993).

Apathy, as characterized by a primary diminution of goal-directed behavior (Brown & Pluck, 2000; Levy & Dubois, 2006), is increasingly recognized as a significant clinical problem in patients with neurological disorders including Alzheimer's disease, stroke, frontotemporal dementia (FTD), and Parkinson's disease (PD). A high prevalence of apathy in PD has been documented, with reported estimates ranging from 17% to 70% depending on population characteristics and assessment procedures (Drijgers, Dujardin, Reijnders, Defebvre, & Leentjens, 2010; Kirsch-Darrow, Fernandez, Marsiske, Okun, & Bowers, 2006; Pluck & Brown, 2002; Starkstein et al., 2009). Apathy

Correspondence should be addressed to: Andrew D. Lawrence, School of Psychology, Cardiff University, Cardiff CF10 3AT, UK. E-mail: LawrenceAD@Cardiff.ac.uk

We thank all volunteers for their participation and the clinical, chemistry, radiography, and instrumentation staff of the Cyclotron Unit, Hammersmith Hospital, for their contributions. This work was supported by Parkinson's UK. I.K.G. was supported by a Parkinson's UK prize studentship and the Richard-Winter-Stiftung. The Wales Institute of Cognitive Neuroscience supports A.D.L.

© 2011 Psychology Press, an imprint of the Taylor & Francis Group, an Informa business
www.psypress.com/socialneuroscience   http://dx.doi.org/10.1080/17470919.2011.556821

in PD can occur independently of other symptoms of depression such as heightened sadness (Kirsch-Darrow et al., 2006; Starkstein et al., 2009), and it impacts significantly on activities of daily living and quality of life (Barone et al., 2009; Pederson, Larsen, Alves, & Aarsland, 2009; Weintraub, Moberg, Duda, Katz, & Stern, 2004).

Despite the high prevalence and clinical significance of apathy in PD, little is known about its pathophysiology. The involvement of the prefrontal-basal ganglia system in apathy in PD has often been hypothesized (Levy & Dubois, 2006), as has involvement of both dopaminergic and nondopaminergic neurotransmission (Rodriguez-Oroz et al., 2009). However, it remains unclear how apathy relates to alterations in the neural circuitry underpinning the affective-motivational and cognitive components of goal-directed behavior.

The motivational control of goal-directed behavior is complex and multifaceted (Berridge, 2001). In addition to a cognitive incentive process, through which individuals evaluate the current reward value (or desirability) of an outcome or goal (based on expected pleasure) (Schoenbaum & Roesch, 2005), reward-predictive environmental cues also contribute importantly to goal-directed behavior. Such cues can increase arousal generally, but also prime specific associated behaviors (Berridge, 2001; Aarts, Custers, & Veltkamp, 2008).

Converging evidence from rodents, primates, and man reveals that several structures, in particular the basolateral amygdala, anterior cingulate cortex, ventromedial prefrontal cortex (vmPFC), and sectors of the striatum are part of a neural circuit that is integral to the representation of reward values and goal-directed behavior (Balleine, Liljeholm, & Ostlund, 2009; Balleine & O'Doherty, 2010; Baxter & Murray, 2002; Rushworth & Behrens, 2008), whereas mesolimbic dopamine appears to be critical for the activating and biasing influence of reward-associated cues on behavior (Berridge, 2007).

Here, we examined the performance of individuals with PD, with or without clinically significant levels of apathy, on a spatial search task during positron emission tomography (PET) $H_2^{15}O$ functional neuroimaging. By manipulating the outcome of the search (money reward vs valueless token), while keeping the actions of the participants constant, we could examine the influence of apathy on the neural coding of money, money being a key player in human social life (Vohs, Mead, & Goode, 2008). In an earlier investigation (Goerendt, Lawrence, & Brooks, 2004), we found that healthy controls showed enhanced spatial search efficiency under conditions of potential monetary

gain, and similar reward responsivity was seen in a group of nonapathetic PD individuals (see also Hall, 1927; Harsay, Buitenweg, Wijnen, Guerreiro, & Ridderinkhof, 2010; Pessiglione et al., 2004; Schmidt et al., 2008).

By contrast, we predicted that apathy in PD would be associated with reduced activity to money in the neural circuitry integral to the representation of reward values, reward-guided behavior, and incentive arousal, namely the amygdala, vmPFC, striatum, and dopaminergic midbrain.

MATERIALS AND METHODS

Participants

There were 20 right-handed PD participants, each of whom was evaluated on the Apathy Scale (AS) (Starkstein et al., 1992), a 14-item scale measuring cognitive, emotional, and behavioral facets of apathy. Sample items include "Do you have plans and goals for the future?" (reverse scored) and "Are you indifferent to things?" Items are rated on a 0–3 Likert scale, and total scores range from 0 to 42, higher scores reflecting increased apathy. The AS has good psychometric properties in PD ($\alpha = .76$, test–retest at 1 week = .90) (Starkstein et al., 1992) and is currently the only recommended scale for the assessment of apathy in PD (Leentjens et al., 2008). We created low- ($n = 10$) and high-apathy ($n = 10$) PD groups (Table 1), based on proposed clinical criteria for apathy (Starkstein et al., 2009). Specifically, apathy was diagnosed whenever patients had poor or no motivation (item 7), interest (items 1 and 2), or effort (items 4 and 9) and had feelings of indifference or lack of emotion most or all of the time (items 10 and 13).

Participants were excluded if they had known current or past psychiatric or neurological disease (including depression) other than PD, or if they were receiving deep brain stimulation treatment. All participants were referred from a neurology clinic and assessed by a neurologist prior to scanning. All fulfilled Queen Square Brain Bank criteria for prospective diagnosis of PD and had a predominantly akinetic-rigid phenotype, except for three patients who had a tremor-predominant phenotype. PD participants had to score >24 on the Mini-Mental Parkinson (Mahieux et al., 1995) and <14 on the Geriatric Depression Scale (GDS) (Ertan, Ertan, Kiziltan, & Uyguçgil, 2006) to exclude coexistence of dementia or clinically significant depression symptoms other than apathy. Premorbid IQ was estimated by the revised National Adult Reading Test (Nelson & Willison,

1991). Severity of disability was evaluated by the Hoehn and Yahr and Unified PD Rating Scales in a practically defined "off" state. The high- and low-apathy PD groups were matched for age, IQ, and disease severity (Table 1). All participants were receiving daily dopamine replacement therapy (levodopa preparations, and/or dopamine receptor agonists), and were stable on their medication doses and responding well. Medication regimes were similar in the two groups. No patient was receiving antidepressant medication. Participants were asked to abstain from taking their medication the night before PET scanning was scheduled to take place. PET scans were performed approximately 12 h after withdrawal of anti-parkinsonian medication.

Written, informed consent was obtained according to the Declaration of Helsinki after the nature and possible risks of the study were fully explained. The Imperial College Research Ethics Committee gave approval for the experiment. Permission to administer radioisotopes was given by the Administration of Radioactive Substances Advisory Committee of the Department of Health, UK.

Task

The task (Figure 1) was written in Microsoft Visual Basic 4 and run on a Dell PC fitted with a touch-sensitive screen. During each of six regional cerebral blood-flow (rCBF) scans, participants had to complete three trials of the task. Each trial consisted of the presentation of four boxes, in one of several configurations, on the screen. A small search array was used to minimize demands on working memory capacity and strategic processing (Robbins et al., 1994). A pacing tone sounded every 3 s. When the participant heard the tone, he touched a box on the screen. The box opened, to reveal what was hidden behind, before it

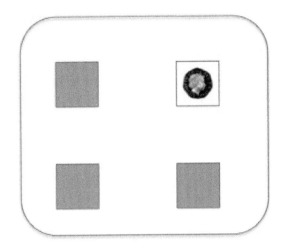

Figure 1. Screen layout.

was covered up again. There were two different experimental conditions: no money scans (box opened to reveal a valueless "chip") or money scans (box opened to reveal a 50p coin). Each box contained a chip or money only once per trial. On each trial, a total of four chips or coins had to be found (equal to the number of boxes per trial). To maximize financial return, participants had to remember which box had previously contained a chip or coin on that trial, and avoid returning to it (as a chip or coin was never hidden in the same box twice on any one trial). Memory load was equivalent across conditions. When all four coins or chips had been found on a trial, a new configuration of boxes appeared on the screen for participants to search.

Each condition (money or chip) was replicated three times. The order of conditions was randomized within and between subjects. Participants were explicitly informed that they would receive the equivalent amount of money to what they accrued in the task (up to a maximum possible reward of £18). Three practice trials were given prior to PET to ensure that participants were able to perform the task successfully.

TABLE 1

Participant details

Variable	Low apathy M (SD)	High apathy M (SD)	Comparison t statistic
Age (in years)	59.6 (6.4)	61.7 (6.1)	$-0.76, p > .05$
NART verbal IQ	119.1 (7.0)	115.1 (6.6)	$1.28, p > .05$
AS	6.2 (2.9)	15.5 (2.3)	$-7.60, p < .001$
GDS*	4.7 (2)	8.2 (3.5)	$-2.68, p < .05$
MMPD*	31.5 (0.7)	29.3 (2.8)	$2.27, p = .05$
H&Y	2.35 (0.6)	2.7 (0.8)	$-1.01, p > .05$
UPDRS III	25.6 (11.0)	28.2 (11.4)	$-0.52, p > .05$

Notes: For both groups, $n = 10$. NART = National Adult Reading Test; AS = Apathy Scale; GDS = Geriatric Depression Scale; MMPD = Mini-Mental Parkinson; H&Y = Hoehn and Yahr; UPDRS III = Unified Parkinson's Disease Rating Scale, Part III (motor score).
*One participant failed to complete these measures.

Both reaction times (RT) and errors were recorded. RT was defined as the time taken to make one response, that is, touch one box, from the onset of the pacing tone. Search errors were defined as a return to a previously successful or unsuccessful box within an individual trial. Data analyses were performed with SPSS for Windows, Version 13.

Image acquisition

PET measurements of regional cerebral blood flow (rCBF) were acquired with an ECAT EXACT HR++ camera (CTI/Siemens 966; Knoxville, TN), with a total axial field of view of 23.4 cm, operating in 3D mode. Prior to data collection, a transmission scan was conducted to correct for attenuation effects. For each PET scan, we administered ~185 MBq (5 mCi) of oxygen-15-labeled water intravenously over 20 s via automatic pump, followed by a 20-s saline flush. rCBF and task-performance data were collected over a 90-s activation period that began 5 s before the rising phase of radioactivity in the head. Six such scans were collected, with an interval of 6 min between PET measurements to allow for radioactivity to decay. Images were reconstructed by 3D filtered back-projection.

Image analysis

Image analysis was performed by statistical parametric mapping (SPM99) (Wellcome Trust Centre for Functional Neuroimaging, London, UK), executed in MATLAB. For each participant, the six scans were realigned, using the mean image as a reference. All images were then spatially normalized into standard anatomic space, using the template developed at the Montreal Neurological Institute (MNI). Images were then smoothed with an isotropic Gaussian kernel (FWHM 12 mm). This accommodated intersubject differences in anatomy, increased the signal to noise ratio and aligned the data more closely to a Gaussian field model.

Subsequent analysis used the general linear model to estimate effects at each voxel in the brain (Friston et al., 1995). An analysis of covariance (ANCOVA) model was fitted to the data at each voxel, with group effects for the two groups (high- and low-apathy), condition effects for the two experimental conditions (reward, no reward), with scan-to-scan differences in global blood flow, scan order, and depression scores (GDS) as confounding covariates.[1] Contrasts of group

and condition effects at each voxel were assessed with Student's t statistic, transformed to a Z statistic to provide statistical parametric maps.

To identify regions of money-related activity, independent of apathy, we first compared all money with non-money scans across participants. To identify money-related activity that was significantly reduced in the high- relative to the low-apathy group, the following contrast was computed: (money − no money)low-apathy group − (money − no money) high-apathy group.

Amygdala, striatal, and anterior cingulate regions of interest (ROIs) were defined with structural templates derived by automated anatomic labeling (AAL) (Tzourio-Mazoyer et al., 2002). The vmPFC was sampled with a spherical ROI (12-mm radius) centered on 4, 52, −10 (Knutson, Fong, Bennett, Adams, & Hommer, 2003). Midbrain was sampled with a 6-mm radius spherical ROI, centered on 0, −20, −28 (Beaver et al., 2006). We used small volume correction (SVC) for multiple comparisons applied at $p < .05$ to each ROI. Other activations are reported if they survived correction for multiple comparisons across the whole brain ($p < .05$, FWE corrected). For anatomic labeling purposes, activation coordinates were transformed into the Talairach and Tournoux coordinate system with an automated nonlinear transform (Laird et al., 2010). For visualizing activations, group maps were overlaid on the ICBM 152 structural template, an average T1-weighted image of 152 individuals co-registered to MNI space. Activations are reported using (x, y, z) coordinates in MNI standardized space.

RESULTS

Performance data

Software failure resulted in a loss of behavioral data for four participants. Error rates were minimal and not amenable to statistical analysis. For RT data, ANOVA revealed a main effect of group, with apathetic patients responding more slowly than nonapathetic individuals, $F(1, 14) = 13.5$, p = .003; no main effect of money, $F(1, 14) = 0.05$, $p = .82$; and no interaction, $F(1, 14) = 0.38$, $p = .54$. Mean RTs for the low-apathy group were 3.02 and 2.95 s in the no-money and money conditions, respectively, and for the high-apathy group 3.75 and 3.78 s.

Imaging data

Across participants, activation in our ROIs for the contrast (money vs no money) failed to reach significance.

---

[1] A similar ANCOVA, controlling for MMPD, produced identical results.

TABLE 2
Regions of reduced money-related activity in apathetic Parkinson's disease participants

Region	MNI coordinates	Z score
Left amygdala	−28, −6, −18	2.53
Left striatum	−28, −10, −2	3.04
vmPFC	4, 32, −2	3.35
vmPFC	8, 56, −8	3.65
Midbrain	0, −24, −30	2.69

Notes: MNI = Montreal Neurological Institute; vmPFC = ventromedial prefrontal cortex.

However, this null result was qualified by a significant interaction between reward and apathy level, with regions in left amygdala, left striatum, vmPFC, and midbrain showing reduced activity in the high-apathy group relative to the low-apathy group in the money relative to the no-money scans at $p_{svc} < .05$ (Table 2, Figure 2). These regions showed a significant increase to money in the low-apathy group, but not in the high-apathy group. There were no other differences at $p_{FWE} < .05$.

DISCUSSION

While previous neuroimaging studies have shown altered money reward processing in the parkinsonian brain (e.g., Künig et al., 2000; Rowe et al., 2008), this is the first study to examine the influence of apathy. Although apathy is broadly recognized as a primary diminution in motivation, there are currently only provisional clinical diagnostic criteria (Starkstein

et al., 2009). To assess apathy, we used the AS, the only scale recommended for the measurement of apathy in PD (Leentjens et al., 2008). It encompasses alterations in the overt manifestation of goal-directed action, in diminished goal-related thought content, and in diminished emotional responses to goal-related events. Thus, the AS represents the current reference standard tool for the assessment of apathy.

We found that apathy, as defined by the AS, was associated with a diminished response to money in several brain regions: the amygdala, vmPFC, striatum, and midbrain, all of which have been previously shown to be integral to the representation of reward value and goal-directed behavior.

Reduced money-related activity was found in two distinct foci in the vmPFC centered on frontal polar area 10p and a region of the ventral medial wall corresponding to the intersection of area 10r/subgenual area 24/area 32, according to the maps of Öngür, Ferry, and Price (2003). Structural alterations in these regions (centered on 9, 57, 3) have previously been linked to apathy in the context of FTD (Rosen et al., 2005); reduced metabolism in medial area 10 is correlated with apathy following frontal lobe lesions (Sarazin et al., 2003); and lesions encompassing these regions of the vmPFC can result in apathy (Fellows & Farah, 2005; Jorge, Starkstein, & Robinson, 2010; Starkstein & Manes, 2000). A number of functional magnetic resonance imaging (fMRI) studies in man have found that these sectors of the vmPFC track the expected value of an associated outcome while participants perform actions in order to obtain reward (e.g., Behrens, Hunt, Woolrich, & Rushworth, 2008; Gläscher, Hampton,

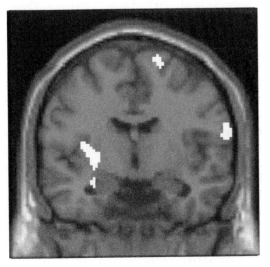
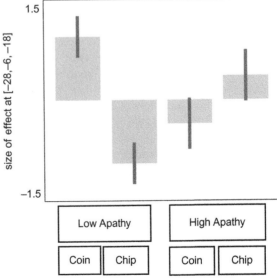

Figure 2. Blunted amygdala reactivity to money in apathetic Parkinson's disease participants.

& O'Doherty, 2009; Hampton, Adolphs, Tyszka, & O'Doherty, 2007; Kim, Shimojo, & O'Doherty, 2010; Knutson et al., 2003) at coordinates very close to the ones reported here (4, 32, −2; 8, 56, −8) (Knutson et al., 4, 52, −10; Hampton et al., 6, 57, −6; Behrens et al., 2, 52, −8; Gläscher et al., 3, 54, −3; Kim et al., −6, 36, −12). The vmPFC appears to play a very flexible role in representing outcome values for both money and other (e.g., food) rewards, and also represents a variety of different value signals, including expected rewards and decision values in addition to outcome values (Kim et al., 2010). Hence apathy appears to be associated with diminished high-level value representations in the vmPFC.

We also found diminished money-related activity in the amygdala in apathetic patients. Previous studies in rodents, monkeys, and man have demonstrated a key role for the amygdala in reward processing – both in encoding stimulus-incentive properties and in encoding the value of chosen actions (Baxter & Murray, 2002; Gläscher et al., 2009; Ostlund & Balleine, 2008). The amygdala also appears to be critical for integrating information about reward with effort costs (Ghods-Sharifi, St. Onge, & Floresco, 2009), and is involved in the biasing effects of reward-associated cues on behavior (Corbit & Balleine, 2005). Moreover, interactions between the amygdala and the vmPFC have been shown to be necessary to establish and maintain expected reward representations in the vmPFC in man: Hampton et al. (2007) found that individuals with lesions in the amygdala showed a profound change in vmPFC activity (centered on 6, 57, −3), associated with reward expectation and behavioral choice. Diminished amygdala activity may be especially relevant to the emotional blunting seen in apathy (Balleine & Killcross, 2006).

Likewise, apathy was associated with reduced striatal response to money. The striatum is a key target of PD-related pathology (Rodriguez-Oroz et al., 2009), and striatal lesions have long been associated with apathy (Jorge et al., 2010; Levy & Dubois, 2006). The striatum is massively interconnected with both the vmPFC and the amygdala (Corbitt & Janak, 2010; Haber & Knutson, 2010), and, like the vmPFC and the amygdala, striatal activity codes the value associated with both stimuli and actions (Croxson, Walton, O'Reilly, Behrens, & Rushworth, 2009; Hori, Minamimoto, & Kimura, 2009; Samejima, Ueda, Doya, & Rimura, 2005). The striatum has been argued to mediate the ability of the incentive value of rewards to influence goal-directed performance (Balleine et al., 2009). The striatum has also been implicated in the biasing effect of reward-predictive cues on action selection (Bray, Rangel, Shimojo, Balleine, &

O'Doherty, 2008; Corbit & Janak, 2010) and in decision effort (Croxson et al., 2009). This broadly integrative role for the striatum in motivated behavior indicates why lesions to this structure have long been implicated in the pathophysiology of apathy.

Of relevance to the complex cognitive manifestations of apathy, the vmPFC has been shown to play a role in the representation of hypothetical, imagined rewards as well as actual rewards (Bray, Shimojo, & O'Doherty, 2010). Furthermore, the vmPFC, amygdala, and striatum are activated when people imagine positive future goals, hopes, and aspirations (Johnson et al., 2006; Sharot, Riccardi, Raio, & Phelps, 2007). Predictions of the emotional impact of potential future experiences ("pre-feelings") (Gilbert & Wilson, 2007) have motivating qualities that initiate plans of action and influence choice. For example, low expectations of how happy or content an event will make one feel will lower one's motivation to pursue that event. The vmPFC and amygdala likely encode the expected value of both present and future rewards including abstract, future goals (Schoenbaum & Roesch, 2005), and diminished activity in vmPFC-related circuitry may underpin the impaired long-term planning and foreshortened future time perspective seen in apathetic individuals (Fellows & Farah, 2005; Weintraub et al., 2005; see also Jiga-Boy, Clark, & Semin, 2010).

Apathy was also associated with reduced money-related activity in the midbrain, in the region of the ventral tegmental area. Previous studies in PD have suggested that apathy is at least partly a dopamine-dependent syndrome. For example, withdrawal of dopamine replacement therapy leads to increases in rated apathy and impaired reward-related performance (Czernecki et al., 2002; Evans, Lawrence, Cresswell, Katzenschlager, & Lees, 2010), and PET neuroreceptor mapping studies show that apathy is correlated with altered indices of dopamine function in the striatum, amygdala, and vmPFC (Remy, Doder, Lees, Turjanski, & Brooks, 2005; Thobois et al., 2010). While the limitations of our scanning technique preclude definitive anatomic and neurochemical localization, our findings are at least consistent with a role for dopamine in the pathophysiology of apathy in PD. In animal models, dopamine has been primarily associated with the activating or biasing influences of reward ("incentive salience") (Berridge, 2007), although dopamine may play a more general role in behavior initiation (Corbit, Janak, & Balleine, 2007), and in the regulation of effort (Salamone, Correa, Farrar, Nunes, & Pardo, 2009) consistent with our behavioral findings of a general prolonging of RT, which was not specific to rewarded scans. However, recent findings also implicate dopamine in pleasure expectation or pre-feeling (Sharot, Shiner, Brown, Fan, & Dolan, 2009). Future studies should examine the influence of dopamine replacement therapy on the neural correlates of reward-related activity in PD apathy. Nondopaminergic cell loss in regions including the amygdala and vmPFC (Harding, Stimson, Henderson, & Halliday, 2002; Silveira-Moriyama et al., 2009) and white matter pathology (Cacciari et al., 2010; but see Gaffan & Eacott, 1995) could also contribute to the reduced money-related activity seen here.

Another area for future investigation will be the influence of apathy on aversive valuation and motivation (see Chase et al., 2010). The vmPFC, amygdala, striatum, and dopamine have all been implicated in both reward- and aversion-related processes (Balleine & Killcross, 2006; Balleine & O'Doherty, 2010; Berridge, 2007; Plassmann, O'Doherty, & Rangel, 2010), and it will be important to determine the extent to which apathy in PD is associated with a diminished response to reward-related events alone, or with a more general emotional "indifference."

Finally, it is notable that the regions showing blunted money-related activity here have been implicated in the pathophysiology of depression (Ressler & Mayberg, 2007). The relationship between apathy and depression is complex. Apathy is a core symptom of depression (Marin et al., 1993), but, in the context of PD, apathy appears to be somewhat independent of other depressive symptoms, such as heightened sadness (Kirsch-Darrow et al., 2006; Starkstein et al., 2009). Our results held when controlling for depression and suggest that a similar blunting of reward-related activity might contribute to apathetic symptoms in the context of major depression. Intriguingly, Canli et al. (2005) found that greater amygdala and striatal reactivity to happy faces predicted subsequent improvement in major depression, and depression outcome is inversely related to apathy (Chaturvedi & Sarmukaddam, 1986). This is an important topic for the future.

Before closing, we must discuss several limitations of our study. We did not find a behavioral effect of money on error rates or RTs. These findings contrast with those of our earlier study (Goerendt et al., 2004), which found that both healthy controls and nonapathetic PD individuals showed increased search efficiency with increasing amounts of money, findings that serve to validate the current task as a measure sensitive to motivation. However, in our earlier study, there were twice the number of scans as here, and a wider range of money values was used, likely leading to differences in the operation of reward expectancies (Watanabe et al., 2001). Furthermore, the use of a pacing tone to control movement onset severely limited

the potential for reward to influence RTs. On balance, however, we do not see the lack of performance effects as a major limitation, since this had the advantage of allowing us to examine neural responses to money valuation and expectation distinct from downstream motivational influences on motor systems (Roesch & Olson, 2007). Though subtle, the distinction between the representation of an event and the response it elicits is nevertheless important. We did not measure subjective hedonic responses to money. However, "reward" should not be equated with subjective reports of desire or pleasure alone. The emotional and motivational components of reward can exist objectively apart from conscious awareness of them; that is, they can occur implicitly (Berridge & Robinson, 2003). Further, a previous study has shown that apathy in PD is not associated with diminished subjective pleasure report (anhedonia) (Isella et al., 2003; see also Brown, Schwartz, & Sweeney, 1978). Moreover, there was a clear differentiation between money and non-financial cue activity in the brains of the low-apathy PD group in regions previously implicated in reward valuation. Nevertheless, it would be interesting to examine the impact of apathy on implicit measures of liking, such as facial affect expressions.

Further, rather than directly reflecting differences in value computations, differences in activity in vmPFC-related circuitry could reflect differences in attention or cognitive control linked to anticipated reward (Pessoa & Engelmann, 2010). However, other research has demonstrated that reward-related activity in the vmPFC does not code saliency or urgency signals linked to motor readiness, attention, and arousal (Litt, Plassman, Shiv, & Rangel, 2011; Roesch & Olson, 2007). Future studies could usefully examine the influence of apathy on the neural correlates of the motivational modulation of performance, in addition to the influence of apathy on cue-related activity. A further limitation is that we did not include a non-PD control group. However, the focus of this study was on the influence of apathy in PD, and so a nonapathetic PD group was deemed the most appropriate control group.

In summary, apathy is a key syndrome associated with PD, linked with reduced activities of daily living and poorer quality of life. Despite its clinical significance, the functional anatomy of apathy is poorly understood. Here, we found that apathy in PD is associated with reduced money-related activity in the valuation circuitry of the vmPFC, amygdala, striatum, and midbrain. Deficits in this circuitry potentially underpin the expression of the various manifestations of apathy in PD, including reduced cognitive, emotional, and behavioral facets of goal-directed behavior. More generally, we suggest that the study of apathy within the framework of theories of valuation and motivation might prove extremely fruitful for both neuropsychiatry and social neuroscience.

REFERENCES

Aarts, H., Custers, R., & Veltkamp, M. (2008). Goal priming and the affective-motivational route to nonconscious goal pursuit. *Social Cognition, 26*, 555–577.

Balleine, B. W., & Killcross, S. (2006). Parallel incentive processing: An integrated view of amygdala function. *Trends in Neurosciences, 29*, 272–279.

Balleine, B. W., Liljeholm, M., & Ostlund, S. B. (2009). The integrative function of the basal ganglia in instrumental conditioning. *Behavioral Brain Research, 199*, 43–52.

Balleine, B. W., & O'Doherty, J. P. (2010). Human and rodent homologies in action control: Corticostriatal determinants of goal-directed and habitual action. *Neuropsychopharmacology, 35*, 48–69.

Barone, P., Antonini, A., Colosimo, C., Marconi, R., Morgante, L., Avarello, T. P., et al., on behalf of the PRIAMO study group (2009). The PRIAMO study: A multicenter assessment of nonmotor symptoms and their impact on quality of life in Parkinson's disease. *Movement Disorders, 24*, 1641–1649.

Baxter, M. G., & Murray, E. A. (2002). The amygdala and reward. *Nature Reviews Neuroscience, 3*, 563–573.

Beaver, J. D., Lawrence, A. D., van Ditzhuijzen, J., Davis, M. H., Woods, A., & Calder, A. J. (2006). Individual differences in reward drive predict neural response to images of food. *Journal of Neuroscience, 26*, 5160–5166.

Behrens, T. E. J., Hunt, L. T., Woolrich, M. W., & Rushworth, M. F. S. (2008). Associative learning of social value. *Nature, 456*, 245–249.

Berridge, K. C. (2001). Reward learning: Reinforcement, incentives, and expectations. In D. L. Medin (Ed.), *The psychology of learning and motivation: Advances in research and theory* (Vol. 40, pp. 223–278). San Diego, CA: Academic Press.

Berridge, K.C. (2007). The debate over dopamine's role in reward: The case for incentive salience. *Psychopharmacology, 191*, 391–431.

Berridge, K. C., & Robinson, T. E. (2003). Parsing reward. *Trends in Neurosciences, 26*, 507–513.

Berrios, G. E., & Grli, M. (2007). Abulia and impulsiveness revisited: A conceptual history. *Acta Psychiatrica Scandinavica, 92*, 161–167.

Bray, S., Rangel, A., Shimojo, S., Balleine, B. W., & O'Doherty, J. P. (2008). The neural mechanisms underlying the influence of Pavlovian cues on human decision-making. *Journal of Neuroscience, 28*, 5861–5866.

Bray, S., Shimojo, S., & O'Doherty, J. P. (2010). Human medial orbitofrontal cortex is recruited during experience of imagined and real rewards. *Journal of Neurophysiology, 103*, 2506–2512.

Brown, R. G., & Pluck, G. (2000). Negative symptoms: The 'pathology' of motivation and goal-directed behaviour. *Trends in Neuroscience, 23*, 412–417.

Brown, S.-L., Schwartz, G. E., & Sweeney, D. R. (1978). Dissociation of self-reported and observed pleasure in depression. *Psychosomatic Medicine, 40*, 536–548.

Cacciari, C., Moraschi, M., Di Paola, M., Cherubini, A., Orfei, M. D., Giove, F., et al. (2010). White matter microstructure and apathy level in amnestic mild cognitive impairment. *Journal of Alzheimer's Disease, 20*, 501–507.

Canli, T., Cooney, R. E., Goldin, P., Shah, M., Sivers, H., Thomason, M. E., et al. (2005). Amygdala reactivity to emotional faces predicts improvement in major depression. *NeuroReport, 16*, 1267–1270.

Chase, H. W., Camille, N., Michael, A., Bullmore, E. T., Robbins, T. W., & Sahakian, B. J. (2010). Regret and the negative evaluation of decision outcomes in major depression. *Cognitive, Affective, & Behavioral Neuroscience, 10*, 406–413.

Chaturvedi, S. K., & Sarmukaddam, S. B. (1986). Prediction of outcome in depression by negative symptoms. *Acta Psychiatrica Scandinavica, 74*, 183–186.

Corbit, L. H., & Balleine, B. W. (2005). Double dissociation of basolateral and central amygdala lesions on the general and outcome-specific forms of Pavlovian-instrumental transfer. *Journal of Neuroscience, 25*, 962–970.

Corbit, L. H., & Janak, P. H. (2010). Posterior dorsomedial striatum is critical for both selective instrumental and Pavlovian reward learning. *European Journal of Neuroscience, 31*, 1312–1321.

Corbit, L. H., Janak, P. H., & Balleine, B. W. (2007). General and outcome-specific forms of Pavlovian-instrumental transfer: The effect of shifts in motivational state and inactivation of the ventral tegmental area. *European Journal of Neuroscience, 26*, 3141–3149.

Croxson, P. L., Walton, M. E., O'Reilly, J. X., Behrens, T. E. J., & Rushworth, M. F. S. (2009). Effort-based cost-benefit valuation and the human brain. *Journal of Neuroscience, 29*, 4531–4541.

Czernecki, V., Pillon, B., Houeto, J. L., Pichon, J. B., Levy, R., & Dubois, B. (2002). Motivation, reward, and Parkinson's disease: Influence of dopatherapy. *Neuropsychologia, 40*, 2257–2267.

Drijgers, R. L., Dujardin, K., Reijnders, J. S. A. M., Defebvre, L., & Leentjens, A. F. G. (2010). Validation of diagnostic criteria for apathy in Parkinson's disease. *Parkinsonism and Related Disorders, 16*, 656–660.

Ertan, F. S., Ertan, T., Kiziltan, G., & Uyguçgil, H. (2006). Reliability and validity of the Geriatric Depression Scale in depression in Parkinson's disease. *Journal of Neurology, Neurosurgery and Psychiatry, 76*, 1445–1447.

Evans, A. H., Lawrence, A. D., Cresswell, S. A., Katzenschlager, R., & Lees, A. J. (2010). Compulsive use of dopaminergic drug therapy in Parkinson's disease: Reward and anti-reward. *Movement Disorders, 25*, 867–876.

Fellows, L. K., & Farah, M. J. (2005). Dissociable elements of human foresight: A role for the ventromedial frontal lobes in framing the future, but not in discounting future rewards. *Neuropsychologia, 43*, 1214–1221.

Friston, K. J., Holmes, A. P., Worlsey, K. J., Poline, J. B., Frith, C. D., & Frackowiak, R. S. (1995). Statistical parametric maps in functional imaging: A general linear approach. *Human Brain Mapping, 2*, 189–210.

Gaffan, D., & Eacott, M. J. (1995). Visual learning for an auditory secondary reinforcer by macaques is intact after uncinate fascicle section: Indirect evidence for the involvement of the corpus striatum. *European Journal of Neuroscience, 7*, 1866–1871.

Ghods-Sharifi, S., St. Onge, J. R., & Floresco, S. B. (2009). Fundamental contribution by the basolateral amygdala to different forms of decision making. *Journal of Neuroscience, 29,* 5251–5259.

Gilbert, D. T., & Wilson, T. D. (2007). Prospection: Experiencing the future. *Science, 317,* 1351–1354.

Gläscher, J., Hampton, A. N., & O'Doherty, J. P. (2010). Determining a role for ventromedial prefrontal cortex in encoding action-based value signals during reward-related decision making. *Cerebral Cortex, 19,* 483–495.

Goerendt, I. K., Lawrence, A. D., & Brooks, D. J. (2004). Reward processing in health and Parkinson's disease: Neural organization and re-organization. *Cerebral Cortex, 14,* 73–80.

Haber, S. N., & Knutson, B. (2010). The reward circuit: Linking primate anatomy and human imaging. *Neuropsychopharmacology, 35,* 4–26.

Hall, A. J. (1927). Rate of movement in post-encephalitic Parkinsonism. *Lancet, 210,* 1009–1012.

Hampton, A. N., Adolphs, R., Tyszka, J. M., & O'Doherty, J. P. (2007). Contributions of the amygdala to reward expectancy and choice signals in human prefrontal cortex. *Neuron, 55,* 545–555.

Harding, A. J., Stimson, E., Henderson, J. M., & Halliday, G. M. (2002). Clinical correlates of selective pathology in the amygdala of patients with Parkinson's disease. *Brain, 125,* 2431–2445.

Harsay, H. A., Buitenweg, J. I., Wijnen, J. G., Guerreiro, M. J., & Ridderinkhof, K. R. (2010). Remedial effects of motivational incentive on declining cognitive control in healthy aging and Parkinson's disease. *Frontiers in Aging Neuroscience, 2,* 144.

Hori, Y., Minamimoto, T., & Kimura, M. (2009). Neuronal encoding of reward value and direction of actions in the primate putamen. *Journal of Neurophysiology, 102,* 3530–3543.

Isella, V., Iurlaro, S., Piolti, R., Ferrarese, L., & Appollonio, I. (2003). Physical anhedonia in Parkinson's disease. *Journal of Neurology, Neurosurgery and Psychiatry, 74,* 1308–1311.

Jiga-Boy, G. M., Clark, A. E., & Semin, G. R. (2010). So much to do and so little time: Effort and perceived temporal distance. *Psychological Science, 21,* 1811–1817.

Johnson, M. K., Raye, C. L., Mitchell, K. J., Touryan, S. R., Greene, E. J., & Nolen-Hoeksema, S. (2006). Dissociating medial frontal and posterior cingulate activity during self-reflection. *Social, Cognitive and Affective Neuroscience, 1,* 56–64.

Jorge, R. E., Starkstein, S. E., & Robinson, R. G. (2010). Apathy following stroke. *Canadian Journal of Psychiatry, 55,* 350–354.

Kim, H., Shimojo, S., & O'Doherty, J. P. (2010). Overlapping responses for the expectation of juice and money rewards in human ventromedial prefrontal cortex. *Cerebral Cortex, Aug 23* [Epub ahead of print].

Kirsch-Darrow, L., Fernandez, H. H., Marsiske, M., Okun, M. S., & Bowers, D. (2006). Dissociating apathy and depression in Parkinson disease. *Neurology, 67,* 33–38.

Knutson, B., Fong, G. W., Bennett, S. M., Adams, C. M., & Hommer, D. (2003). A region of mesial prefrontal cortex tracks monetarily rewarding outcomes: Characterization with rapid event-related fMRI. *NeuroImage, 18,* 263–272.

Künig, G., Leenders, K. L., Martin-Sölch, C., Missimer, J., Magyar, S., & Schultz, W. (2000). Reduced reward processing in the brains of Parkinsonian patients. *NeuroReport, 11,* 3681–3687.

Laird, A. R., Robinson, J. L., McMillan, K. M., Tordesillas-Gutiérrez, D., Moran, S. T., Gonzales, S. M., et al. (2010). Comparison of the disparity between Talairach and MNI coordinates in functional neuroimaging data: Validation of the Lancaster transform. *NeuroImage, 51,* 677–683.

Leentjens, A. F. G., Dujardin, K., Marsh, L., Martinez-Martin, P., Richard, I. H., Starkstein, S. E., et al. (2008). Apathy and anhedonia rating scales in Parkinson's disease: Critiques and recommendations. *Movement Disorders, 23,* 2004–2014.

Levy, R., & Dubois, B. (2006). Apathy and the functional anatomy of the prefrontal cortex-basal ganglia circuits. *Cerebral Cortex, 16,* 916–928.

Litt, A., Plassmann, H., Shiv, B., & Rangel, A. (2011). Dissociating valuation and saliency signals during decision-making. *Cerebral Cortex, 21,* 95–102.

Mahieux, F., Michelet, D., Manifacier, M. J., Boller, F., Fermanian, J., & Guillard, A. (1995). Mini-Mental Parkinson: First validation study of a new bedside test constructed for Parkinson's disease. *Behavioral Neurology, 8,* 15–22.

Marin, R. S. (1991). Apathy: A neuropsychiatric syndrome. *Journal of Neuropsychiatry and Clinical Neurosciences, 3,* 243–254.

Marin, R. S., Firinciogullari, S., & Biedrzycki, R. C. (1993). The sources of convergence between measures of apathy and depression. *Journal of Affective Disorders, 28,* 117–124.

Nelson, H. E., & Willison, J. R. (1991). *The Revised National Adult Reading Test: Manual.* Windsor, UK: NFER-Nelson.

Öngür, S., Ferry, A. T., & Price, J. L. (2003). Architectonic subdivision of the human orbital and medial prefrontal cortex. *Journal of Comparative Neurology, 460,* 425–449.

Ostlund, S. B., & Balleine, B. W. (2008). Differential involvement of the basolateral amygdala and mediodorsal thalamus in instrumental action selection. *Journal of Neuroscience, 28,* 4398–4405.

Pederson, K. F., Larsen, J. P., Alves, G., & Aarsland, D. (2009). Prevalence and clinical correlates of apathy in Parkinson's disease: A Community-based study. *Parkinsonism and Related Disorders, 15,* 295–299.

Pessiglione, M., Guehl, D., Jan, C., François, C., Hirsch, E. C., Féger, J., et al. (2004). Disruption of self-organized actions in monkeys with progressive MPTP-induced parkinsonism: II. Effects of reward preference. *European Journal of Neuroscience, 19,* 437–446.

Pessoa, L., & Engelmann, J. B. (2010). Embedding reward signals into perception and cognition. *Frontiers in Neuroscience, 4,* 17.

Plassmann, H., O'Doherty, J. P., & Rangel, A. (2010). Appetitive and aversive goal values are encoded in the medial orbitofrontal cortex at the time of decision making. *Journal of Neuroscience, 30,* 10799–10808.

Pluck, G. C., & Brown, R. G. (2002). Apathy in Parkinson's disease. *Journal of Neurology, Neurosurgery and Psychiatry, 73,* 636–642.

Remy, P., Doder, M., Lees, A. J., Turjanski, N., & Brooks, D. J. (2005). Depression in Parkinson's disease: Loss of dopamine and noradrenaline innervation in the limbic system. *Brain, 128,* 1314–1322.

Ressler, K. J., & Mayberg, H. S. (2007). Targeting abnormal neural circuits in mood and anxiety disorders: From the laboratory to the clinic. *Nature Neuroscience, 10,* 1116–1124.

Robbins, T. W., James, M., Owen, A. M., Lange, K. W., Lees, A. J., Leigh, P. N., et al. (1994). Cognitive deficits in progressive supranuclear palsy, Parkinson's disease, and multiple system atrophy in tests sensitive to frontal lobe dysfunction. *Journal of Neurology, Neurosurgery, and Psychiatry, 57,* 79–88.

Rodriguez-Oroz, M. C., Jahanshahi, M., Krack, P., Litvan, I., Macias, R., Bezard, E., et al. (2009). Initial clinical manifestations of Parkinson's disease: Features and pathophysiological mechanisms. *Lancet Neurology, 8,* 1128–1139.

Roesch, M. R., & Olson, C. R. (2007). Neuronal activity related to anticipated reward in frontal cortex: Does it represent value or reflect motivation? *Annals of the New York Academy of Sciences, 1121,* 431–446.

Rosen, H. J., Allison, S. C., Schauer, G. F., Gorno-Tempini, M. L., Weiner, M. W., & Miller, B. W. (2005). Neuroanatomical correlates of behavioral disorders in dementia. *Brain, 128,* 2612–2625.

Rowe, J. B., Hughes, L., Ghosh, B. C., Eckstein, D., Williams-Gray, C. H., Fallon, S., et al. (2008). Parkinson's disease and dopaminergic therapy – differential effects on movement, reward and cognition. *Brain, 131,* 2094–2105.

Rushworth, M. F., & Behrens, T. E. (2008). Choice, uncertainty and value in prefrontal and cingulate cortex. *Nature Neuroscience, 11,* 389–397.

Salamone, J. D., Correa, M., Farrar, A. M., Nunes, E. J., & Pardo, M. (2009). Dopamine, behavioral economics, and effort. *Frontiers in Behavioral Neuroscience, 3,* 13.

Samejima, K., Ueda, Y., Doya, K., & Rimura, M. (2005). Representation of action-specific reward values in the striatum. *Science, 310,* 1337–1340.

Sarazin, M., Michon, A., Pillon, B., Samson, Y., Canuto, A., Gold, G., et al. (2003). Metabolic correlates of behavioral and affective disturbances in frontal lobe pathologies. *Journal of Neurology, 250,* 827–833.

Schmidt, L., d'Arc, B. F., Lafargue, G., Galanaud, D., Czernecki, V., Grabli, D., et al. (2008). Disconnecting force from money: Effects of basal ganglia damage on incentive motivation. *Brain, 131,* 1303–1310.

Schoenbaum, G., & Roesch, M. (2005). Orbitofrontal cortex, associative learning, and expectancies. *Neuron, 47,* 633–636.

Sharot, T., Riccardi, A. M., Raio, C. M., & Phelps, E. A. (2007). Neural mechanisms mediating optimism bias. *Nature, 450,* 102–106.

Sharot, T., Shiner, T., Brown, A. C., Fan, J., & Dolan, R. J. (2009). Dopamine enhances expectation of pleasure in humans. *Current Biology, 19,* 2077–2080.

Silveira-Moriyama, L., Holton, J. L., Kinsbury, A., Ayling, H., Petrie, A., Sterlacci, W., et al. (2009). Regional differences in the severity of Lewy body pathology across the olfactory cortex. *Neuroscience Letters, 453,* 77–80.

Sorabji, R. (2000). *Emotion and peace of mind: From Stoic agitation to Christian temptation.* Oxford, UK: Oxford University Press.

Starkstein, S. E., & Manes, F. (2000). Apathy and depression following stroke. *CNS Spectrums, 5,* 43–50.

Starkstein, S. E., Mayberg, H. S., Preziosi, T. J., Andrezejewski, P., Leiguarda, R., & Robinson, R. G. (1992). Reliability, validity, and clinical correlates of apathy in Parkinson's disease. *Journal of Neuropsychiatry and Clinical Neurosciences, 4,* 134–139.

Starkstein, S. E., Merello, M., Jorge, R., Brockman S., Bruce, D., & Power, B. (2009). The syndromal validity and nosological position of apathy in Parkinson's disease. *Movement Disorders, 24,* 1211–1216.

Thobois, S., Ardouin, C., Lhommée, E., Klinger, H., Lagrange, C., Xie, J., et al. (2010). Non-motor dopamine withdrawal syndrome after surgery for Parkinson's disease: Predictors and underlying mesolimbic denervation. *Brain, 133,* 1111–1127.

Tzourio-Mazoyer, N., Landeau, B., Papathanassiou D, Crivello, F., Etard, O., Delcroix, N., et al. (2002). Automated anatomical labeling of activations in SPM using a macroscopic anatomical parcellation of the MNI MRI single-subject brain. *NeuroImage, 15,* 273–289.

Vohs, K. D., Mead, N. L., & Goode, M. R. (2008). Merely activating the concept of money changes personal and interpersonal behavior. *Current Directions in Psychological Science, 17,* 208–212.

Watanabe, M., Cromwell, H. C., Tremblay, L., Hollerman, J., Hikosaka, K., & Schultz, W. (2001). Behavioral reactions reflecting differential reward expectations in monkeys. *Experimental Brain Research, 140,* 511–518.

Weintraub, D., Moberg, P. J., Culbertson, W. C., Duda, J. E., Katz, I. R., & Stein, M. B. (2005). Dimensions of executive function in Parkinson's disease. *Dementia and Geriatric Cognitive Disorders, 20,* 140–144.

Weintraub, D., Moberg, P. J., Duda, J. E., Katz, I. R., & Stern, M. B. (2004). Effects of psychiatric and other non-motor symptoms on disability in Parkinson's disease. *Journal of the American Geriatrics Society, 52,* 784–788.

Index

Page numbers in *Italics* represent tables.
Page numbers in **Bold** represent figures.

Printed and bound by CPI Group (UK) Ltd, Croydon, CR0 4YY

18/10/2024

01776253-0016